design

test

NASA SYSTEMS ENGINEERING HANDBOOK

integrate

fly

NASA SP-2016-6105 Rev2 supersedes SP-2007-6105 Rev 1 dated December, 2007.

Cover photos: *Top left:* In this photo, engineers led by researcher Greg Gatlin have sprayed fluorescent oil on a 5.8 percent scale model of a futuristic hybrid wing body during tests in the 14- by 22-Foot Subsonic Wind Tunnel at NASA's Langley Research Center in Hampton, VA. The oil helps researchers "see" the flow patterns when air passes over and around the model. (NASA Langley/ Preston Martin) *Top right:* Water impact test of a test version of the Orion spacecraft took place on August 24, 2016, at NASA Langley Research Center (NASA) *Bottom left:* two test mirror segments are placed onto the support structure that will hold them. (NASA/Chris Gunn) *Bottom right:* This self-portrait of NASA's Curiosity Mars rover shows the vehicle at the "Mojave" site, where its drill collected the mission's second taste of Mount Sharp. (NASA/JPL-Caltech/MSSS)

Table of Contents

Table of Figures

Table of Tables

Table of Boxes

Preface

Since the initial writing of NASA/SP-6105 in 1995 and the following revision (Rev 1) in 2007, systems engineering as a discipline at the National Aeronautics and Space Administration (NASA) has undergone rapid and continued evolution. Changes include using Model-Based Systems Engineering to improve development and delivery of products, and accommodating updates to NASA Procedural Requirements (NPR) 7123.1. Lessons learned on systems engineering were documented in reports such as those by the NASA Integrated Action Team (NIAT), the Columbia Accident Investigation Board (CAIB), and the follow-on Diaz Report. Other lessons learned were garnered from the robotic missions such as Genesis and the Mars Reconnaissance Orbiter as well as from mishaps from ground operations and the commercial space flight industry. Out of these reports came the NASA Office of the Chief Engineer (OCE) initiative to improve the overall Agency systems engineering infrastructure and capability for the efficient and effective engineering of NASA systems, to produce quality products, and to achieve mission success. This handbook update is a part of that OCE-sponsored Agency-wide systems engineering initiative.

In 1995, SP-6105 was initially published to bring the fundamental concepts and techniques of systems engineering to NASA personnel in a way that recognized the nature of NASA systems and the NASA environment. This revision (Rev 2) of SP-6105 maintains that original philosophy while updating the Agency's systems engineering body of knowledge, providing guidance for insight into current best Agency practices, and maintaining the alignment of the handbook with the Agency's systems engineering policy.

The update of this handbook continues the methodology of the previous revision: a top-down compatibility with higher level Agency policy and a bottom-up infusion of guidance from the NASA practitioners in the field. This approach provides the opportunity to obtain best practices from across NASA and bridge the information to the established NASA systems engineering processes and to communicate principles of good practice as well as alternative approaches rather than specify a particular way to accomplish a task. The result embodied in this handbook is a top-level implementation approach on the practice of systems engineering unique to NASA. Material used for updating this handbook has been drawn from many sources, including NPRs, Center systems engineering handbooks and processes, other Agency best practices, and external systems engineering textbooks and guides.

This handbook consists of six chapters: (1) an introduction, (2) a systems engineering fundamentals discussion, (3) the NASA program/project life cycles, (4) systems engineering processes to get from a concept to a design, (5) systems engineering processes to get from a design to a final product, and (6) crosscutting management processes in systems engineering. The chapters are supplemented by appendices that provide outlines, examples, and further information to illustrate topics in the chapters. The handbook makes extensive use of boxes and figures to define, refine, illustrate, and extend concepts in the chapters.

Finally, it should be noted that this handbook provides top-level guidance for good systems engineering practices; it is not intended in any way to be a directive.

NASA/SP-2016-6105 Rev2 supersedes SP-2007-6105 Rev 1 dated December, 2007.

Acknowledgments

The following individuals are recognized as contributing practitioners to the content of this expanded guidance:

Alexander, Michael, NASA/Langley Research Center

Allen, Martha, NASA/Marshall Space Flight Center

Baumann, Ethan, NASA/Armstrong Flight Research Center

Bixby, CJ, NASA/Armstrong Flight Research Center

Boland, Brian, NASA/Langley Research Center

Brady, Timothy, NASA/NASA Engineering and Safety Center

Bromley, Linda, NASA/Headquarters/Bromley SE Consulting

Brown, Mark, NASA/Jet Propulsion Laboratory

Brumfield, Mark, NASA/Goddard Space Flight Center

Campbell, Paul, NASA/Johnson Space Center

Carek, David, NASA/Glenn Research Center

Cox, Renee, NASA/Marshall Space Flight Center

Crable, Vicki, NASA/Glenn Research Center

Crocker, Alan, NASA/Ames Research Center

DeLoof, Richard, NASA/Glenn Research Center

Demo, Andrew/Ames Research Center

Dezfuli, Homayoon, NASA/HQ

Diehl, Roger, NASA/Jet Propulsion Laboratory

DiPietro, David, NASA/Goddard Space Flight Center

Doehne, Thomas, NASA/Glenn Research Center

Duarte, Alberto, NASA/Marshall Space Flight Center

Durham, David, NASA/Jet Propulsion Laboratory

Epps, Amy, NASA/Marshall Space Flight Center

Fashimpaur, Karen, Vantage Partners

Feikema, Douglas, NASA/Glenn Research Center

Fitts, David, NASA/Johnson Space Flight Center

Foster, Michele, NASA/Marshall Space Flight Center

Fuller, David, NASA/Glenn Research Center

Gati, Frank, NASA/Glenn Research Center

Gefert, Leon, NASA/Glenn Research Center

Ghassemieh, Shakib, NASA/Ames Research Center

Grantier, Julie, NASA/Glenn Research Center

Hack, Kurt, NASA/Glenn Research Center

Hall, Kelly, NASA/Glenn Research Center

Hamaker, Franci, NASA/Kennedy Space Center

Hange, Craig, NASA/Ames Research Center

Henry, Thad, NASA/Marshall Space Flight Center

Hill, Nancy, NASA/Marshall Space Flight Center

Hirshorn, Steven, NASA/Headquarters

Holladay, Jon, NASA/NASA Engineering and Safety Center

Hyatt, Mark, NASA/Glenn Research Center

Killebrew, Jana, NASA/Ames Research Center

Jannette, Tony, NASA/Glenn Research Center

Jenks, Kenneth, NASA/Johnson Space Center

Jones, Melissa, NASA/Jet Propulsion Laboratory

Jones, Ross, NASA/Jet Propulsion Laboratory

Killebrew, Jana, NASA/Ames Research Center

Leitner, Jesse, NASA/Goddard Space Flight Center

Lin, Chi, NASA/Jet Propulsion Laboratory

Mascia, Anne Marie, Graphic Artist

McKay, Terri, NASA/Marshall Space Flight Center

McNelis, Nancy, NASA/Glenn Research Center

Mendoza, Donald, NASA/Ames Research Center

Miller, Scott, NASA/Ames Research Center

Montgomery, Patty, NASA/Marshall Space Flight Center

Mosier, Gary, NASA/Goddard Space Flight Center

Noble, Lee, NASA/Langley Research Center

Oleson, Steven, NASA/Glenn Research Center

Parrott, Edith, NASA/Glenn Research Center

Powell, Christine, NASA/Stennis Space Center

Powell, Joseph, NASA/Glenn Research Center

Price, James, NASA/Langley Research Center

Rawlin, Adam, NASA/Johnson Space Center

Rochlis-Zumbado, Jennifer, NASA/Johnson Space Center

Rohn, Dennis, NASA/Glenn Research Center

Rosenbaum, Nancy, NASA/Goddard Space Flight Center

Ryan, Victoria, NASA/Jet Propulsion Laboratory
Sadler, Gerald, NASA/Glenn Research Center
Salazar, George, NASA/Johnson Space Center
Sanchez, Hugo, NASA/Ames Research Center
Schuyler, Joseph, NASA/Stennis Space Center
Sheehe, Charles, NASA/Glenn Research Center
Shepherd, Christena, NASA/Marshall Space Flight Center
Shull, Thomas, NASA/Langley Research Center
Singer, Bart, NASA/Langley Research Center
Slywczak, Richard, NASA/Glenn Research Center
Smith, Scott, NASA/Goddard Space Flight Center
Smith, Joseph, NASA/Headquarters
Sprague, George, NASA/Jet Propulsion Laboratory
Trase, Kathryn, NASA/Glenn Research Center
Trenkle, Timothy, NASA/Goddard Space Flight Center
Vipavetz, Kevin, NASA/Langley Research Center
Voss, Linda, Dell Services
Walters, James Britton, NASA/Johnson Space Center
Watson, Michael, NASA/Marshall Space Flight Center
Weiland, Karen, NASA/Glenn Research Center
Wiedeman, Grace, Dell Services
Wiedenmannott, Ulrich, NASA/Glenn Research Center
Witt, Elton, NASA/Johnson Space Center
Woytach, Jeffrey, NASA/Glenn Research Center
Wright, Michael, NASA/Marshall Space Flight Center
Yu, Henry, NASA/Kennedy Space Center

1.1 Purpose

This handbook is intended to provide general guidance and information on systems engineering that will be useful to the NASA community. It provides a generic description of Systems Engineering (SE) as it should be applied throughout NASA. A goal of the handbook is to increase awareness and consistency across the Agency and advance the practice of SE. This handbook provides perspectives relevant to NASA and data particular to NASA.

This handbook should be used as a companion for implementing NPR 7123.1, Systems Engineering Processes and Requirements, as well as the Center-specific handbooks and directives developed for implementing systems engineering at NASA. It provides a companion reference book for the various systems engineering-related training being offered under NASA's auspices.

1.2 Scope and Depth

This handbook describes systems engineering best practices that should be incorporated in the development and implementation of large and small NASA programs and projects. The engineering of NASA systems requires a systematic and disciplined set of processes that are applied recursively and iteratively for the design, development, operation, maintenance, and closeout of systems throughout the life cycle of the programs and projects. The scope of this handbook includes systems engineering functions regardless of whether they are performed by a manager or an engineer, in-house or by a contractor.

There are many Center-specific handbooks and directives as well as textbooks that can be consulted for in-depth tutorials. For guidance on systems engineering for information technology projects, refer to Office of Chief Information Officer *Information Technology Systems Engineering Handbook Version 2.0.* For guidance on entrance and exit criteria for milestone reviews of software projects, refer to *NASA-HDBK-2203, NASA Software Engineering Handbook.* A NASA systems engineer can also participate in the NASA Engineering Network (NEN) Systems Engineering Community of Practice, located at *https://nen.nasa.gov/web/se*. This Web site includes many resources useful to systems engineers, including document templates for many of the work products and milestone review presentations required by the NASA SE process.

This handbook is applicable to NASA space flight projects of all sizes and to research and development programs and projects. While all 17 processes are applicable to all projects, the amount of formality, depth of documentation, and timescales are varied as appropriate for the type, size, and complexity of the project. References to "documents" are intended to include not only paper or digital files but also models, graphics, drawings, or other appropriate forms that capture the intended information.

For a more in-depth discussion of the principles provided in this handbook, refer to the NASA Expanded Guidance for SE document which can be found at *https://nen.nasa.gov/web/se/doc-repository*. This handbook is an abridged version of that reference.

2.0 Fundamentals of Systems Engineering

At NASA, "systems engineering" is defined as a methodical, multi-disciplinary approach for the design, realization, technical management, operations, and retirement of a system. A "system" is the combination of elements that function together to produce the capability required to meet a need. The elements include all hardware, software, equipment, facilities, personnel, processes, and procedures needed for this purpose; that is, all things required to produce system-level results. The results include system-level qualities, properties, characteristics, functions, behavior, and performance. The value added by the system as a whole, beyond that contributed independently by the parts, is primarily created by the relationship among the parts; that is, how they are interconnected.[1] It is a way of looking at the "big picture" when making technical decisions. It is a way of achieving stakeholder functional, physical, and operational performance requirements in the intended use environment over the planned life of the system within cost, schedule, and other constraints. It is a methodology that supports the containment of the life cycle cost of a system. In other words, systems engineering is a logical way of thinking.

1 Eberhardt Rechtin, *Systems Architecting of Organizations: Why Eagles Can't Swim.*

Systems engineering is the art and science of developing an operable system capable of meeting requirements within often opposed constraints. Systems engineering is a holistic, integrative discipline, wherein the contributions of structural engineers, electrical engineers, mechanism designers, power engineers, human factors engineers, and many more disciplines are evaluated and balanced, one against another, to produce a coherent whole that is not dominated by the perspective of a single discipline.[2]

Systems engineering seeks a safe and balanced design in the face of opposing interests and multiple, sometimes conflicting constraints. The systems engineer should develop the skill for identifying and focusing efforts on assessments to optimize the overall design and not favor one system/subsystem at the expense of another while constantly validating that the goals of the operational system will be met. The art is in knowing when and where to probe. Personnel with these skills are usually tagged as "systems engineers." They may have other titles—lead systems engineer,

2 Comments on systems engineering throughout Chapter 2.0 are extracted from the speech "System Engineering and the Two Cultures of Engineering" by Michael D. Griffin, NASA Administrator.

technical manager, chief engineer—but for this document, the term "systems engineer" is used.

The exact role and responsibility of the systems engineer may change from project to project depending on the size and complexity of the project and from phase to phase of the life cycle. For large projects, there may be one or more systems engineers. For small projects, the project manager may sometimes perform these practices. But whoever assumes those responsibilities, the systems engineering functions should be performed. The actual assignment of the roles and responsibilities of the named systems engineer may also therefore vary. The lead systems engineer ensures that the system technically fulfills the defined needs and requirements and that a proper systems engineering approach is being followed. The systems engineer oversees the project's systems engineering activities as performed by the technical team and directs, communicates, monitors, and coordinates tasks. The systems engineer reviews and evaluates the technical aspects of the project to ensure that the systems/subsystems engineering processes are functioning properly and evolves the system from concept to product. The entire technical team is involved in the systems engineering process.

The systems engineer usually plays the key role in leading the development of the concept of operations (ConOps) and resulting system architecture, defining boundaries, defining and allocating requirements, evaluating design tradeoffs, balancing technical risk between systems, defining and assessing interfaces, and providing oversight of verification and validation activities, as well as many other tasks. The systems engineer typically leads the technical planning effort and has the prime responsibility in documenting many of the technical plans, requirements and specification documents, verification and validation documents, certification packages, and other technical documentation.

In summary, the systems engineer is skilled in the art and science of balancing organizational, cost, and technical interactions in complex systems. The systems engineer and supporting organization are vital to supporting program and Project Planning and Control (PP&C) with accurate and timely cost and schedule information for the technical activities. Systems engineering is about tradeoffs and compromises; it uses a broad crosscutting view of the system rather than a single discipline view. Systems engineering is about looking at the "big picture" and not only ensuring that they get the design right (meet requirements) but that they also get the right design (enable operational goals and meet stakeholder expectations).

Systems engineering plays a key role in the project organization. Managing a project consists of three main objectives: managing the technical aspects of the project, managing the project team, and managing the cost and schedule. As shown in FIGURE 2.0-1, these three functions are interrelated. Systems engineering is focused on the technical characteristics of decisions including technical, cost, and schedule and on providing these to the project manager. The Project Planning and Control (PP&C) function is responsible for identifying and controlling the cost and schedules of the project. The project manager has overall responsibility for managing the project team and ensuring that the project delivers a technically correct system within cost and schedule. Note that there are areas where the two cornerstones of project management, SE and PP&C, overlap. In these areas, SE provides the technical aspects or inputs whereas PP&C provides the programmatic, cost, and schedule inputs.

This document focuses on the SE side of the diagram. The practices/processes are taken from NPR 7123.1, NASA Systems Engineering Processes and Requirements. Each process is described in much greater detail in subsequent chapters of this

PROJECT MANAGEMENT

PROJECT MANAGEMENT ACTIVITIES

- Setting up Project Team
- Programmatic Stakeholders (non-technical, non-business)
- Programmatic Planning (non-technical, non-business)
- Identifying Programmatic (non-technical) requirements
- Identifying Programmatic Risks
- Technology Transfer and Commercialization
- Integration of technical and non-technical activities
- Overall Approver/Decider

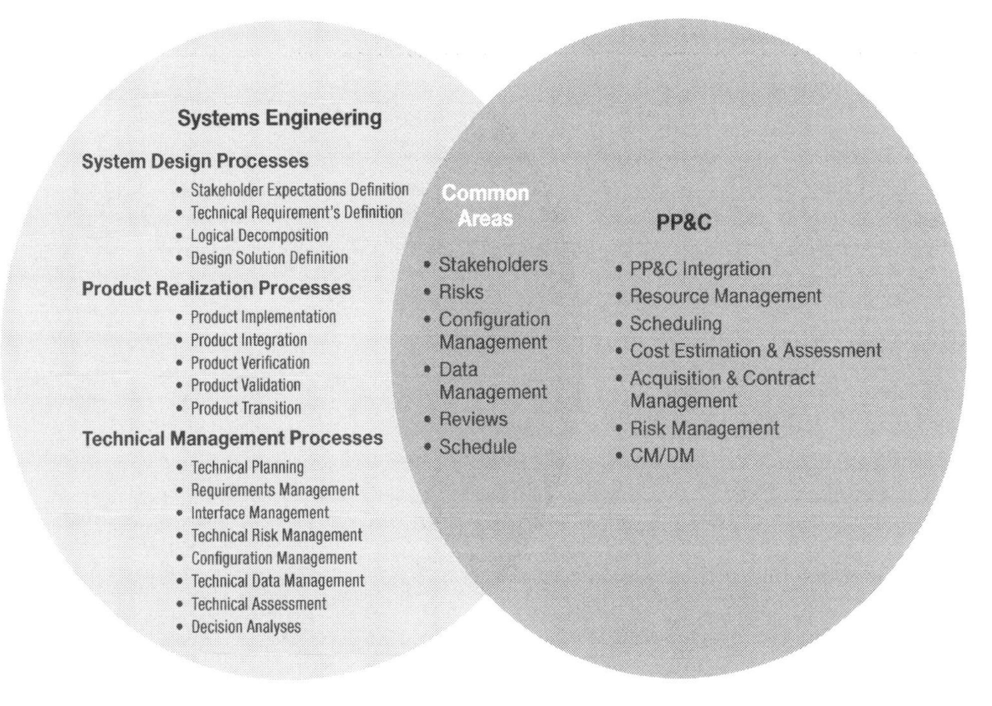

FIGURE 2.0-1 SE in Context of Overall Project Management

document, but an overview is given in the following subsections of this chapter.

2.1 The Common Technical Processes and the SE Engine

There are three sets of common technical processes in NPR 7123.1, NASA Systems Engineering Processes and Requirements: system design, product realization, and technical management. The processes in each set and their interactions and flows are illustrated by the NPR systems engineering "engine" shown in FIGURE 2.1-1. The processes of the SE engine are used to develop and realize the end products. This chapter provides the application context of the 17 common technical processes required in NPR7123.1. The system design processes, the product realization processes, and the technical management processes are discussed in more detail in *Chapters 4.0*, *5.0*, and *6.0*, respectively. Processes 1 through 9 indicated in FIGURE 2.1-1 represent the tasks in the execution of a project. Processes 10 through17 are crosscutting tools for carrying out the processes.

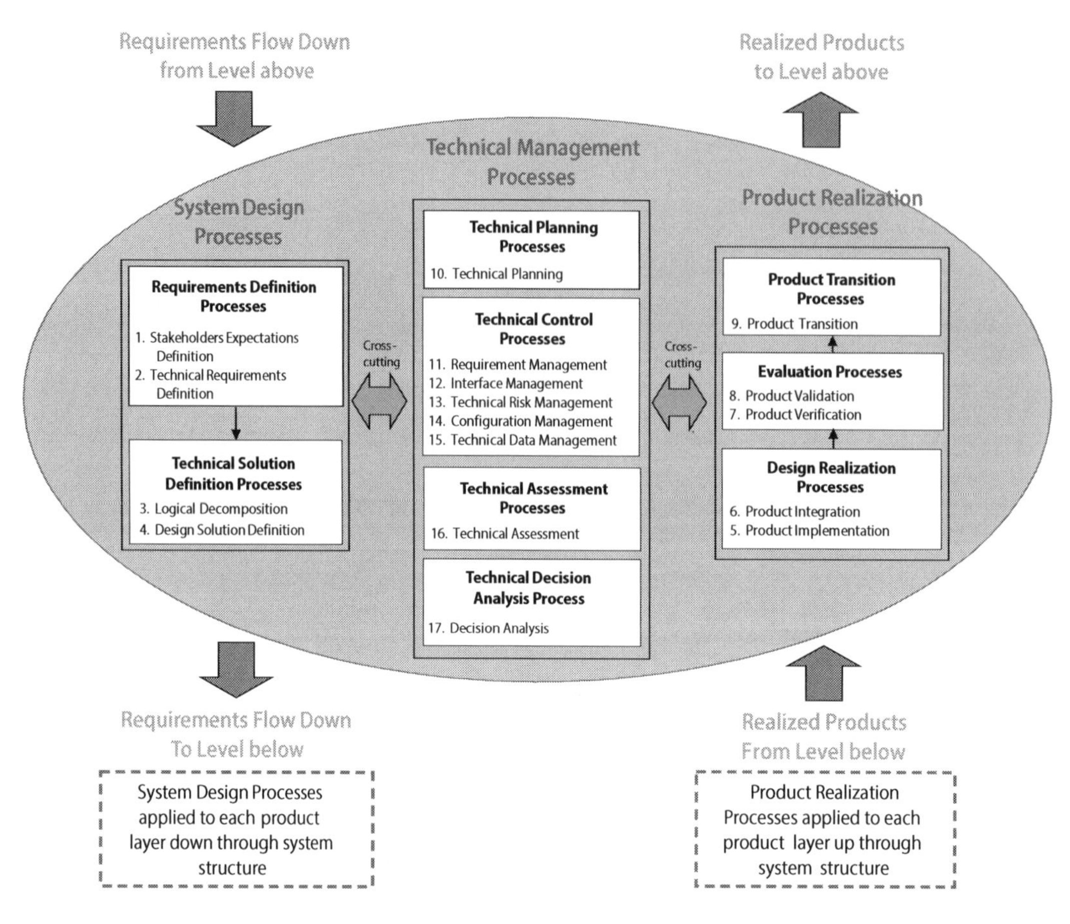

FIGURE 2.1-1 The Systems Engineering Engine (NPR 7123.1)

- **System Design Processes:** The four system design processes shown in FIGURE 2.1-1 are used to define and baseline stakeholder expectations, generate and baseline technical requirements, decompose the requirements into logical and behavioral models, and convert the technical requirements into a design solution that will satisfy the baselined stakeholder expectations. These processes are applied to each product of the system structure from the top of the structure to the bottom until the lowest products in any system structure branch are defined to the point where they can be built, bought, or reused. All other products in the system structure are realized by implementation or integration.

- **Product Realization Processes:** The product realization processes are applied to each operational/mission product in the system structure starting from the lowest level product and working up to higher level integrated products. These processes are used to create the design solution for each product (through buying, coding, building, or reusing) and to verify, validate, and transition up to the next hierarchical level those products that satisfy their design solutions and meet stakeholder expectations as a function of the applicable life cycle phase.

- **Technical Management Processes:** The technical management processes are used to establish and evolve technical plans for the project, to manage communication across interfaces, to assess progress against the plans and requirements for the system products or services, to control technical execution of the project through to completion, and to aid in the decision-making process.

The processes within the SE engine are used both iteratively and recursively. As defined in NPR 7123.1, "iterative" is the "application of a process to the same product or set of products to correct a discovered discrepancy or other variation from requirements," whereas "recursive" is defined as adding value to the system "by the repeated application of processes to design next lower layer system products or to realize next upper layer end products within the system structure. This also applies to repeating application of the same processes to the system structure in the next life cycle phase to mature the system definition and satisfy phase success criteria." The technical processes are applied recursively and iteratively to break down the initializing concepts of the system to a level of detail concrete enough that the technical team can implement a product from the information. Then the processes are applied recursively and iteratively to

TABLE 2.1-1 Alignment of the 17 SE Processes to AS9100

SE Process	AS9100 Requirement
Stakeholder Expectations	Customer Requirements
Technical Requirements Definition	Planning of Product Realization
Logical Decomposition	Design and Development Input
Design Solution Definition	Design and Development Output
Product Implementation	Control of Production
Product Integration	Control of Production
Product Verification	Verification
Product Validation	Validation
Product Transition	Control of Work Transfers; Post Delivery Support, Preservation of Product
Technical Planning	Planning of Product Realization; Review of Requirements; Measurement, Analysis and Improvement
Requirements Management	Design and Development Planning; Purchasing
Interface Management	Configuration Management
Technical Risk Management	Risk Management
Configuration Management	Configuration Management; Identification and Traceability; Control of Nonconforming Product
Technical Data Management	Control of Documents; Control of Records; Control of Design and Development Changes
Technical Assessment	Design and Development Review
Decision Analysis	Measurement, Analysis and Improvement; Analysis of Data

integrate the smallest product into greater and larger systems until the whole of the system or product has been assembled, verified, validated, and transitioned.

For a detailed example of how the SE Engine could be used, refer to the NASA Expanded Guidance for SE document at *https://nen.nasa.gov/web/se/doc-repository.*

AS9100 is a widely adopted and standardized quality management system developed for the commercial aerospace industry. Some NASA Centers have chosen to certify to the AS9100 quality system and may require their contractors to follow NPR 7123.1. TABLE 2.1-1 shows how the 17 NASA SE processes align with AS9100.

2.2 An Overview of the SE Engine by Project Phase

FIGURE 2.2-1 conceptually illustrates how the SE engine is used during each phase of a project (Pre-Phase A through Phase F). The life cycle phases are described in TABLE 2.2-1. FIGURE 2.2-1 is a *conceptual* diagram. For full details, refer to the poster version of this figure, which is located at *https://nen.nasa.gov/web/se/doc-repository.*

The uppermost horizontal portion of this chart is used as a reference to project system maturity, as the project progresses from a feasible concept to an as-deployed system; phase activities; Key Decision Points (KDPs); and major project reviews. The next major horizontal band shows the technical development

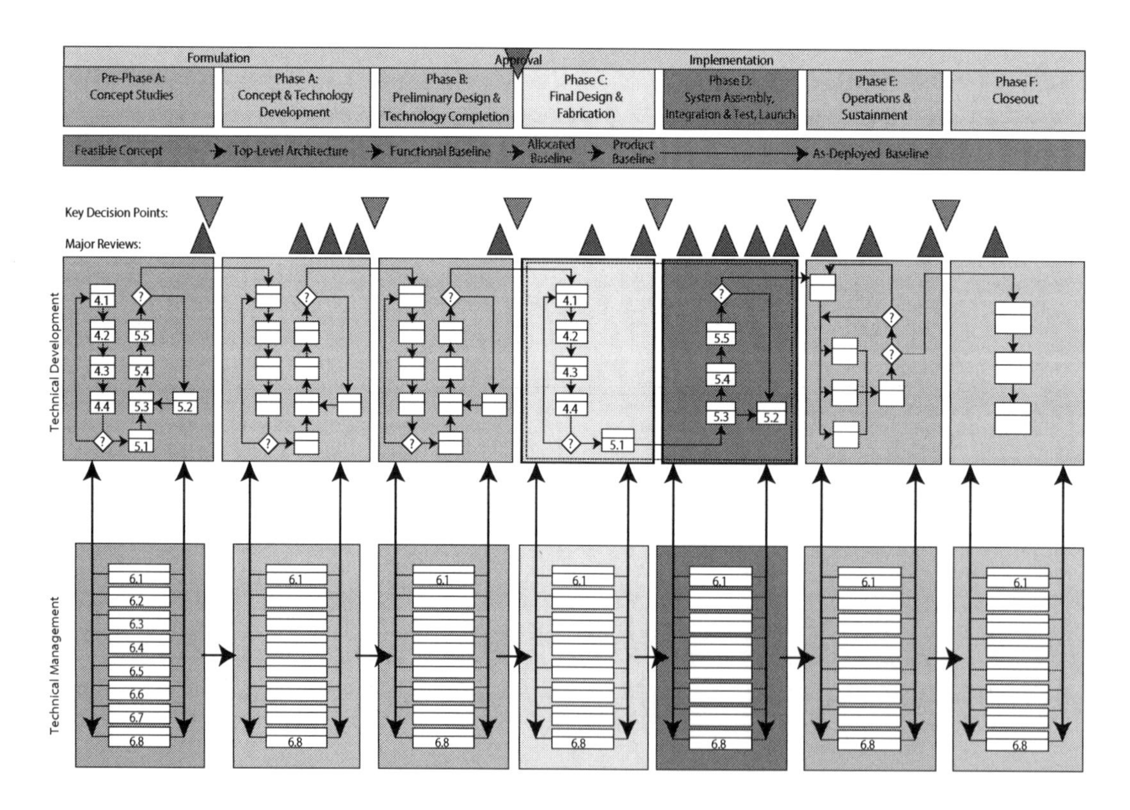

FIGURE 2.2-1 Miniature Version of the Poster-Size NASA Project Life Cycle Process Flow for Flight and Ground Systems Accompanying this Handbook

TABLE 2.2-1 Project Life Cycle Phases

Phase		Purpose	Typical Outcomes
Pre-Formulation	**Pre-Phase A** Concept Studies	To produce a broad spectrum of ideas and alternatives for missions from which new programs/projects can be selected. Determine feasibility of desired system, develop mission concepts, draft system-level requirements, assess performance, cost, and schedule feasibility; identify potential technology needs, and scope.	Feasible system concepts in the form of simulations, analysis, study reports, models, and mock-ups
Formulation	**Phase A** Concept and Technology Development	To determine the feasibility and desirability of a suggested new system and establish an initial baseline compatibility with NASA's strategic plans. Develop final mission concept, system-level requirements, needed system technology developments, and program/project technical management plans.	System concept definition in the form of simulations, analysis, engineering models and mock-ups, and trade study definition
Formulation	**Phase B** Preliminary Design and Technology Completion	To define the project in enough detail to establish an initial baseline capable of meeting mission needs. Develop system structure end product (and enabling product) requirements and generate a preliminary design for each system structure end product.	End products in the form of mock-ups, trade study results, specification and interface documents, and prototypes
Implementation	**Phase C** Final Design and Fabrication	To complete the detailed design of the system (and its associated subsystems, including its operations systems), fabricate hardware, and code software. Generate final designs for each system structure end product.	End product detailed designs, end product component fabrication, and software development
Implementation	**Phase D** System Assembly, Integration and Test, Launch	To assemble and integrate the system (hardware, software, and humans), meanwhile developing confidence that it is able to meet the system requirements. Launch and prepare for operations. Perform system end product implementation, assembly, integration and test, and transition to use.	Operations-ready system end product with supporting related enabling products
Implementation	**Phase E** Operations and Sustainment	To conduct the mission and meet the initially identified need and maintain support for that need. Implement the mission operations plan.	Desired system
Implementation	**Phase F** Closeout	To implement the systems decommissioning/disposal plan developed in Phase E and perform analyses of the returned data and any returned samples.	Product closeout

processes (steps 1 through 9) in each project phase. The SE engine cycles five times from Pre-Phase A through Phase D. Note that NASA's management has structured Phases C and D to "split" the technical development processes in half in Phases C and D to ensure closer management control. The engine is bound by a dashed line in Phases C and D. Once a project enters into its operational state (Phase E) and closes out (Phase F), the technical work shifts to activities commensurate with these last two project phases. The next major horizontal band shows the eight technical management processes (steps 10 through 17) in each project phase. The SE engine cycles the technical management processes seven times from Pre-Phase A through Phase F.

2.3 Example of Using the SE Engine

In Pre-Phase A, the SE engine is used to develop the initial concepts; clearly define the unique roles of humans, hardware, and software in performing the missions objectives; establish the system functional and performance boundaries; develop/identify a preliminary/draft set of key high-level requirements, define one or more initial Concept of Operations (ConOps) scenarios; realize these concepts through iterative modeling, mock-ups, simulation, or other means; and verify and validate that these concepts and products would be able to meet the key high-level requirements and ConOps. The operational concept must include scenarios for all significant operational situations, including known off-nominal situations. To develop a useful and complete set of scenarios, important malfunctions and degraded-mode operational situations must be considered. The importance of early ConOps development cannot be underestimated. As system requirements become more detailed and contain more complex technical information, it becomes harder for the stakeholders and users to understand what the requirements are conveying; i.e., it may become more difficult to visualize the end product. The ConOps can serve as a check in identifying missing or conflicting requirements.

Note that this Pre-Phase A initial concepts development work is not the formal verification and validation program that is performed on the final product, but rather it is a methodical run through ensuring that the concepts that are being developed in this Pre-Phase A are able to meet the likely requirements and expectations of the stakeholders. Concepts are developed to the lowest level necessary to ensure that they are feasible and to a level that reduces the risk low enough to satisfy the project. Academically, this process could proceed down to the circuit board level for every system; however, that would involve a great deal of time and money. There may be a higher level or tier of product than circuit board level that would enable designers to accurately determine the feasibility of accomplishing the project, which is the purpose of Pre-Phase A.

During Phase A, the recursive use of the SE engine is continued, this time taking the concepts and draft key requirements that were developed and validated during Pre-Phase A and fleshing them out to become the set of baseline system requirements and ConOps. During this phase, key areas of high risk might be simulated to ensure that the concepts and requirements being developed are good ones and to identify verification and validation tools and techniques that will be needed in later phases.

During Phase B, the SE engine is applied recursively to further mature requirements and designs for all products in the developing product tree and perform verification and validation of concepts to ensure that the designs are able to meet their requirements. Operational designs and mission scenarios are evaluated and feasibility of execution within design capabilities and cost estimates are assessed.

Phase C again uses the left side of the SE engine to finalize all requirement updates, finalize the ConOps validation, develop the final designs to the lowest level of the product tree, and begin fabrication.

Phase D uses the right side of the SE engine to recursively perform the final implementation, integration, verification, and validation of the end product, and at the final pass, transition the end product to the user.

The technical management processes of the SE engine are used in Phases E and F to monitor performance; control configuration; and make decisions associated with the operations, sustaining engineering, and closeout of the system. Any new capabilities or upgrades of the existing system reenter the SE engine as new developments.

2.4 Distinctions between Product Verification and Product Validation

From a process perspective, the Product Verification and Product Validation processes may be similar in nature, but the objectives are fundamentally different:

- Verification of a product shows proof of compliance with requirements—that the product can meet each "shall" statement as proven though performance of a test, analysis, inspection, or demonstration (or combination of these).
- Validation of a product shows that the product accomplishes the intended purpose in the intended environment—that it meets the expectations of the customer and other stakeholders as shown through performance of a test, analysis, inspection, or demonstration.

Verification testing relates back to the approved requirements set and can be performed at different stages in the product life cycle. The approved specifications, drawings, parts lists, and other configuration documentation establish the configuration baseline of that product, which may have to be modified at a later time. Without a verified baseline and appropriate configuration controls, later modifications could be costly or cause major performance problems.

Validation relates back to the ConOps document. Validation testing is conducted under realistic conditions (or simulated conditions) on end products for the purpose of determining the effectiveness and suitability of the product for use in mission operations by typical users. Validation can be performed in each development phase using phase products (e.g., models) and not only at delivery using end products.

It is appropriate for verification and validation methods to differ between phases as designs advance. The ultimate success of a program or project may relate to the frequency and diligence of validation efforts during the design process, especially in Pre-Phase A and Phase A during which corrections in the direction of product design might still be made cost-effectively. The question should be continually asked, "Are we building the right product for our users and other stakeholders?" The selection of the verification or validation method is based on engineering judgment as to which is the most effective way to reliably show the product's conformance to requirements or that it will operate as intended and described in the ConOps.

2.5 Cost Effectiveness Considerations

The objective of systems engineering is to see that the system is designed, built, and can be operated so that it accomplishes its purpose safely in the most cost-effective way possible considering performance, cost, schedule, and risk. A cost-effective and safe system should provide a particular kind of balance between effectiveness and cost. This causality is an indefinite one because there are usually many designs that meet the cost-effective condition.

Design trade studies, an important part of the systems engineering process, often attempt to find designs that provide the best combination of cost and effectiveness. At times there are alternatives that either reduce costs without reducing effectiveness or increase effectiveness without increasing cost. In such "win-win" cases, the systems engineer's decision is easy. When the alternatives in a design trade study require trading cost for effectiveness, the decisions become harder.

THE SYSTEMS ENGINEER'S DILEMMA

At each cost-effective solution:

- To reduce cost at constant risk, performance must be reduced.
- To reduce risk at constant cost, performance must be reduced.
- To reduce cost at constant performance, higher risks must be accepted.
- To reduce risk at constant performance, higher costs must be accepted.

In this context, time in the schedule is often a critical resource, so that *schedule* behaves like a kind of *cost*.

FIGURE 2.5-1 shows that the life cycle costs of a program or project tend to get "locked in" early in design and development. The cost curves clearly show that late identification of and fixes to problems cost considerably more later in the life cycle. Conversely, descopes taken later versus earlier in the project life cycle result in reduced cost savings. This figure, obtained from the Defense Acquisition University, is an example of how these costs are determined by the early concepts and designs. The numbers will vary from project to project, but the general shape of the curves and the message they send will be similar. For example, the figure shows that during design, only about 15% of the costs might be expended, but the design itself will commit about 75% of the life cycle costs. This is because the way the system is designed will determine how expensive it will be to test, manufacture, integrate, operate, and sustain. If these factors have not been considered during design, they pose significant cost risks later in the life cycle. Also note that the cost to change the design increases as you get later in the life cycle. If the project waits until verification to do any type of test or analysis, any problems found will have a significant cost impact to redesign and reverify.

The technical team may have to choose among designs that differ in terms of numerous attributes. A variety of methods have been developed that can be used to help uncover preferences between attributes and to quantify subjective assessments of relative value. When this can be done, trades between attributes can be assessed quantitatively. Often, however, the attributes are incompatible. In the end, decisions need to be made in spite of the given variety of attributes. There are several decision analysis techniques (*Section 6.8*) that can aid in complex decision analysis. The systems engineer should always keep in mind the information that needs to be available to help the decision-makers choose the most cost-effective option.

2.6 Human Systems Integration (HSI) in the SE Process

As noted at the beginning of NPR 7123.1, the "systems approach is applied to all elements of a system (i.e., hardware, software, human systems integration. In short, the systems engineering approach must equally address and integrate these three key elements: hardware, software, and human systems

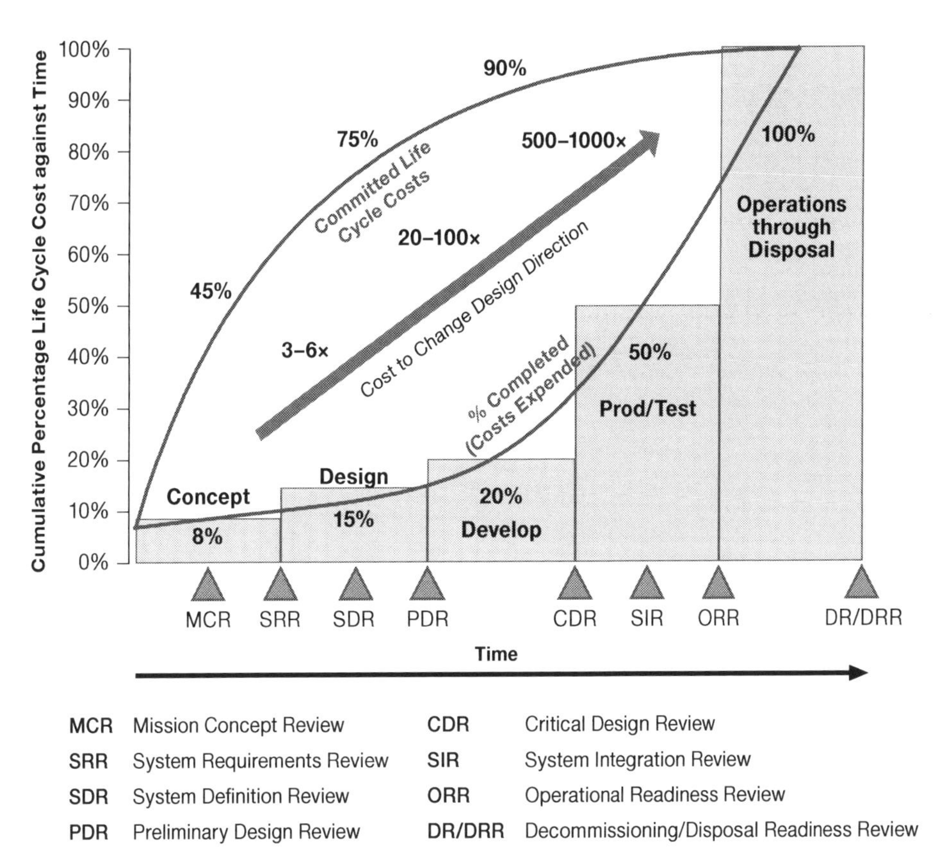

MCR	Mission Concept Review	CDR	Critical Design Review
SRR	System Requirements Review	SIR	System Integration Review
SDR	System Definition Review	ORR	Operational Readiness Review
PDR	Preliminary Design Review	DR/DRR	Decommissioning/Disposal Readiness Review

Adapted from INCOSE-TP-2003-002-04, 2015

FIGURE 2.5-1 Life-Cycle Cost Impacts from Early Phase Decision-Making

integration. Therefore, the human element is something that integration and systems engineering processes must address. The definition of "system" in NPR 7123.1 is inclusive; i.e., a system is "the combination of elements that function together to produce the capability required to meet a need. The elements include all hardware, software, equipment, facilities, personnel, processes, and procedures needed for this purpose. For additional information and guidance on his, refer to Section 2.6 of the NASA Expanded Guidance for Systems Engineering at *https://nen.nasa.gov/web/se/doc-repository*.

2.7 Competency Model for Systems Engineers

TABLE 2.7-1 provides a summary of the Competency Model for Systems Engineering. For more information on the NASA SE Competency model refer to: *http://appel.nasa.gov/competency-model/*.

There are four levels of proficiencies associated with each of these competencies:

- Team Practitioner/Technical Engineer
- Team Lead/Subsystem Lead
- Project Systems Engineer
- Chief Engineer

TABLE 2.7-1 NASA System Engineering Competency Model

Competency Area	Competency	Description
SE 1.0 System Design	SE 1.1 Stakeholder Expectation Definition & Management	Eliciting and defining use cases, scenarios, concept of operations and stakeholder expectations. This includes identifying stakeholders, establishing support strategies, establishing a set of Measures of Effectiveness (MOEs), validating stakeholder expectation statements, and obtaining commitments from the customer and other stakeholders, as well as using the baselined stakeholder expectations for product validation during product realization
	SE 1.2 Technical Requirements Definition	Transforming the baseline stakeholder expectations into unique, quantitative, and measurable technical requirements expressed as "shall" statements that can be used for defining the design solution. This includes analyzing the scope of the technical problems to be solved, defining constraints affecting the designs, defining the performance requirements, validating the resulting technical requirement statements, defining the Measures of Performance (MOPs) for each MOE, and defining appropriate Technical Performance Measures (TPMs) by which technical progress will be assessed.
	SE 1.3 Logical Decomposition	Transforming the defined set of technical requirements into a set of logical decomposition models and their associated set of derived technical requirements for lower levels of the system, and for input to the design solution efforts. This includes decomposing and analyzing by function, time, behavior, data flow, object, and other models. It also includes allocating requirements to these decomposition models, resolving conflicts between derived requirements as revealed by the models, defining a system architecture for establishing the levels of allocation, and validating the derived technical requirements.
	SE 1.4 Design Solution Definition	Translating the decomposition models and derived requirements into one or more design solutions, and using the Decision Analysis process to analyze each alternative and for selecting a preferred alternative that will satisfy the technical requirements. A full technical data package is developed describing the selected solution. This includes generating a full design description for the selected solution; developing a set of 'make-to,' 'buy-to,' 'reuse-to,' specifications; and initiating the development or acquisition of system products and enabling products.
SE 2.0 Product Realization	SE 2.1 Product Implementation	Generating a specific product through buying, making, or reusing so as to satisfy the design requirements. This includes preparing the implementation strategy; building or coding the produce; reviewing vendor technical information; inspecting delivered, built, or reused products; and preparing product support documentation for integration.
	SE 2.2 Product Integration	Assembling and integrating lower-level validated end products into the desired end product of the higher-level product. This includes preparing the product integration strategy, performing detailed planning, obtaining products to integrate, confirming that the products are ready for integration, preparing the integration environment, and preparing product support documentation.
	SE 2.3 Product Verification	Proving the end product conforms to its requirements. This includes preparing for the verification efforts, analyzing the outcomes of verification (including identifying anomalies and establishing recommended corrective actions), and preparing a product verification report providing the evidence of product conformance with the applicable requirements.

(continued)

Competency Area	Competency	Description
SE 2.0 Product Realization	SE 2.4 Product Validation	Confirming that a verified end product satisfies the stakeholder expectations for its intended use when placed in its intended environment and ensuring that any anomalies discovered during validation are appropriately resolved prior to product transition. This includes preparing to conduct product validation, performing the product validation, analyzing the results of validation (including identifying anomalies and establishing recommended corrective actions), and preparing a product validation report providing the evidence of product conformance with the stakeholder expectations baseline.
	SE 2.5 Product Transition	Transitioning the verified and validated product to the customer at the next level in the system structure. This includes preparing to conduct product transition, evaluating the product and enabling product readiness for product transition, preparing the product for transition (including handling, storing, and shipping preparation), preparing sites, and generating required documentation to accompany the product
SE 3.0 Technical Management	SE 3.1 Technical Planning	Planning for the application and management of each common technical process, as well as identifying, defining, and planning the technical effort necessary to meet project objectives. This includes preparing or updating a planning strategy for each of the technical processes, and determining deliverable work products from technical efforts; identifying technical reporting requirements; identifying entry and success criteria for technical reviews; identifying product and process measures to be used; identifying critical technical events; defining cross domain interoperability and collaboration needs; defining the data management approach; identifying the technical risks to be addressed in the planning effort; identifying tools and engineering methods to be employed; and defining the approach to acquire and maintain technical expertise needed. This also includes preparing the Systems Engineering Management Plan (SEMP) and other technical plans; obtaining stakeholder commitments to the technical plans; and issuing authorized technical work directives to implement the technical work
	SE 3.2 Requirements Management	Managing the product requirements, including providing bidirectional traceability, and managing changes to establish requirement baselines over the life cycle of the system products. This includes preparing or updating a strategy for requirements management; selecting an appropriate requirements management tool; training technical team members in established requirement management procedures; conducting expectation and requirements traceability audits; managing expectation and requirement changes; and communicating expectation and requirement change information
	SE 3.3 Interface Management	Establishing and using formal interface management to maintain internal and external interface definition and compliance among the end products and enabling products. This includes preparing interface management procedures, identifying interfaces, generating and maintaining interface documentation, managing changes to interfaces, disseminating interface information, and conducting interface control
	SE 3.4 Technical Risk Management	Examining on a continual basis the risks of technical deviations from the plans, and identifying potential technical problems before they occur. Planning, invoking, and performing risk-handling activities as needed across the life of the product or project to mitigate impacts on meeting technical objectives. This includes developing the strategy for technical risk management, identifying technical risks, and conducting technical risk assessment; preparing for technical risk mitigation, monitoring the status of each technical risk, and implementing technical risk mitigation and contingency action plans when applicable thresholds have been triggered.

(continued)

Competency Area	Competency	Description
SE 3.0 Technical Management	SE 3.5 Configuration Management	Identifying the configuration of the product at various points in time, systematically controlling changes to the configuration of the product, maintaining the integrity and traceability of product configuration, and preserving the records of the product configuration throughout its life cycle. This includes establishing configuration management strategies and policies, identifying baselines to be under configuration control, maintaining the status of configuration documentation, and conducting configuration audits
	SE 3.6 Technical Data Management	Identifying and controlling product-related data throughout its life cycle; acquiring, accessing, and distributing data needed to develop, manage, operate, support, and retire system products; managing and disposing data as records; analyzing data use; obtaining technical data feedback for managing the contracted technical efforts; assessing the collection of appropriate technical data and information; maintaining the integrity and security of the technical data, effectively managing authoritative data that defines, describes, analyzes, and characterizes a product life cycle; and ensuring consistent, repeatable use of effective Product Data and Life-cycle Management processes, best practices, interoperability approaches, methodologies, and traceability. This includes establishing technical data management strategies and policies; maintaining revision, status, and history of stored technical data and associated metadata; providing approved, published technical data; providing technical data to authorized parties; and collecting and storing required technical data.
	SE 3.7 Technical Assessment	Monitoring progress of the technical effort and providing status information for support of the system design, product realization, and technical management efforts. This includes developing technical assessment strategies and policies, assessing technical work productivity, assessing product quality, tracking and trending technical metrics, and conducting technical, peer, and life cycle reviews.
	SE 3.8 Technical Decision Analysis	Evaluating technical decision issues, identifying decision criteria, identifying alternatives, analyzing alternatives, and selecting alternatives. Performed throughout the system life cycle to formulate candidate decision alternatives, and evaluate their impacts on health and safety, technical, cost, and schedule performance. This includes establishing guidelines for determining which technical issues are subject to formal analysis processes; defining the criteria for evaluating alternative solutions; identifying alternative solutions to address decision issues; selecting evaluation methods; selecting recommended solutions; and reporting the results and findings with recommendations, impacts, and corrective actions.

3.0 NASA Program/Project Life Cycle

One of the fundamental concepts used within NASA for the management of major systems is the program/project life cycle, which categorizes everything that should be done to accomplish a program or project into distinct phases that are separated by Key Decision Points (KDPs). *KDPs are the events at which the decision authority determines the readiness of a program/project to progress to the next phase of the life cycle (or to the next KDP).* Phase boundaries are defined so that they provide natural points for "go" or "no-go" decisions. Decisions to proceed may be qualified by liens that should be removed within an agreed-to time period. A program or project that fails to pass a KDP may be allowed to try again later after addressing deficiencies that precluded passing the KDP, or it may be terminated.

All systems start with the recognition of a need or the discovery of an opportunity and proceed through various stages of development to the end of the project. While the most dramatic impacts of the analysis and optimization activities associated with systems engineering are obtained in the early stages, decisions that affect cost continue to be amenable to the systems approach even as the end of the system lifetime approaches.

Decomposing the program/project life cycle into phases organizes the entire process into more manageable pieces. The program/project life cycle should provide managers with incremental visibility into the progress being made at points in time that fit with the management and budgetary environments.

For NASA projects, the life cycle is defined in the applicable governing document:

- **For space flight projects:** NPR 7120.5, NASA Space Flight Program and Project Management Requirements
- **For information technology:** NPR 7120.7, NASA Information Technology and Institutional Infrastructure Program and Project Management Requirements
- **For NASA research and technology:** NPR 7120.8, NASA Research and Technology Program and Project Management Requirements
- **For software:** NPR 7150.2 NASA Software Engineering Requirements

For example, NPR 7120.5 defines the major NASA life cycle phases as Formulation and Implementation. For space flight systems projects, the NASA life cycle phases of Formulation and Implementation divide into the following seven incremental pieces. The phases of the project life cycle are:

Program Pre-Formulation:

- **Pre-Phase A:** Concept Studies

Program Formulation

- **Phase A:** Concept and Technology Development
- **Phase B:** Preliminary Design and Technology Completion

Program Implementation:

- **Phase C:** Final Design and Fabrication
- **Phase D:** System Assembly, Integration and Test, Launch
- **Phase E:** Operations and Sustainment
- **Phase F:** Closeout

FIGURE 3.0-1 is taken from NPR 7120.5 and provides the life cycle for NASA space flight projects and identifies the KDPs and reviews that characterize the phases. More information concerning life cycles can be found in the NASA Expanded Guidance for SE document at *https://nen.nasa.gov/web/se/doc-repository* and in the *SP-2014-3705, NASA Space Flight Program and Project Management Handbook.*

TABLE 3.0-1 is taken from NPR 7123.1 and represents the product maturity for the major SE products developed and matured during the product life cycle.

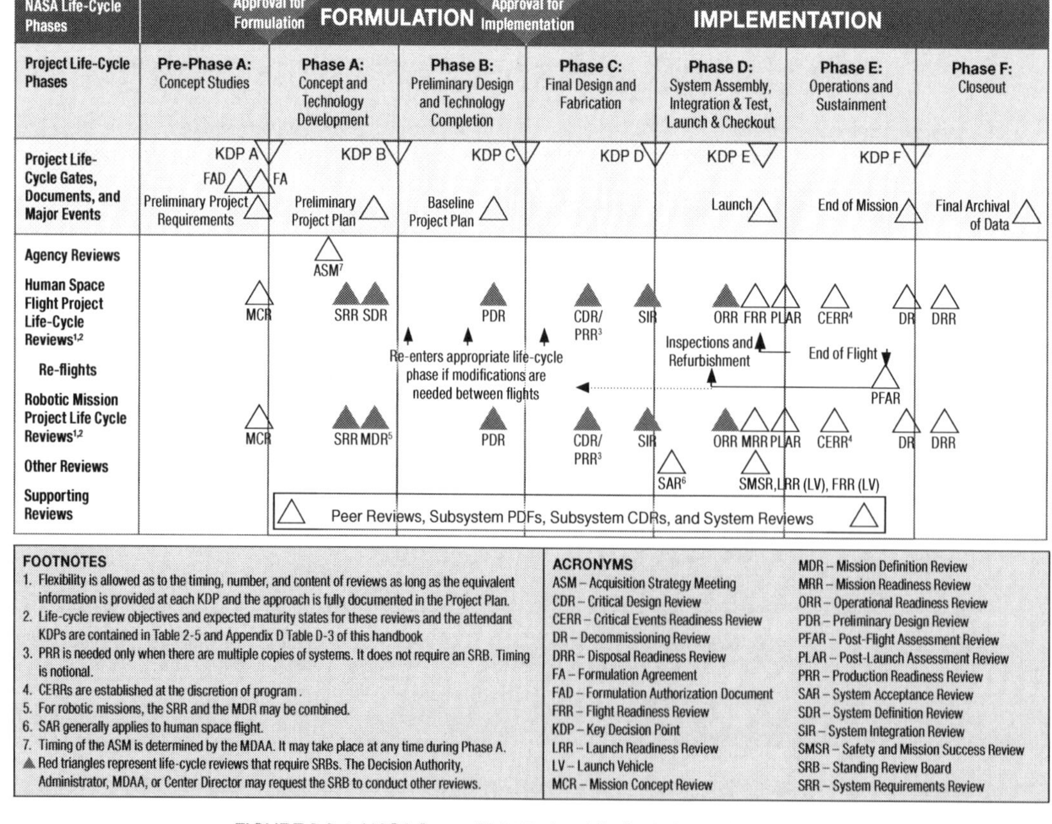

FIGURE 3.0-1 NASA Space Flight Project Life Cycle from NPR 7120.5E

TABLE 3.0-1 SE Product Maturity from NPR 7123.1

Products	Formulation			Implementation						
Uncoupled/ Loosely Coupled	KDP 0		KDP I	Periodic KDPs						
Tightly Coupled Programs	KDP 0			KDP I	KDP II		KDP III		Periodic KDPs	
Projects and Single Project Programs	Pre-Phase A	Phase A		Phase B	Phase C		Phase D		Phase E	Phase F
	KDP A	KDP B		KDP C	KDP D		KDP E		KDP F	
	MCR	SRR	MDR/SDR	PDR	CDR	SIR	ORR	FRR	DR	DRR
Stakeholder identification and	**Baseline	Update	Update	Update						
Concept definition	**Baseline	Update	Update	Update	Update					
Measure of effectiveness definition	**Approve									
Cost and schedule for technical	Initial	Update	Update		Update	Update	Update	Update	Update	Update
SEMP[1]	Preliminary	**Baseline	**Baseline	Update	Update	Update				
Requirements	Preliminary	**Baseline	Update	Update	Update					
Technical Performance Measures definition			**Approve							
Architecture definition			**Baseline							
Allocation of requirements to next lower level			**Baseline							
Required leading indicator trends			**Initial	Update	Update	Update				
Design solution definition			Preliminary	**Preliminary	**Baseline	Update	Update			
Interface definition(s)			Preliminary	Baseline	Update	Update				
Implementation plans (Make/ code, buy, reuse)			Preliminary	Baseline	Update					
Integration plans			Preliminary	Baseline	Update	**Update				
Verification and validation plans	Approach		Preliminary	Baseline	Update	Update				
Verification and validation results						**Initial	**Preliminary	**Baseline		
Transportation criteria and instructions					Initial	Final	Update			
Operations plans				Baseline	Update	Update	**Update			
Operational procedures					Preliminary	Baseline	**Update	Update		
Certification (flight/use)							Preliminary	**Final		
Decommissioning plans				Preliminary	Preliminary	Preliminary	**Baseline	Update	**Update	
Disposal plans				Preliminary	Preliminary	Preliminary	**Baseline	Update	Update	**Update

** Item is a required product for that review

1 SEMP is baselined at SRR for projects, tightly coupled programs and single-project programs, and at MDR/SDR for uncoupled, and loosely coupled programs.

3.1 Program Formulation

The program Formulation Phase establishes a cost-effective program that is demonstrably capable of meeting Agency and mission directorate goals and objectives. The program Formulation Authorization Document (FAD) authorizes a Program Manager (PM) to initiate the planning of a new program and to perform the analyses required to formulate a sound program plan. The lead systems engineer provides the technical planning and concept development or this phase of the program life cycle. Planning includes identifying the major technical reviews that are needed and associated entrance and exit criteria. Major reviews leading to approval at KDP I are the SRR, SDR, PDR, and governing Program Management Council (PMC) review. A summary of the required gate products for the program Formulation Phase can be found in the governing NASA directive (e.g., NPR 7120.5 for space flight programs, NPR 7120.7 for IT projects, NPR 7120.8 for research and technology projects). Formulation for all program types is the same, involving one or more program reviews followed by KDP I where a decision is made approving a program to begin implementation.

3.2 Program Implementation

During the program Implementation phase, the PM works with the Mission Directorate Associate Administrator (MDAA) and the constituent project

SPACE FLIGHT PROGRAM FORMULATION

Purpose

To establish a cost-effective program that is demonstrably capable of meeting Agency and mission directorate goals and objectives

Typical Activities and Their Products for Space Flight Programs

- Identify program stakeholders and users
- Develop program requirements based on user expectations and allocate them to initial projects
- Identify NASA risk classification
- Define and approve program acquisition strategies
- Develop interfaces to other programs
- Start developing technologies that cut across multiple projects within the program
- Derive initial cost estimates and approve a program budget based on the project's life cycle costs
- Perform required program Formulation technical activities defined in NPR 7120.5
- Satisfy program Formulation reviews' entrance/success criteria detailed in NPR 7123.1
- Develop a clear vision of the program's benefits and usage in the operational era and document it in a ConOps

Reviews

- MCR (pre-Formulation)
- SRR
- SDR

SPACE FLIGHT PROGRAM IMPLEMENTATION

Purpose

To execute the program and constituent projects and ensure that the program continues to contribute to Agency goals and objectives within funding constraints

Typical Activities and Their Products

- Initiate projects through direct assignment or competitive process (e.g., Request for Proposal (RFP), Announcement of Opportunity (AO)
- Monitor project's formulation, approval, implementation, integration, operation, and ultimate decommissioning
- Adjust program as resources and requirements change
- Perform required program Implementation technical activities from NPR 7120.5
- Satisfy program Implementation reviews' entrance/success criteria from NPR 7123.1

Reviews

- PSR/PIR (uncoupled and loosely coupled programs only)
- Reviews synonymous (not duplicative) with the project reviews in the project life cycle (see FIGURE 3.0-4) through Phase D (single-project and tightly coupled programs only)

managers to execute the program plan cost-effectively. Program reviews ensure that the program continues to contribute to Agency and mission directorate goals and objectives within funding constraints. A summary of the required gate products for the program Implementation Phase can be found in the governing NASA directive; e.g., NPR 7120.5 for space flight programs. The program life cycle has two different implementation paths, depending on program type. Each implementation path has different types of major reviews. It is important for the systems engineer to know what type of program a project falls under so that the appropriate scope of the technical work, documentation requirements, and set of reviews can be determined.

3.3 Project Pre-Phase A: Concept Studies

The purpose of Pre-Phase A is to produce a broad spectrum of ideas and alternatives for missions from which new programs/projects can be selected. During Pre-Phase A, a study or proposal team analyses a broad range of mission concepts that can fall within technical, cost, and schedule constraints and that contribute to program and Mission Directorate goals and objectives. Pre-Phase A effort could include focused examinations on high-risk or high technology development areas. These advanced studies, along with interactions with customers and other potential stakeholders, help the team to identify promising mission concept(s). The key stakeholders (including the customer) are determined and

SPACE FLIGHT PRE-PHASE A: CONCEPT STUDIES

Purpose

To produce a broad spectrum of ideas and alternatives for missions from which new programs and projects can be selected. Determine feasibility of desired system; develop mission concepts; draft system-level requirements; assess performance, cost, and schedule feasibility; identify potential technology needs and scope.

Typical Activities and Products

- Review/identify any initial customer requirements or scope of work, which may include:
  - Mission
  - Science
  - Top-level system
- Identify and involve users and other stakeholders
  - Identify key stakeholders for each phase of the life cycle
  - Capture and baseline expectations as Needs, Goals, and Objectives (NGOs)
  - Define measures of effectiveness
- Develop and baseline the Concept of Operations
  - Identify and perform trade-offs and analyses of alternatives (AoA)
  - Perform preliminary evaluations of possible missions
- Identify risk classification
- Identify initial technical risks
- Identify the roles and responsibilities in performing mission objectives (i.e., technical team, flight, and ground crew) including training
- Develop plans
  - Develop preliminary SEMP
  - Develop and baseline Technology Development Plan
  - Define preliminary verification and validation approach
- Prepare program/project proposals, which may include:
  - Mission justification and objectives;
  - A ConOps that exhibits clear understanding of how the program's outcomes will cost-effectively satisfy mission objectives;
  - High-level Work Breakdown Structures (WBSs);
  - Life cycle rough order of magnitude (ROM) cost, schedule, and risk estimates; and
  - Technology assessment and maturation strategies.
- Satisfy MCR entrance/success criteria from NPR 7123.1

Reviews

- MCR
- Informal proposal review

expectations for the project are gathered from them. If feasible concepts can be found, one or more may be selected to go into Phase A for further development. Typically, the system engineers are heavily involved in the development and assessment of the concept options. In projects governed by NPR 7120.5, the descope options define what the system can accomplish if the resources are not available to accomplish the entire mission. This could be in the form of fewer instruments, a less ambitious mission profile, accomplishing only a few goals, or using cheaper, less capable technology. Descope options can also reflect what the mission can accomplish in case a hardware failure results in the loss of a portion of the spacecraft architecture; for example, what an orbiter can accomplish after the loss of a lander. The success criteria are reduced to correspond with a descoped mission.

Descope options are developed when the NGOs or other stakeholder expectation documentation is developed. The project team develops a preliminary set of mission descope options as a gate product for the MCR, but these preliminary descope options are not baselined or maintained. They are kept in the documentation archive in case they are needed later in the life cycle.

It is important in Pre-Phase A to define an accurate group of stakeholders and users to help ensure that mission goals and operations concepts meet the needs and expectations of the end users. In addition, it is important to estimate the composition of the technical team and identify any unique facility or personnel requirements.

Advanced studies may extend for several years and are typically focused on establishing mission goals and formulating top-level system requirements and ConOps. Conceptual designs may be developed to demonstrate feasibility and support programmatic estimates. The emphasis is on establishing feasibility and desirability rather than optimality. Analyses and designs are accordingly limited in both depth and number of options, but each option should be evaluated for its implications through the full life cycle, i.e., through Operations and Disposal. It is important in Pre-Phase A to develop and mature a clear vision of what problems the proposed program will address, how it will address them, and how the solution will be feasible and cost-effective.

3.4 Project Phase A: Concept and Technology Development

The purpose of Phase A is to develop a proposed mission/system architecture that is credible and responsive to program expectations, requirements, and constraints on the project, including resources. During Phase A, activities are performed to fully develop a baseline mission concept, begin or assume responsibility for the development of needed technologies, and clarify expected reliance on human elements to achieve full system functionality or autonomous system development. This work, along with interactions with stakeholders, helps mature the mission concept and the program requirements on the project. Systems engineers are heavily involved during this phase in the development and assessment of the architecture and the allocation of requirements to the architecture elements.

In Phase A, a team—often associated with a program or informal project office—readdresses the mission concept first developed in Pre-Phase A to ensure that the project justification and practicality are sufficient to warrant a place in NASA's budget. The team's effort focuses on analyzing mission requirements and establishing a mission architecture. Activities become formal, and the emphasis shifts toward optimizing the concept design. The effort addresses more depth and considers many alternatives. Goals and objectives are

SPACE FLIGHT PHASE A: CONCEPT AND TECHNOLOGY DEVELOPMENT

Purpose

To determine the feasibility and desirability of a suggested new system and establish an initial baseline compatibility with NASA's strategic plans. Develop final mission concept, system-level requirements, needed system technology developments, and program/project technical management plans.

Typical Activities and Their Products

- Review and update documents baselined in Pre-Phase A if needed
- Monitor progress against plans
- Develop and baseline top-level requirements and constraints including internal and external interfaces, integrated logistics and maintenance support, and system software functionality
- Allocate system requirements to functions and to next lower level
- Validate requirements
- Baseline plans
    - > Systems Engineering Management Plan
    - > Human Systems Integration Plan
    - > Control plans such as the Risk Management Plan, Configuration Management Plan, Data Management Plan, Safety and Mission Assurance Plan, and Software Development or Management Plan (See NPR 7150.2)
    - > Other crosscutting and specialty plans such as environmental compliance documentation, acquisition surveillance plan, contamination control plan, electromagnetic interference/ electromagnetic compatibility control plan, reliability plan, quality control plan, parts management plan, logistics plan
- Develop preliminary Verification and Validation Plan
- Establish human rating plan and perform initial evaluations
- Develop and baseline mission architecture
    - > Develop breadboards, engineering units or models identify and reduce high risk concepts
    - > Demonstrate that credible, feasible design(s) exist
    - > Perform and archive trade studies
    - > Initiate studies on human systems interactions
- Initiate environmental evaluation/National Environmental Policy Act process
- Develop initial orbital debris assessment (NASA-STD-8719.14)
- Perform technical management
    - > Provide technical cost estimate and range and develop system-level cost-effectiveness model
    - > Define the WBS
    - > Develop SOWs
    - > Acquire systems engineering tools and models

(continued)

    - > Establish technical resource estimates
- Identify, analyze and update risks
- Perform required Phase A technical activities from NPR 7120.5 as applicable
- Satisfy Phase A reviews' entrance/success criteria from NPR 7123.1

Reviews
- SRR
- MDR/SDR

solidified, and the project develops more definition in the system requirements, top-level system architecture, and ConOps. Conceptual designs and analyses (including engineering units and physical models, as appropriate) are developed and exhibit more engineering detail than in Pre-Phase A. Technical risks are identified in more detail, and technology development needs become focused. A Systems Engineering Management Plan (SEMP) is baselined in Phase A to document how NASA systems engineering requirements and practices of NPR 7123.1 will be addressed throughout the program life cycle.

In Phase A, the effort focuses on allocating functions to particular items of hardware, software, and to humans. System functional and performance requirements, along with architectures and designs, become firm as system tradeoffs and subsystem tradeoffs iterate back and forth, while collaborating with subject matter experts in the effort to seek out more cost-effective designs. A method of determining life cycle cost (i.e., system-level cost-effectiveness model) is refined in order to compare cost impacts for each of the different alternatives. (Trade studies should precede—rather than follow—system design decisions.) Major products to this point include an accepted functional baseline for the system and its major end items. The project team conducts the security categorization of IT systems required by NPR 2810.1 and Federal Information Processing Standard Publication (FIPS PUB) 199. The effort also produces various engineering and management plans to prepare for managing the project's downstream processes such as verification and operations.

3.5 Project Phase B: Preliminary Design and Technology Completion

The purpose of Phase B is for the project team to complete the technology development, engineering prototyping, heritage hardware and software assessments, and other risk-mitigation activities identified in the project Formulation Agreement (FA) and the preliminary design. The project demonstrates that its planning, technical, cost, and schedule baselines developed during Formulation are complete and consistent; that the preliminary design complies with its requirements; that the project is sufficiently mature to begin Phase C; and that the cost and schedule are adequate to enable mission success with acceptable risk. It is at the conclusion of this phase that the project and the Agency commit to accomplishing the project's objectives for a given cost and schedule. For projects with a Life Cycle Cost (LCC) greater than $250 million, this commitment is made with the Congress and the U.S. Office of Management and Budget (OMB). This external commitment is the Agency Baseline Commitment (ABC). Systems

SPACE FLIGHT PHASE B: PRELIMINARY DESIGN AND TECHNOLOGY COMPLETION

Purpose

To define the project in enough detail to establish an initial baseline capable of meeting mission needs. Develop system structure end product (and enabling product) requirements and generate a preliminary design for each system structure end product.

Typical Activities and Their Products

- Review and update documents baselined in previous phases
- Monitor progress against plans
- Develop the preliminary design
  - Identify one or more feasible preliminary designs including internal and external interfaces
  - Perform analyses of candidate designs and report results
  - Conduct engineering development tests as needed and report results
  - Perform human systems integration assessments
  - Select a preliminary design solution
- Develop operations plans based on matured ConOps
  - Define system operations as well as Principal Investigator (PI)/contract proposal management, review, and access and contingency planning
- Report technology development results
- Update cost range estimate and schedule data (Note that after PDR changes are incorporated and costed, at KDP C this will turn into the Agency Baseline Commitment)
- Improve fidelity of models and prototypes used in evaluations
- Identify and update risks
- Develop appropriate level safety data package and security plan
- Develop preliminary plans
  - Orbital Debris Assessment
  - Decommissioning Plan
  - Disposal Plan
- Perform required Phase B technical activities from NPR 7120.5 as applicable
- Satisfy Phase B reviews' entrance/success criteria from NPR 7123.1

Reviews

- PDR
- Safety review

engineers are involved in this phase to ensure the preliminary designs of the various systems will work together, are compatible, and are likely to meet the customer expectations and applicable requirements.

During Phase B, activities are performed to establish an initial project baseline, which (according to NPR 7120.5 and NPR 7123.1) includes "a formal flow down of the project-level performance requirements to a complete set of system and subsystem design specifications for both flight and ground elements" and "corresponding preliminary designs." The technical requirements should be sufficiently detailed to establish firm schedule and cost estimates for the project. It also should be noted, especially for AO-driven projects, that Phase B is where the top-level requirements and the requirements flowed down to the next level are finalized and placed under configuration control. While the requirements should be baselined in Phase A, changes resulting from the trade studies and analyses in late Phase A and early Phase B may result in changes or refinement to system requirements.

It is important in Phase B to validate design decisions against the original goals and objectives and ConOps. All aspects of the life cycle should be considered, including design decisions that affect training, operations resource management, human factors, safety, habitability and environment, and maintainability and supportability.

The Phase B baseline consists of a collection of evolving baselines covering technical and business aspects of the project: system (and subsystem) requirements and specifications, designs, verification and operations plans, and so on in the technical portion of the baseline, and schedules, cost projections, and management plans in the business portion. Establishment of baselines implies the implementation of configuration management procedures. (See *Section 6.5*.)

Phase B culminates in a series of PDRs, containing the system-level PDR and PDRs for lower level end items as appropriate. The PDRs reflect the successive refinement of requirements into designs. Design issues uncovered in the PDRs should be resolved so that final design can begin with unambiguous design-to specifications. From this point on, almost all changes to the baseline are expected to represent successive refinements, not fundamental changes. As noted in FIGURE 2.5-1, significant design changes at and beyond Phase B become increasingly expensive.

3.6 Project Phase C: Final Design and Fabrication

The purpose of Phase C is to complete and document the detailed design of the system that meets the detailed requirements and to fabricate, code, or otherwise realize the products. During Phase C, activities are performed to establish a complete design (product baseline), fabricate or produce hardware, and code software in preparation for integration. Trade studies continue and results are used to validate the design against project goals, objectives, and ConOps. Engineering test units more closely resembling actual hardware are built and tested to establish confidence that the design will function in the expected environments. Human subjects representing the user population participate in operations evaluations of the design, use, maintenance, training procedures, and interfaces. Engineering specialty and crosscutting analysis results are integrated into the design, and the manufacturing process and controls are defined and valid. Systems engineers are involved in this phase to ensure the final detailed designs of the various systems will work together, are compatible, and are likely to meet the customer expectations and applicable requirements. During fabrication, the systems engineer is available to answer questions and work any interfacing issues that might arise.

SPACE FLIGHT PHASE C: FINAL DESIGN AND FABRICATION

Purpose

To complete the detailed design of the system (and its associated subsystems, including its operations systems), fabricate hardware, and code software. Generate final designs for each system structure end product.

Typical Activities and Their Products

- Review and update documents baselined in previous phases
- Monitor progress against plans
- Develop and document hardware and software detailed designs
  > Fully mature and define selected preliminary designs
  > Add remaining lower level design specifications to the system architecture
  > Perform and archive trade studies
  > Perform development testing at the component or subsystem level
  > Fully document final design and develop data package
- Develop/refine and baseline plans
  > Interface definitions
  > Implementation plans
  > Integration plans
  > Verification and validation plans
  > Operations plans
- Develop/refine preliminary plans
  > Decommissioning and disposal plans, including human capital transition
  > Spares
  > Communications (including command and telemetry lists)
- Develop/refine procedures for
  > Refine integration
  > Manufacturing and assembly
  > Verification and validation
- Fabricate (or code) the product
- Identify and update risks
- Monitor project progress against project plans
- Prepare launch site checkout and post launch activation and checkout
- Finalize appropriate level safety data package and updated security plan
- Identify opportunities for preplanned product improvement
- Refine orbital debris assessment
- Perform required Phase C technical activities from NPR 7120.5 as applicable
- Satisfy Phase C review entrance/success criteria from NPR 7123.1

(continued)

Reviews
- CDR
- PRR
- SIR
- Safety review

All the planning initiated back in Phase A for the testing and operational equipment, processes and analysis, integration of the crosscutting and engineering specialty analysis, and manufacturing processes and controls is implemented. Configuration management continues to track and control design changes as detailed interfaces are defined. At each step in the successive refinement of the final design, corresponding integration and verification activities are planned in greater detail. During this phase, technical parameters, schedules, and budgets are closely tracked to ensure that undesirable trends (such as an unexpected growth in spacecraft mass or increase in its cost) are recognized early enough to take corrective action. These activities focus on preparing for the CDR, Production Readiness Review (PRR) (if required), and the SIR.

Phase C contains a series of CDRs containing the system-level CDR and CDRs corresponding to the different levels of the system hierarchy. A CDR for each end item should be held prior to the start of fabrication/production for hardware and prior to the start of coding of deliverable software products. Typically, the sequence of CDRs reflects the integration process that will occur in the next phase; that is, from lower level CDRs to the system-level CDR. Projects, however, should tailor the sequencing of the reviews to meet the needs of the project. If there is a production run of products, a PRR will be performed to ensure the production plans, facilities, and personnel are ready to begin production. Phase C culminates with an SIR. Training requirements and preliminary mission operations procedures are created and baselined. The final product of this phase is a product ready for integration.

3.7 Project Phase D: System Assembly, Integration and Test, Launch

The purpose of Phase D is to assemble, integrate, verify, validate, and launch the system. These activities focus on preparing for the Flight Readiness Review (FRR)/Mission Readiness Review (MRR). Activities include assembly, integration, verification, and validation of the system, including testing the flight system to expected environments within margin. Other activities include updating operational procedures, rehearsals and training of operating personnel and crew members, and implementation of the logistics and spares planning. For flight projects, the focus of activities then shifts to prelaunch integration and launch. System engineering is involved in all aspects of this phase including answering questions, providing advice, resolving issues, assessing results of the verification and validation tests, ensuring that the V&V results meet the customer expectations and applicable requirements, and providing information to decision makers for go/no-go decisions.

The planning for Phase D activities was initiated in Phase A. For IT projects, refer to the *IT Systems Engineering Handbook*. The planning for the activities should be performed as early as possible since

SPACE FLIGHT PHASE D: SYSTEM ASSEMBLY, INTEGRATION AND TEST, LAUNCH

Purpose

To assemble and integrate the system (hardware, software, and humans), meanwhile developing confidence that it will be able to meet the system requirements. Launch and prepare for operations. Perform system end product implementation, assembly, integration and test, and transition to use.

Typical Activities and Their Products

- Update documents developed and baselined in previous phases
- Monitor project progress against plans
- Identify and update risks
- Integrate/assemble components according to the integration plans
- Perform verification and validation on assemblies according to the V&V Plan and procedures
  - Perform system qualification verifications, including environmental verifications
  - Perform system acceptance verifications and validation(s) (e.g., end-to-end tests encompassing all elements; i.e., space element, ground system, data processing system)
  - Assess and approve verification and validation results
  - Resolve verification and validation discrepancies
  - Archive documentation for verifications and validations performed
  - Baseline verification and validation report
- Prepare and baseline
  - Operator's manuals
  - Maintenance manuals
  - Operations handbook
- Prepare launch, operations, and ground support sites including training as needed
  - Train initial system operators and maintainers
  - Train on contingency planning
  - Confirm telemetry validation and ground data processing
  - Confirm system and support elements are ready for flight
  - Provide support to the launch and checkout of the system
  - Perform planned on-orbit operational verification(s) and validation(s)
- Document lessons learned. Perform required Phase D technical activities from NPR 7120.5
- Satisfy Phase D reviews' entrance/success criteria from NPR 7123.1

Reviews

- Test Readiness Reviews (TRRs)
- System Acceptance Review (SAR) or pre-Ship Review
- ORR

(continued)

- FRR
- System functional and physical configuration audits
- Safety review

changes at this point can become costly. Phase D concludes with a system that has been shown to be capable of accomplishing the purpose for which it was created.

3.8 Project Phase E: Operations and Sustainment

The purpose of Phase E is to conduct the prime mission to meet the initially identified need and to maintain support for that need. The products of the phase are the results of the mission and performance of the system.

Systems engineering personnel continue to play a role during this phase since integration often overlaps with operations for complex systems. Some programs have repeated operations/flights which require configuration changes and new mission objectives with each occurrence. And systems with complex sustainment needs or human involvement will likely require evaluation and adjustments that may be beyond the scope of operators to perform. Specialty engineering disciplines, like maintainability and logistics servicing, will be performing tasks during this phase as well. Such tasks may require reiteration and/or recursion of the common systems engineering processes.

Systems engineering personnel also may be involved in in-flight anomaly resolution. Additionally, software development may continue well into Phase E. For example, software for a planetary probe may be developed and uplinked while in-flight. Another example would be new hardware developed for space station increments.

This phase encompasses the evolution of the system only insofar as that evolution does not involve major changes to the system architecture. Changes of that scope constitute new "needs," and the project life cycle starts over. For large flight projects, there may be an extended period of cruise, orbit insertion, on-orbit assembly, and initial shakedown operations. Near the end of the prime mission, the project may apply for a mission extension to continue mission activities or attempt to perform additional mission objectives.

For additional information on systems engineering in Phase E, see *Appendix T*.

3.9 Project Phase F: Closeout

The purpose of Phase F is to implement the systems decommissioning and disposal planning and analyze any returned data and samples. The products of the phase are the results of the mission. The system engineer is involved in this phase to ensure all technical information is properly identified and archived, to answer questions, and to resolve issues as they arise.

Phase F deals with the final closeout of the system when it has completed its mission; the time at which this occurs depends on many factors. For a flight system that returns to Earth after a short mission duration, closeout may require little more than de-integrating the hardware and returning it to its owner. On flight projects of long duration, closeout may proceed according to established plans or may begin as a result of unplanned events, such as failures. Refer to NASA Policy Directive (NPD) 8010.3, Notification of Intent to Decommission or Terminate Operating Space Systems and Terminate Missions,

SPACE FLIGHT PHASE E: OPERATIONS AND SUSTAINMENT

Purpose

To conduct the mission and meet the initially identified need and maintain support for that need. Implement the mission operations plan.

Typical Activities and Their Products

- Conduct launch vehicle performance assessment. Commission and activate science instruments
- Conduct the intended prime mission(s)
- Provide sustaining support as planned
    - > Implement spares plan
    - > Collect engineering and science data
    - > Train replacement operators and maintainers
    - > Train the flight team for future mission phases (e.g., planetary landed operations)
    - > Maintain and approve operations and maintenance logs
    - > Maintain and upgrade the system
    - > Identify and update risks
    - > Address problem/failure reports
    - > Process and analyze mission data
    - > Apply for mission extensions, if warranted
- Prepare for deactivation, disassembly, decommissioning as planned (subject to mission extension)
- Capture lessons learned
- Complete post-flight evaluation reports
- Develop final mission report
- Perform required Phase E technical activities from NPR 7120.5
- Satisfy Phase E reviews' entrance/success criteria from NPR 7123.1

Reviews

- Post-Launch Assessment Review (PLAR)
- Critical Event Readiness Review (CERR)
- Post-Flight Assessment Review (PFAR) (human space flight only)
- DR
- System upgrade review
- Safety review

PHASE F: CLOSEOUT

Purpose

To implement the systems decommissioning/disposal plan developed in Phase E and perform analyses of the returned data and any returned samples.

Typical Activities and Their Products

- Dispose of the system and supporting processes
- Document lessons learned
- Baseline mission final report
- Archive data
- Capture lessons learned
- Perform required Phase F technical activities from NPR 7120.5
- Satisfy Phase F reviews' entrance/success criteria from NPR 7123.1

Reviews

- DRR

for terminating an operating mission. Alternatively, technological advances may make it uneconomical to continue operating the system either in its current configuration or an improved one.

To limit space debris, NPR 8715.6, NASA Procedural Requirements for Limiting Orbital Debris, provides requirements for removing Earth-orbiting robotic satellites from their operational orbits at the end of their useful life. For Low Earth Orbit (LEO) missions, the satellite is usually deorbited. For small satellites, this is accomplished by allowing the orbit to slowly decay until the satellite eventually burns up in Earth's atmosphere. Larger, more massive satellites and observatories should be designed to demise or deorbit in a controlled manner so that they can be safely targeted for impact in a remote area of the ocean. The Geostationary (GEO) satellites at 35,790 km above the Earth cannot be practically deorbited, so they are boosted to a higher orbit well beyond the crowded operational GEO orbit.

In addition to uncertainty about when this part of the phase begins, the activities associated with safe closeout of a system may be long and complex and may affect the system design. Consequently, different options and strategies should be considered during the project's earlier phases along with the costs and risks associated with the different options.

3.10 Funding: The Budget Cycle

For a description of the NASA Budget Cycle, refer to the NASA Expanded Guidance for Systems Engineering document found at *https://nen.nasa.gov/web/se/doc-repository*. See also Section 5.8 of *NASA/SP-2014-3705, NASA Space Flight Program and Project Management Handbook.*

3.11 Tailoring and Customization of NPR 7123.1 Requirements

In this section, the term *requirements* refers to the "shall" statements imposed from Agency directives. This discussion focuses on the tailoring of the requirements contained in NPR 7123.1.

3.11.1 Introduction

NASA policy recognizes the need to accommodate the unique aspects of each program or project to achieve mission success in an efficient and economical manner. Tailoring is a process used to accomplish this.

NPR 7123.1 defines *tailoring* as "the process used to seek relief from SE NPR requirements consistent with program or project objectives, allowable risk, and constraints." Tailoring results in deviations or waivers (see NPR 7120.5, Section 3.5) to SE requirements and is documented in the next revision of the SEMP (e.g., via the Compliance Matrix).

Since NPR 7123.1 was written to accommodate programs and projects regardless of size or complexity, the NPR requirements leave considerable latitude for interpretation. Therefore, the term "customization" is introduced and is defined as "the modification of recommended SE practices that are used to accomplish the SE requirements." Customization does not require waivers or deviations, but significant customization should be documented in the SEMP.

Tailoring and customization are essential systems engineering tools that are an accepted and expected part of establishing the proper SE NPR requirements for a program or project. Although tailoring is expected for all sizes of projects and programs, small projects present opportunities and challenges that are different from those of large, traditional projects such as the Shuttle, International Space Station, Hubble Space Telescope, and Mars Science Laboratory.

While the technical aspects of small projects are generally narrower and more focused, they can also be challenging when their objectives are to demonstrate advanced technologies or provide "one of a kind" capabilities. At the same time, their comparatively small budgets and restricted schedules dictate lean and innovative implementation approaches to project management and systems engineering. Tailoring and customization allow programs and projects to be successful in achieving technical objectives within cost and schedule constraints. The key is effective tailoring that reflects lessons learned and best practices. Tailoring the SE requirements and customizing the SE best practices to the specific needs of the project helps to obtain the desired benefits while eliminating unnecessary overhead. To accomplish this, an acceptable risk posture must be understood and agreed upon by the project, customer/stakeholder, Center management, and independent reviewers. Even with this foundation, however, the actual process of appropriately tailoring SE requirements and customizing NPR 7123.1 practices to a specific project can be complicated and arduous. Effective approaches and experienced mentors make the tailoring process for any project more systematic and efficient.

Chapter 6 of the *NASA Software Engineering Handbook* provides guidance on tailoring SE requirements for software projects.

3.11.2 Criteria for Tailoring

NPR 8705.4, Risk Classification for NASA Payloads, is intended for assigning a risk classification to projects and programs. It establishes baseline criteria that enable users to define the risk classification level for NASA payloads on human or non-human-rated launch systems or carrier vehicles. It is also a starting point for understanding and defining criteria for tailoring.

The extent of acceptable tailoring depends on several characteristics of the program/project such as the following:

1. **Type of mission.** For example, the requirements for a human space flight mission are much more rigorous than those for a small robotic mission.

2. **Criticality of the mission** in meeting the Agency Strategic Plan. Critical missions that absolutely must be successful may not be able to get relief from NPR requirements.

3. **Acceptable risk level.** If the Agency and the customer are willing to accept a higher risk of failure, some NPR requirements may be waived.

4. **National significance.** A project that has great national significance may not be able to get relief from NPR requirements.

5. **Complexity.** Highly complex missions may require more NPR requirements in order to keep systems compatible, whereas simpler ones may not require the same level of rigor.

6. **Mission lifetime.** Missions with a longer lifetime need to more strictly adhere to NPR requirements than short-lived programs/projects.

7. **Cost of mission.** Higher cost missions may require stricter adherence to NPR requirements to ensure proper program/project control.

8. **Launch constraints.** If there are several launch constraints, a project may need to be more fully compliant with Agency requirements.

3.11.3 Tailoring SE NPR Requirements Using the Compliance Matrix

NPR 7123.1 includes a Compliance Matrix (Appendix H.2) to assist programs and projects in verifying that they meet the specified NPR requirements. The Compliance Matrix documents the program/project's compliance or intent to comply with the requirements of the NPR or justification for tailoring. The Compliance Matrix can be used to assist in identifying where major customization of the way (e.g., formality and rigor) the NPR requirements will be accomplished and to communicate that customization to the stakeholders. The tailoring process (which can occur at any time in the program or project's life cycle) results in deviations or waivers to the NPR requirements depending on the timing of the request. Deviations and waivers of the requirements can be submitted separately to the Designated Governing Authority or via the Compliance Matrix. The Compliance Matrix is attached to the Systems Engineering Management Plan (SEMP) when submitted for approval. Alternatively, if there is no stand-alone SEMP and the contents of the SEMP are incorporated into another document such as the project plan, the Compliance Matrix can be captured within that plan.

FIGURE 3.11-1 illustrates a notional tailoring process for a space flight project. Project management (such as the project manager/the Principal Investigator/the task lead, etc.) assembles a project team to tailor the NPR requirements codified in the Compliance Matrix. To properly classify the project, the team (chief engineer, lead systems engineer, safety and mission assurance, etc.) needs to understand the building blocks of the project such as the needs, goals, and objectives as well as the appropriate risk posture.

Through an iterative process, the project team goes through the NPR requirements in the Compliance Matrix to tailor the requirements. A tailoring tool with suggested guidelines may make the tailoring process easier if available. Several NASA Centers including LaRC and MSFC have developed tools for use at their Centers which could be adapted for other Centers. Guidance from Subject Matter Experts (SMEs) should be sought to determine the appropriate amount of tailoring for a specific project.

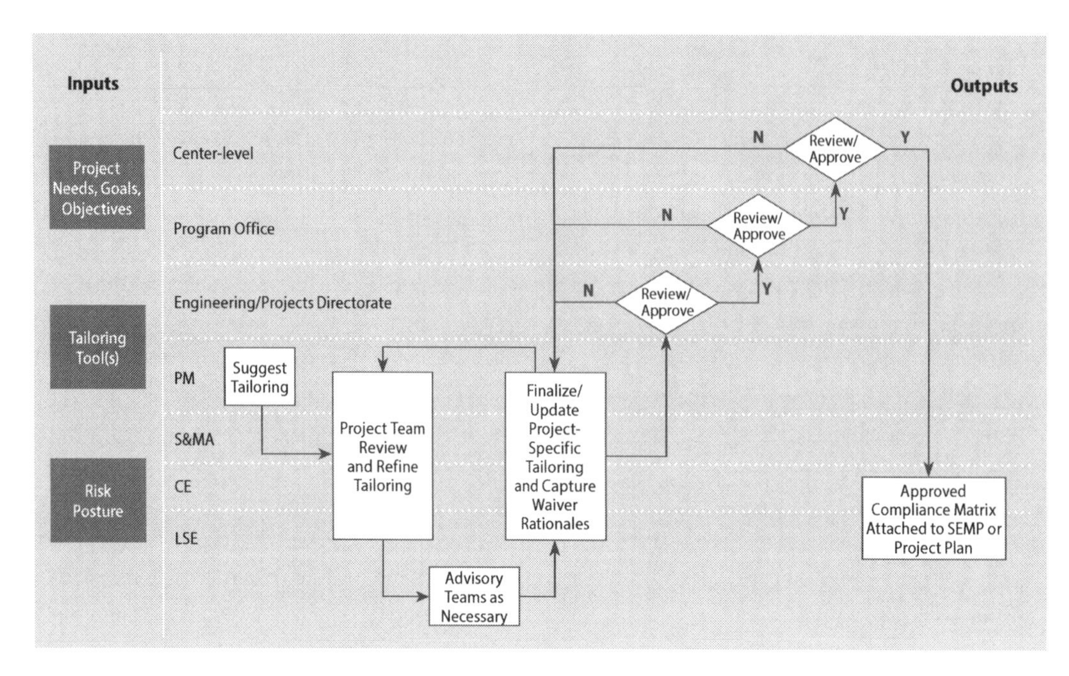

FIGURE 3.11-1 Notional Space Flight Products Tailoring Process

The Compliance Matrix provides rationales for each of the NPR requirements to assist in understanding. Once the tailoring is finalized and recorded in the Compliance Matrix with appropriate rationales, the requested tailoring proceeds through the appropriate governance model for approval.

3.11.4 Ways to Tailor a SE Requirement

Tailoring often comes in three areas:

1. Eliminating a requirement that does not apply to the specific program/project.
2. Eliminating a requirement that is overly burdensome (i.e., when the cost of implementing the requirement adds more risk to the project by diverting resources than the risk of not complying with the requirement).
3. Scaling the requirement in a manner that better balances the cost of implementation and the project risk.

Customizing SE practices can include the following:

1. Adjusting the way each of the 17 SE processes is implemented.
2. Adjusting the formality and timing of reviews.

3.11.4.1 *Non-Applicable NPR Requirements*

Each requirement in NPR 7123.1 is assessed for applicability to the individual project or program. For example, if the project is to be developed completely in-house, the requirements of the NPR's Chapter 4 on contracts would not be applicable. If a system does not contain software, then none of the NPR requirements for developing and maintaining software would be applicable.

3.11.4.2 *Adjusting the Scope*

Depending on the project or program, some relief on the scope of a requirement may be appropriate. For example, although the governing project management directive (e.g., NPR 7120.5, 7150.2, 7120.7, 7120.8) for a program/project may require certain documents to be standalone, the SE NPR does not require any additional stand-alone documents. For small projects, many of the plans can be described in just a few paragraphs or pages. In these types of projects, any NPR requirements stating that the plans need to be stand-alone document would be too burdensome. In these cases, the information can simply be written and included as part of the project plan or SEMP. If the applicable project management directive (e.g., NPR 7120.5 or NPR 7120.8) requires documents to be stand-alone, a program/project waiver/deviation is needed. However, if there is no requirement or Center expectation for a stand-alone document, a project can customize where that information is recorded and no waiver or deviation is required. Capturing where this information is documented within the systems engineering or project management Compliance Matrix would be useful for clarity.

3.11.4.3 *Formality and Timing of Reviews*

The governing project management directive identifies the required or recommended life cycle for the specific type of program/project. The life cycle defines the number and timing of the various reviews; however, there is considerable discretion concerning the formality of the review and how to conduct it. NPR 7123.1, Appendix G, provides extensive guidance for suggested review entrance and success criteria. It is expected that the program/project will customize these criteria in a manner that makes sense for their program/project. The SE NPR does not require a waiver/deviation for this customization; however, departures from review elements required by other NPRs need to be addressed by tailoring those documents.

If a program/project decides it does not need one of the required reviews, a waiver or deviation is needed. However, the SE NPR does not specify a minimum amount of spacing for these reviews. A small project may decide to combine the SRR and the SDR (or Mission Definition Review (MDR)) for example. As long as the intent for *both* reviews is accomplished, the SE NPR does not require a waiver or deviation. (Note that even though the SE NPR does not require it, a waiver or deviation may still be required in the governing project management NPR.) This customization and/or tailoring should be documented in the Compliance Matrix and/or the review plan or SEMP.

Unless otherwise required by the governing project management directives, the formality of the review can be customized as appropriate for the type of program/project. For large projects, it might be appropriate to conduct a very formal review with a formal Review Item Discrepancy (RID)/Request for Action (RFA) process, a summary, and detailed presentations to a wide audience including boards and pre-boards over several weeks. For small projects, that same review might be done in a few hours across a tabletop with a few stakeholders and with issues and actions simply documented in a word or PowerPoint document.

The NASA Engineering Network Systems Engineering Community of Practice, located at https://nen.nasa.gov/web/se includes document templates for milestone review presentations required by the NASA SE process.

3.11.5 Examples of Tailoring and Customization

TABLE 3.11-1 shows an example of the types of missions that can be defined based on a system that breaks projects into various types ranging from a very complex type A to a much simpler type F. When tailoring a project, the assignment of specific projects to

TABLE 3.11-1 Example of Program/Project Types

Criteria	Type A	Type B	Type C	Type D	Type E	Type F
Description of the Types of Mission	Human Space Flight or Very Large Science/ Robotic Missions	Non-Human Space Flight or Science/Robotic Missions	Small Science or Robotic Missions	Smaller Science or Technology Missions (ISS payload)	Suborbital or Aircraft or Large Ground based Missions	Aircraft or Ground based technology demonstrations
Priority (Criticality to Agency Strategic Plan) and Acceptable Risk Level	High priority, very low (minimized) risk	High priority, low risk	Medium priority, medium risk	Low priority, high risk	Low priority, high risk	Low to very low priority, high risk
National Significance	Very high	High	Medium	Medium to Low	Low	Very Low
Complexity	Very high to high	High to Medium	Medium to Low	Medium to Low	Low	Low to Very Low
Mission Lifetime (Primary Baseline Mission)	Long. >5 years	Medium. 2–5 years	Short. <2 years	Short. <2 years	N/A	N/A
Cost Guidance (estimate LCC)	High (greater than ~$1B)	High to Medium (~$500M–$1B)	Medium to Low (~$100M–$500M)	Low (~$50M–$100M)	(~$10–50M)	(less than $10–15M)
Launch Constraints	Critical	Medium	Few	Few to none	Few to none	N/A
Alternative Research Opportunities or Re-flight Opportunities	No alternative or re-flight opportunities	Few or no alternative or re-flight opportunities	Some or few alternative or re-flight opportunities	Significant alternative or re-flight opportunities	Significant alternative or re-flight opportunities	Significant alternative or re-flight opportunities
Achievement of Mission Success Criteria	All practical measures are taken to achieve minimum risk to mission success. The highest assurance standards are used.	Stringent assurance standards with only minor compromises in application to maintain a low risk to mission success.	Medium risk of not achieving mission success may be acceptable. Reduced assurance standards are permitted.	Medium or significant risk of not achieving mission success is permitted. Minimal assurance standards are permitted.	Significant risk of not achieving mission success is permitted. Minimal assurance standards are permitted.	Significant risk of not achieving mission success is permitted. Minimal assurance standards are permitted.
Examples	HST, Cassini, JIMO, JWST, MPCV, SLS, ISS	MER, MRO, Discovery payloads, ISS Facility Class payloads, Attached ISS payloads	ESSP, Explorer payloads, MIDES, ISS complex subrack payloads, PA-1, ARES 1-X, MEDLI, CLARREO, SAGE III, Calipso	SPARTAN, GAS Can, technology demonstrators, simple ISS, express middeck and subrack payloads, SMEX, MISSE-X, EV-2	IRVE-2, IRVE-3, HiFIRE, HyBoLT, ALHAT, STORRM, Earth Venture I	DAWNAir, InFlame, Research, technology demonstrations

particular types should be viewed as guidance, not as rigid characterization. Many projects will have characteristics of multiple types, so the tailoring approach may permit more tailoring for those aspects of the project that are simpler and more open to risk and less tailoring for those aspects of the project where complexity and/or risk aversion dominate. These tailoring criteria and definitions of project "types" may vary from Center to Center and from Mission Directorate to Mission Directorate according to what is appropriate for their missions. TABLE 3.11-2 shows an example of how the documentation required of

TABLE 3.11-2 Example of Tailoring NPR 7120.5 Required Project Products

	Type A	Type B	Type C	Type D	Type E	Type F
Example Project Technical Products						
Concept Documentation	Fully Compliant	Fully Compliant	Fully Compliant	Tailor	Tailor	Tailor
Mission, Spacecraft, Ground, and Payload Architectures	Fully Compliant	Fully Compliant	Fully Compliant	Tailor	Tailor	Tailor
Project-Level, System and Subsystem Requirements	Fully Compliant	Fully Compliant	Fully Compliant	Fully Compliant	Tailor	Tailor
Design Documentation	Fully Compliant	Fully Compliant	Fully Compliant	Fully Compliant	Tailor	Tailor
Operations Concept	Fully Compliant	Fully Compliant	Fully Compliant	Tailor	Tailor	Tailor
Technology Readiness Assessment Documentation	Fully Compliant	Fully Compliant	Fully Compliant	Tailor	Tailor	Tailor
Human Systems Integration Plan	Fully Compliant	Fully Compliant	Fully Compliant	Tailor	Tailor	Tailor
Heritage Assessment Documentation	Fully Compliant	Fully Compliant	Fully Compliant	Tailor	Tailor	Tailor
Safety Data Packages	Fully Compliant	Fully Compliant	Fully Compliant	Fully Compliant	Tailor	Tailor
ELV Payload Safety Process Deliverables	Fully Compliant	Fully Compliant	Fully Compliant	Fully Compliant	Fully Compliant	Not Applicable
Verification and Validation Report	Fully Compliant	Fully Compliant	Fully Compliant	Tailor	Tailor	Tailor
Operations Handbook	Fully Compliant	Fully Compliant	Fully Compliant	Tailor	Tailor	Not Applicable
End of Mission Plans	Fully Compliant	Fully Compliant	Fully Compliant	Tailor	Tailor	Tailor
Mission Report	Fully Compliant	Fully Compliant	Tailor	Tailor	Tailor	Tailor

(continued)

	Type A	Type B	Type C	Type D	Type E	Type F
Example Project Plan Control Plans						
Risk Management Plan	Fully Compliant	Fully Compliant	Fully Compliant	Tailor	Tailor	Not Applicable
Technology Development plan	Fully Compliant	Fully Compliant	Fully Compliant	Fully Compliant	Not Applicable	Not Applicable
Systems Engineering Management Plan	Fully Compliant	Fully Compliant	Fully Compliant	Tailor	Tailor	Tailor
Software Management plan	Fully Compliant	Fully Compliant	Tailor	Tailor	Tailor	Tailor
Verification and Validation Plan	Fully Compliant	Fully Compliant	Tailor	Tailor	Tailor	Tailor
Review Plan	Fully Compliant	Fully Compliant	Fully Compliant	Tailor	Tailor	Tailor
Integrated Logistics Support Plan	Fully Compliant	Fully Compliant	Fully Compliant	Tailor	Tailor	Not Applicable
Science Data Management Plan	Fully Compliant	Fully Compliant	Fully Compliant	Tailor	Tailor	Not Applicable
Integration Plan	Fully Compliant	Fully Compliant	Fully Compliant	Fully Compliant	Tailor	Tailor
Configuration Management Plan	Fully Compliant	Fully Compliant	Fully Compliant	Fully Compliant	Tailor	Tailor
Technology Transfer (formerly Export) Control Plan	Fully Compliant	Fully Compliant	Fully Compliant	Fully Compliant	Tailor	Tailor
Lessons Learned Plan	Fully Compliant	Fully Compliant	Fully Compliant	Fully Compliant	Tailor	Tailor
Human Rating Certification Package	Fully Compliant	Not Applicable	Not Applicable	Not Applicable	Not Applicable	Not Applicable

a program/project might also be tailored or customized. The general philosophy is that the simpler, less complex projects should require much less documentation and fewer formal reviews. Project products should be sensibly scaled.

3.11.6 Approvals for Tailoring

Deviations and waivers of the requirements for the SE NPR can be submitted separately to the requirements owners or in bulk using the appropriate Compliance Matrix found in NPR 7123.1 Appendix H. If it is a Center that is requesting tailoring of the NPR requirements for standard use at the Center, Appendix H.1 is completed and submitted to the OCE for approval upon request or as changes to the Center processes occur. If a program/project whose responsibility has been delegated to a Center is seeking a waiver/deviation from the NPR requirements, the Compliance Matrix in Appendix H.2 is used. In these cases, the Center Director or designee will approve the waiver/deviation.

The result of this tailoring, whether for a Center or for a program/project, should also be captured in the next revision of the SEMP along with supporting rationale and documented approvals from the requirement owner. This allows communication of the approved waivers/deviations to the entire project team as well as associated managers. If an independent assessment is being conducted on the program/project, this also allows appropriate modification of expectations and assessment criteria. TABLE 3.11-3 provides some examples of tailoring captured within the H.2 Compliance Matrix.

TABLE 3.11-3 Example Use of a Compliance Matrix

Req ID	SE NPR Section	Requirement Statement	Rationale	Req. Owner	Comply?	Justification
SE-05	2.1.5.2	For those requirements owned by Center Directors, the technical team shall complete the Compliance Matrix in *Appendix H.2* and include it in the SEMP.	For programs and projects, the Compliance Matrix in *Appendix H.2* is filled out showing that the program/project is compliant with the requirements of this NPR (or a particular Center's implementation of NPR 7123.1, whichever is applicable) or any tailoring thereof is identified and approved by the Center Director or designee as part of the program/project SEMP.	CD	Fully Compliant	
SE-06	2.1.6.1	The DGA shall approve the SEMP, waiver authorizations, and other key technical documents to ensure independent assessment of technical content.	The DGA, who is often the TA, provides an approval of the SEMPs, waivers to technical requirements and other key technical document to provide assurance of the applicability and technical quality of the products.	CD	Fully Complaint	
SE-24	4.2.1	The NASA technical team shall define the engineering activities for the periods before contract award, during contract performance, and upon contract completion in the SEMP.	It is important for both the government and contractor technical teams to understand what activities will be handled by which organization throughout the product life cycle. The contractor(s) will typically develop a SEMP or its equivalent to describe the technical activities in their portion of the project, but an overarching SEMP is needed that will describe all technical activities across the life cycle whether contracted or not.	CD	Not Applicable	Project is conducted entirely in-house and therefore there are no contracts involved

4.0 System Design Processes

This chapter describes the activities in the system design processes listed in FIGURE 2.1-1. The chapter is separated into sections corresponding to processes 1 to 4 listed in FIGURE 2.1-1. The tasks within each process are discussed in terms of inputs, activities, and outputs. Additional guidance is provided using examples that are relevant to NASA projects.

The system design processes are interdependent, highly iterative and recursive processes resulting in a validated set of requirements and a design solution that satisfies a set of stakeholder expectations. There are four system design processes: developing stakeholder expectations, technical requirements, logical decompositions, and design solutions.

FIGURE 4.0-1 illustrates the recursive relationship among the four system design processes. These processes start with a study team collecting and clarifying the stakeholder expectations, including the mission objectives, constraints, design drivers, operational objectives, and criteria for defining mission success. This set of stakeholder expectations and high-level requirements is used to drive an iterative design loop where a straw man architecture/design, the concept of operations, and derived requirements are developed. These three products should be consistent with each other and will require iterations and design decisions to achieve this consistency. Once consistency is achieved, analyses allow the project team to validate the proposed design against the stakeholder expectations. A simplified validation asks the questions: Will the system work as expected? Is the system achievable within budget and schedule constraints? Does the system provide the functionality and fulfill the operational needs that drove the project's funding approval? If the answer to any of these questions is no, then changes to the design or stakeholder expectations will be required, and the process starts again. This process continues until the system—architecture, ConOps, and requirements—meets the stakeholder expectations.

The depth of the design effort should be sufficient to allow analytical verification of the design to the requirements. The design should be feasible and credible when judged by a knowledgeable independent review team and should have sufficient depth to support cost modeling and operational assessment.

Once the system meets the stakeholder expectations, the study team baselines the products and prepares for the next phase. Often, intermediate levels of decomposition are validated as part of the process. In

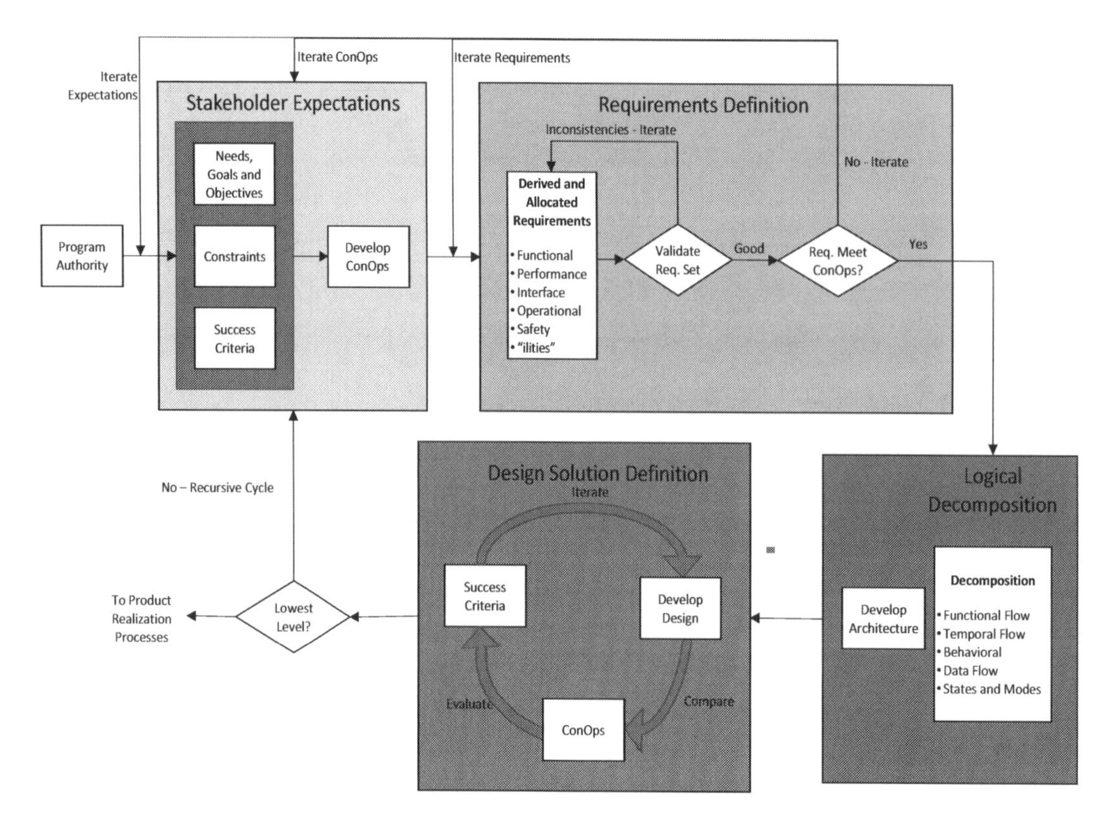

FIGURE 4.0-1 Interrelationships among the System Design Processes

SYSTEM DESIGN KEYS

- Successfully understanding and defining the mission objectives and the concept of operations are keys to capturing the stakeholder expectations, which will translate into quality requirements and operational efficiencies over the life cycle of the project.
- Complete and thorough requirements traceability is a critical factor in successful validation of requirements.
- Clear and unambiguous requirements will help avoid misunderstanding when developing the overall system and when making major or minor changes.
- Document all decisions made during the development of the original design concept in the technical data package. This will make the original design philosophy and negotiation results available to assess future proposed changes and modifications against.
- The validation of a design solution is a continuing recursive and iterative process during which the design solution is evaluated against stakeholder expectations.

the next level of decomposition, the baselined derived (and allocated) requirements become the set of high-level requirements for the decomposed elements and the process begins again. These system design processes are primarily applied in Pre-Phase A and continue through Phase C.

The system design processes during Pre-Phase A focus on producing a feasible design that will lead to Formulation approval. During Phase A, alternative designs and additional analytical maturity are pursued to optimize the design architecture. Phase B results in a preliminary design that satisfies the approval criteria. During Phase C, detailed, build-to designs are completed.

This is a simplified description intended to demonstrate the recursive relationship among the system design processes. These processes should be used as guidance and tailored for each study team depending on the size of the project and the hierarchical level of the study team. The next sections describe each of the four system design processes and their associated products for a given NASA mission.

4.1 Stakeholder Expectations Definition

The Stakeholder Expectations Definition Process is the initial process within the SE engine that establishes the foundation from which the system is designed and the product is realized. The main purpose of this process is to identify who the stakeholders are and how they intend to use the product. This is usually accomplished through use-case scenarios (sometimes referred to as Design Reference Missions (DRMs)) and the ConOps.

4.1.1 Process Description

FIGURE 4.1-1 provides a typical flow diagram for the Stakeholder Expectations Definition Process and identifies typical inputs, outputs, and activities to consider in defining stakeholder expectations.

4.1.1.1 *Inputs*

Typical inputs needed for the Stakeholder Expectations Definition Process include the following:

- **Initial Customer Expectations:** These are the needs, goals, objectives, desires, capabilities, and other constraints that are received from the customer for the product within the product layer. For the top-tier products (final end item), these are the expectations of the originating customer who requested the product. For an end product within the product layer, these are the expectations of the recipient of the end item when transitioned.

- **Other Stakeholder Expectations:** These are the expectations of key stakeholders other than the customer. For example, such stakeholders may be the test team that will be receiving the transitioned product (end product and enabling products) or the trainers that will be instructing the operators or managers that are accountable for the product at this layer.

- **Customer Flow-down Requirements:** These are any requirements that are being flowed down or allocated from a higher level (i.e., parent requirements). They are helpful in establishing the expectations of the customer at this layer.

4.1.1.2 *Process Activities*

4.1.1.2.1 *Identify Stakeholders*

A "stakeholder" is a group or individual that is affected by or has a stake in the product or project. The key players for a project/product are called the key stakeholders. One key stakeholder is always the

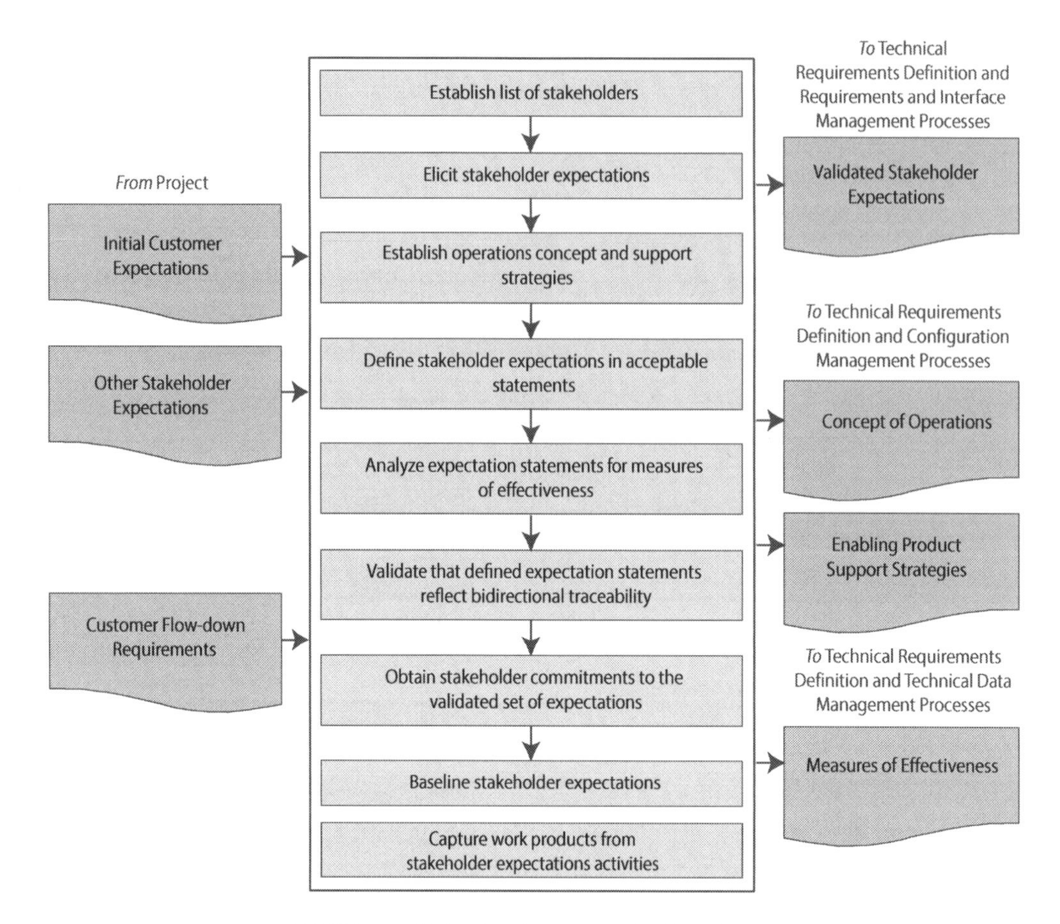

FIGURE 4.1-1 Stakeholder Expectations Definition Process

"customer." The customer may vary depending on where the systems engineer is working in the PBS. For example, at the topmost level, the customer may be the person or organization that is purchasing the product. For a systems engineer working three or four levels down in the PBS, the customer may be the leader of the team that takes the element and integrates it into a larger assembly. Regardless of where the systems engineer is working within the PBS, it is important to understand what is expected by the customer.

Other interested parties are those who affect the project by providing broad, overarching constraints within which the customers' needs should be achieved. These parties may be affected by the resulting product, the manner in which the product is used, or have a responsibility for providing life cycle support services. Examples include Congress, advisory planning teams, program managers, maintainers, and mission partners. It is important that the list of stakeholders be identified early in the process, as well as the primary stakeholders who will have the most significant influence over the project.

The customer and users of the system are usually easy to identify. The other key stakeholders may be more difficult to identify and they may change depending on the type of the project and the phase the project is in. TABLE 4.1-1 provides some

TABLE 4.1-1 Stakeholder Identification throughout the Life Cycle

Life-Cycle Stage	Example Stakeholders
Pre-Phase A	NASA Headquarters, NASA Centers, Presidential Directives, NASA advisory committees, the National Academy of Sciences
Phase A	Mission Directorate, customer, potential users, engineering disciplines, safety organization
Phase B	Customer, engineering disciplines, safety, crew, operations, logistics, production facilities, suppliers, principle investigators
Phase C	Customer, engineering disciplines, safety, crew, operations, logistics, production facilities, suppliers, principle investigators
Phase D	Customer, engineering disciplines, safety, crew, operations, training, logistics, verification team, Flight Readiness Board members
Phase E	Customer, system managers, operations, safety, logistics, sustaining team, crew, principle investigators, users
Phase F	Customer, NASA Headquarters, operators, safety, planetary protection, public

examples of stakeholders in the life cycle phase that should be considered.

4.1.1.2.2 *Understand Stakeholder Expectations*

Thoroughly understanding the customer and other key stakeholders' expectations for the project/product is one of the most important steps in the systems engineering process. It provides the foundation upon which all other systems engineering work depends. It helps ensure that all parties are on the same page and that the product being provided will satisfy the customer. When the customer, other stakeholders, and the systems engineer mutually agree on the functions, characteristics, behaviors, appearance, and performance the product will exhibit, it sets more realistic expectations on the customer's part and helps prevent significant requirements creep later in the life cycle.

Through interviews/discussions, surveys, marketing groups, e-mails, a Statement of Work (SOW), an initial set of customer requirements, or some other means, stakeholders specify what is desired as an end state or as an item to be produced and put bounds on the achievement of the goals. These bounds may encompass expenditures (resources), time to deliver, life cycle support expectations, performance objectives, operational constraints, training goals, or other less obvious quantities such as organizational needs or geopolitical goals. This information is reviewed, summarized, and documented so that all parties can come to an agreement on the expectations.

FIGURE 4.1-2 shows the type of information needed when defining stakeholder expectations and depicts how the information evolves into a set of high-level requirements. The yellow lines depict validation paths. Examples of the types of information that would be defined during each step are also provided.

Defining stakeholder expectations begins with the *mission authority* and *strategic objectives* that the mission is meant to achieve. Mission authority changes depending on the category of the mission. For example, science missions are usually driven by NASA Science Mission Directorate strategic plans, whereas the exploration missions may be driven by a Presidential directive. Understanding the objectives of the mission helps ensure that the project team is working toward a common vision. These goals and

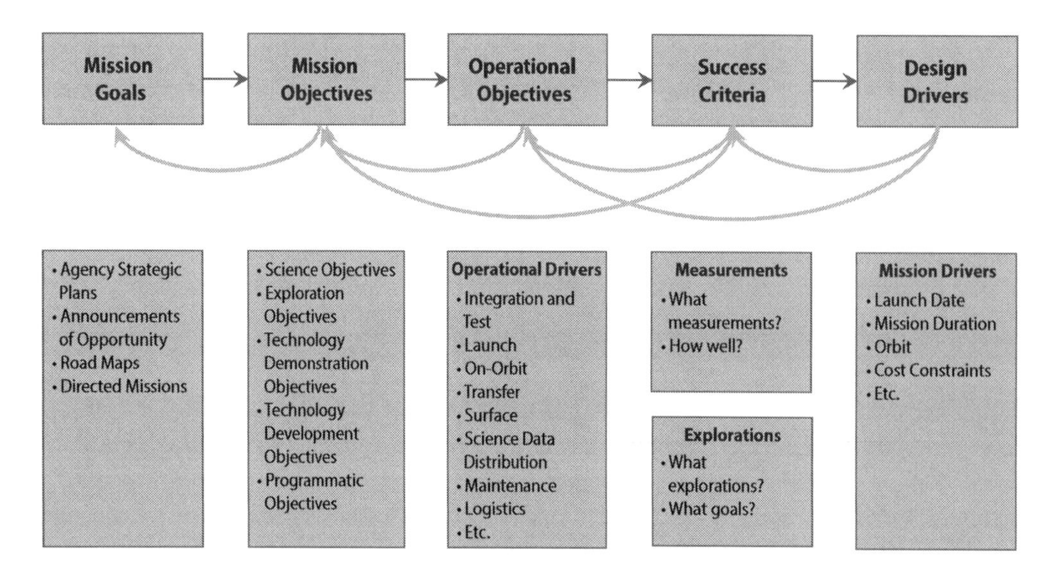

FIGURE 4.1-2 Information Flow for Stakeholder Expectations

objectives form the basis for developing the mission, so they need to be clearly defined and articulated.

The project team should also identify the *constraints* that may apply. A "constraint" is a condition that is to be met. Sometimes a constraint is dictated by external factors such as orbital mechanics, an existing system that must be utilized (external interface), a regulatory restriction, or the state of technology; sometimes constraints are the result of the overall budget environment. Concepts of operation and constraints also need to be included in defining the stakeholder expectations. These identify how the system should be operated to achieve the mission objectives.

NOTE: It is extremely important to involve stakeholders in all phases of a project. Such involvement should be built in as a self-correcting feedback loop that will significantly enhance the chances of mission success. Involving stakeholders in a project builds confidence in the end product and serves as a validation and acceptance with the target audience.

In identifying the full set of expectations, the systems engineer will need to interact with various communities, such as those working in the areas of orbital debris, space asset protection, human systems integration, quality assurance, and reliability. Ensuring that a complete set of expectations is captured will help prevent "surprise" features from arising later in the life cycle. For example, space asset protection may require additional encryption for the forward link commands, additional shielding or filtering for RF systems, use of a different frequency, or other design changes that might be costly to add to a system that has already been developed.

4.1.1.2.3 *Identify Needs, Goals, and Objectives*

In order to define the goals and objectives, it is necessary to elicit the needs, wants, desires, capabilities, external interfaces, assumptions, and constraints from the stakeholders. Arriving at an agreed-to set of goals and objectives can be a long and arduous task. Proactive iteration with the stakeholders throughout the systems engineering process is the way that all parties can come to a true understanding of what should be done and what it takes to do the job. It is important to know who the primary stakeholders are and who has the decision authority to help resolve conflicts.

Needs, Goals, and Objectives (NGOs) provide a mechanism to ensure that everyone (implementer, customer, and other stakeholders) is in agreement at the beginning of a project in terms of defining the problem that needs to be solved and its scope. NGOs are not contractual requirements or designs.

Needs are defined in the answer to the question "What problem are we trying to solve?" Goals address what must be done to meet the needs; i.e., what the customer wants the system to do. Objectives expand on the goals and provide a means to document specific expectations. (Rationale should be provided where needed to explain why the need, goal, or objective exists, any assumptions made, and any other information useful in understanding or managing the NGO.)

Well-written NGOs provide clear traceability from the needs, then to the goals, and then to objectives. For example, if a given goal does not support a need, or an objective does not support a goal, it should not be part of the integrated set of NGOs. This traceability helps ensure that the team is actually providing what is needed.

The following definitions (source: *Applied Space Systems Engineering* edited by Larson, Kirkpatrick, Sellers, Thomas, and Verma) are provided to help the reader interpret the NGOs contained in this product.

- **Need:** A single statement that drives everything else. It should relate to the problem that the system is supposed to solve but not be the solution. The need statement is singular. Trying to satisfy more than one need requires a trade between the two, which could easily result in failing to meet at least one, and possibly several, stakeholder expectations.
- **Goals:** An elaboration of the need, which constitutes a specific set of expectations for the system. Goals address the critical issues identified during the problem assessment. Goals need not be in a quantitative or measurable form, but they should allow us to assess whether the system has achieved them.
- **Objectives:** Specific target levels of outputs the system must achieve. Each objective should relate to a particular goal. Generally, objectives should meet four criteria. (1) They should be specific enough to provide clear direction, so developers, customers, and testers will understand them. They should aim at results and reflect what the system needs to do but not outline how to implement the solution. (2) They should be measurable, quantifiable, and verifiable. The project needs to monitor the system's success in achieving each objective. (3) They should be aggressive but attainable, challenging but reachable, and targets need to be realistic. Objectives "To Be Determined" (TBD) may be included until trade studies occur, operations concepts solidify, or technology matures. Objectives need to be feasible before requirements are written and systems designed. (4) They should be results-oriented focusing on desired outputs and outcomes, not on the methods used to achieve the target (what, not how). It is important to always remember that objectives are not requirements. Objectives are identified during pre-Phase A development and help with the eventual formulation of a requirements set, but it is the requirements themselves that are contractually binding and will be verified against the "as-built" system design.

These stakeholder expectations are captured and are considered as initial until they can be further refined through development of the concept of operations and final agreement by the stakeholders.

4.1.1.2.4 *Establish Concept of Operations and Support Strategies*

After the initial stakeholder expectations have been established, the development of a Concept of Operations (ConOps) will further ensure that the technical team fully understands the expectations and how they may be satisfied by the product, and that understanding has been agreed to by the stakeholders. This may lead to further refinement of the initial set of stakeholder expectations if gaps or ambiguous statements are discovered. These scenarios and concepts of how the system will behave provide an implementation-free understanding of the stakeholders' expectations by defining what is expected without addressing how (the design) to satisfy the need. It captures required behavioral characteristics and the manner in which people will interact with the system. Support strategies include provisions for fabrication, test, deployment, operations, sustainment, and disposal.

The ConOps is an important component in capturing stakeholder expectations and is used in defining requirements and the architecture of a project. It stimulates the development of the requirements and architecture related to the user elements of the system. It serves as the basis for subsequent definition documents such as the operations plan, launch and early orbit plan, and operations handbook, and it provides the foundation for the long-range operational planning activities such as operational facilities, staffing, and network scheduling.

The ConOps is an important driver in the system requirements and therefore should be considered early in the system design processes. Thinking through the ConOps and use cases often reveals requirements and design functions that might otherwise be overlooked. For example, adding system requirements to allow for communication during a particular phase of a mission may require an additional antenna in a specific location that may not be required during the nominal mission. The ConOps should include scenarios for all significant operational situations, including known off-nominal situations. To develop a useful and complete set of scenarios, important malfunctions and degraded-mode operational situations should be considered. The ConOps is also an important aide to characterizing life cycle staffing goals and function allocation between humans and systems. In walking through the accomplishment of mission objectives, it should become clear when decisions need to be made as to what the human operators are contributing vs. what the systems are responsible for delivering.

The ConOps should consider all aspects of operations including nominal and off-nominal operations during integration, test, and launch through disposal. Typical information contained in the ConOps includes a description of the major phases; operation timelines; operational scenarios and/or DRM (see FIGURE 4.1-3 for an example of a DRM); fault management strategies, description of human interaction and required training, end-to-end communications strategy; command and data architecture; operational facilities; integrated logistic support (resupply, maintenance, and assembly); staffing levels and required skill sets; and critical events. The operational scenarios describe the dynamic view of the systems' operations and include how the system is perceived to function throughout the various modes and mode transitions, including interactions with external interfaces, response to anticipated hazard and faults, and during failure mitigations. For exploration missions, multiple DRMs make up a ConOps. The design and performance analysis leading to the requirements should satisfy all of them.

Additional information on the development of the ConOps is discussed in Section 4.1.2.1 of the NASA Expanded Guidance for Systems Engineering document found ***https://nen.nasa.gov/web/se/doc-repository***.

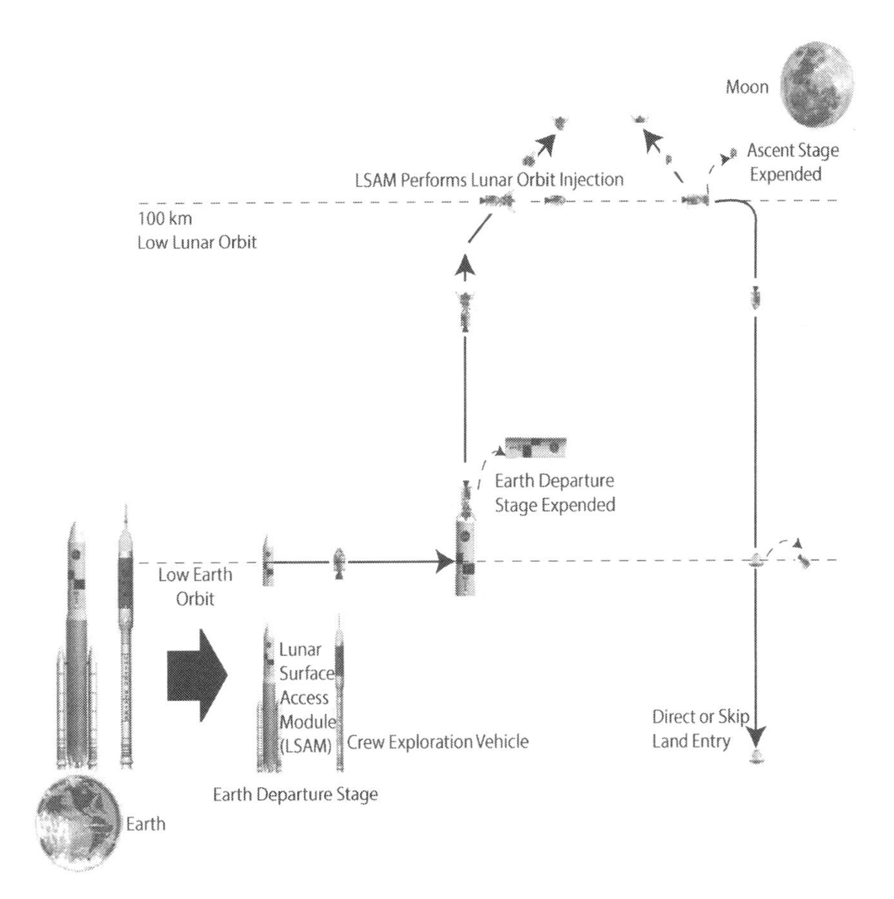

FIGURE 4.1-3 Example of a Lunar Sortie DRM Early in the Life Cycle

CONCEPT OF OPERATIONS VS. OPERATIONS CONCEPT

Concept of Operations

Developed early in Pre-Phase A by the technical team, describes the overall high-level concept of how the system will be used to meet stakeholder expectations, usually in a time sequenced manner. It describes the system from an operational perspective and helps facilitate an understanding of the system goals. It stimulates the development of the requirements and architecture related to the user elements of the system. It serves as the basis for subsequent definition documents and provides the foundation for the long-range operational planning activities.

Operations Concept

A description of how the flight system and the ground system are used together to ensure that the concept of operation is reasonable. This might include how mission data of interest, such as engineering or scientific data, are captured, returned to Earth, processed, made available to users, and archived for future reference. It is typically developed by the operational team. (See NPR 7120.5.)

Appendix S contains one possible outline for developing a ConOps. The specific sections of the ConOps will vary depending on the scope and purpose of the project.

4.1.1.2.5 *Define Stakeholder Expectations in Acceptable Statements*

Once the ConOps has been developed, any gaps or ambiguities have been resolved, and understanding between the technical team and stakeholders about what is expected/intended for the system/product has been achieved, the expectations can be formally documented. This often comes in the form of NGOs, mission success criteria, and design drivers. These may be captured in a document, spreadsheet, model, or other form appropriate to the product.

The *design drivers* will be strongly dependent upon the ConOps, including the operational environment, orbit, and mission duration requirements. For science missions, the design drivers include, at a minimum, the mission launch date, duration, and orbit, as well as operational considerations. If alternative orbits are to be considered, a separate concept is needed for each orbit. Exploration missions should consider the destination, duration, operational sequence (and system configuration changes), crew interactions, maintenance and repair activities, required training, and in situ exploration activities that allow the exploration to succeed.

4.1.1.2.6 *Analyze Expectations Statements for Measures of Effectiveness*

The *mission success criteria* define what the mission needs to accomplish to be successful. This could be in the form of science missions, exploration concept for human exploration missions, or a technological goal for technology demonstration missions. The success criteria also define how well the concept measurements or exploration activities should be accomplished. The success criteria capture the stakeholder expectations and, along with programmatic requirements and constraints, are used within the high-level requirements.

Measures of Effectiveness (MOEs) are the measures of success that are designed to correspond to accomplishment of the system objectives as defined by the stakeholder's expectations. They are stated from the stakeholder's point of view and represent criteria that are to be met in order for the stakeholder to consider the project successful. As such, they can be synonymous with mission/project success criteria. MOEs are developed when the NGOs or other stakeholder expectation documentation is developed. Additional information on MOEs is contained in Section 6.7.2.4 of the NASA Expanded Guidance for SE document at *https://nen.nasa.gov/web/se/doc-repository.*

4.1.1.2.7 *Validate That Defined Expectation Statements Reflect Bidirectional Traceability*

The NGOs or other stakeholder expectation documentation should also capture the source of the expectation. Depending on the location within the product layer, the expectation may be traced to an NGO or a requirement of a higher layer product, to organizational strategic plans, or other sources. Later functions and requirements will be traced to these NGOs. The use of a requirements management tool or model or other application is particularly useful in capturing and tracing expectations and requirements.

4.1.1.2.8 *Obtain Stakeholder Commitments to the Validated Set of Expectations*

Once the stakeholder and the technical team are in agreement with the expressed stakeholder expectations and the concept of operations, signatures or other forms of commitment are obtained. In order to obtain these commitments, a concept review is typically held on a formal or informal basis depending on the scope and complexity of the system (see *Section 6.7*). The stakeholder expectations (e.g., NGOs), MOEs, and concept of operations are presented, discussed, and refined as necessary to achieve final agreement. This agreement shows that both sides have committed to the development of this product.

4.1.1.2.9 *Baseline Stakeholder Expectations*
The set of stakeholder expectations (e.g., NGOs and MOEs) and the concept of operations that are agreed upon are now baselined. Any further changes will be required to go through a formal or informal (depending on the nature of the product) approval process involving both the stakeholder and the technical team.

4.1.1.2.10 *Capture Work Products*
In addition to developing, documenting, and baselining stakeholder expectations, the ConOps and MOEs discussed above and other work products from this process should be captured. These may include key decisions made, supporting decision rationale and assumptions, and lessons learned in performing these activities.

4.1.1.3 *Outputs*
Typical outputs for capturing stakeholder expectations include the following:

- **Validated Stakeholder Expectations:** These are the agreed-to set of expectations for this product layer. They are typically captured in the form of needs, goals, and objectives with constraints and assumptions identified. They may also be in the form of models or other graphical forms.

- **Concept of Operations:** The ConOps describes how the system will be operated during the life cycle phases that will meet stakeholder expectations. It describes the system characteristics from an operational perspective and helps facilitate an understanding of the system goals and objectives and other stakeholder expectations. Examples would be the ConOps document, model, or a Design Reference Mission (DRM).

- **Enabling Product Support Strategies:** These include any special provisions that might be needed for fabrication, test, deployment, operations sustainment, and disposal of the end product. They identify what support will be needed and any enabling products that will need to be developed in order to generate the end product.

- **Measures of Effectiveness:** A set of MOEs is developed based on the stakeholder expectations. These are measures that represent expectations that are critical to the success of the system, and failure to satisfy these measures will cause the stakeholder to deem the system unacceptable.

Other outputs that might be generated:

- **Human/Systems Function Allocation:** This describes the interaction of the hardware and software systems with all personnel and their supporting infrastructure. In many designs (e.g., human space flight) human operators are a critical total-system component and the roles and responsibilities of the humans-in-the-system should be clearly understood. This should include all human/system interactions required for a mission including assembly, ground operations, logistics, in-flight and ground maintenance, in-flight operations, etc.

4.1.2 Stakeholder Expectations Definition Guidance

Refer to Section 4.1.2 in the NASA Expanded Guidance for Systems Engineering at *https://nen.nasa.gov/web/se/doc-repository* for additional guidance on:

- Concept of Operations (including examples),
- protection of space assets, and
- identification of stakeholders for each phase.

4.2 Technical Requirements Definition

The Technical Requirements Definition Process transforms the stakeholder expectations into a definition of the problem and then into a complete set of validated technical requirements expressed as "shall" statements that can be used for defining a design solution for the Product Breakdown Structure (PBS) and related enabling products. The process of requirements definition is a recursive and iterative one that develops the stakeholders' requirements, product requirements, and lower level product/component requirements. The requirements should enable the description of all inputs, outputs, and required relationships between inputs and outputs, including constraints, and system interactions with operators, maintainers, and other systems. The requirements documents organize and communicate requirements to the customer and other stakeholders and the technical community.

NOTE: It is important to note that the team must not rely solely on the requirements received to design and build the system. Communication and iteration with the relevant stakeholders are essential to ensure a mutual understanding of each requirement. Otherwise, the designers run the risk of misunderstanding and implementing an unwanted solution to a different interpretation of the requirements. This iterative stakeholder communication is a critically important part of project validation. Always confirm that the right products and results are being developed.

Technical requirements definition activities apply to the definition of all technical requirements from the program, project, and system levels down to the lowest level product/component requirements document.

4.2.1 Process Description

FIGURE 4.2-1 provides a typical flow diagram for the Technical Requirements Definition Process and identifies typical inputs, outputs, and activities to consider in addressing technical requirements definition.

4.2.1.1 *Inputs*

Typical inputs needed for the requirements process include the following:

- **Baselined Stakeholder Expectations:** This is the agreed-to set of stakeholder expectations (e.g., needs, goals, objectives, assumptions, constraints, external interfaces) for the product(s) of this product layer.

- **Baselined Concept of Operations:** This describes how the system will be operated during the life cycle phases to meet stakeholder expectations. It describes the system characteristics from an operational perspective and helps facilitate an understanding of the system goals, objectives, and constraints. It includes scenarios, use cases, and/or Design Reference Missions (DRMs) as appropriate for the project. It may be in the form of a document, graphics, videos, models, and/or simulations.

- **Baselined Enabling Support Strategies:** These describe the enabling products that were identified in the Stakeholder Expectations Definition Process as needed to develop, test, produce, operate, or dispose of the end product. They also include descriptions of how the end product will be supported throughout the life cycle.

- **Measures of Effectiveness:** These MOEs were identified during the Stakeholder Expectations Definition Process as measures that the stakeholders deemed necessary to meet in order for the project to be considered a success (i.e., to meet success criteria).

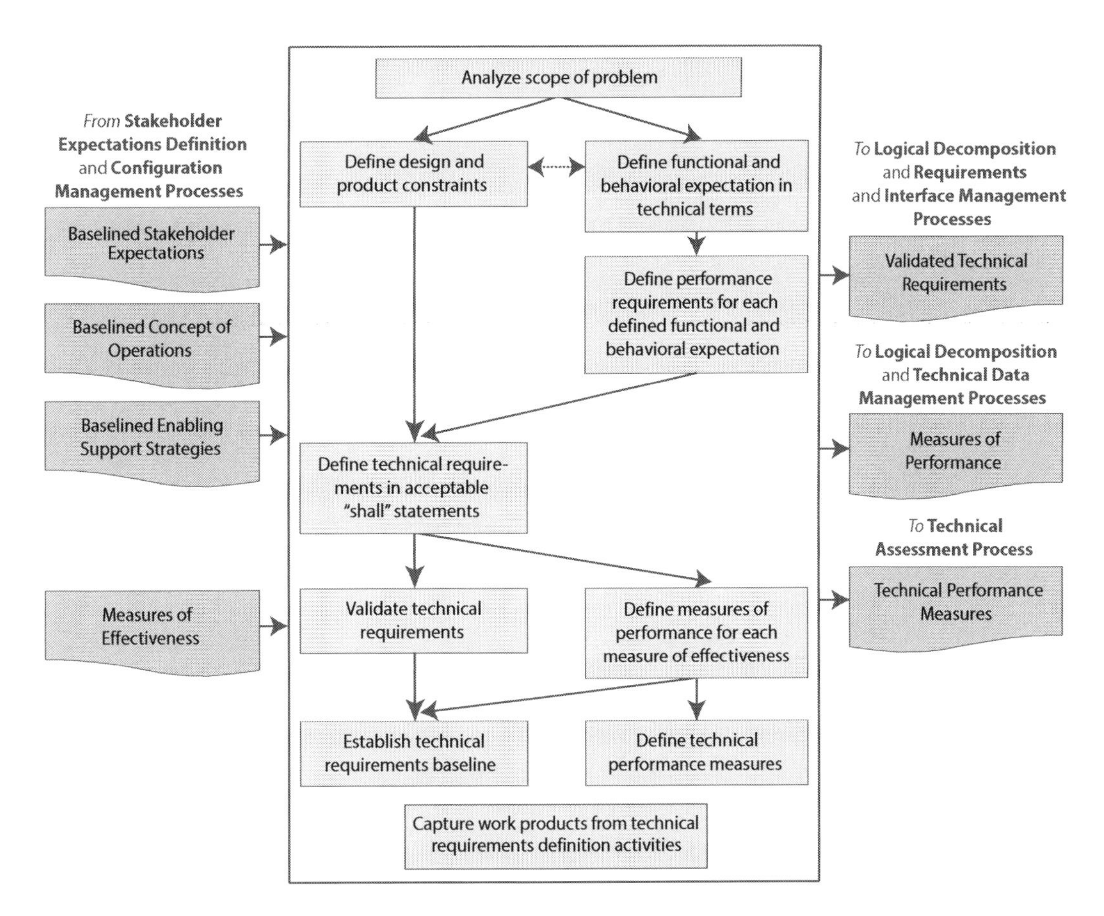

FIGURE 4.2-1 Technical Requirements Definition Process

Other inputs that might be useful in determining the technical requirements:

- **Human/Systems Function Allocation:** This describes the interaction of the hardware and software systems with all personnel and their supporting infrastructure. When human operators are a critical total-system component, the roles and responsibilities of the humans-in-the-system should be clearly understood. This should include all human/system interactions required for a mission including assembly, ground operations, logistics, in-flight and ground maintenance, in-flight operations, etc.

4.2.1.2 *Process Activities*

4.2.1.2.1 *Define Constraints, Functional and Behavioral Expectations*

The top-level requirements and expectations are initially assessed to understand the technical problem to be solved (scope of the problem) and establish the design boundary. This boundary is typically established by performing the following activities:

- Defining constraints that the design needs to adhere to or that limit how the system will be used. The constraints typically cannot be changed based on trade-off analyses.

- Identifying those elements that are already under design control and cannot be changed. This helps establish those areas where further trades will be made to narrow potential design solutions.

- Identifying external and enabling systems with which the system should interact and establishing physical and functional interfaces (e.g., mechanical, electrical, thermal, human, etc.).

- Defining functional and behavioral expectations for the range of anticipated uses of the system as identified in the ConOps. The ConOps describes how the system will be operated and the possible use-case scenarios.

4.2.1.2.2 *Define Requirements*

A complete set of project requirements includes those that are decomposed and allocated down to design elements through the PBS and those that cut across product boundaries. Requirements allocated to the PBS can be functional requirements (what functions need to be performed), performance requirements (how well these functions should be performed), and interface requirements (product to product interaction requirements). Crosscutting requirements include environmental, safety, human factors, and those that originate from the "-ilities" and from Design and Construction (D&C) standards. FIGURE 4.2-2 is a general overview on the flow of requirements, what they are called, and who is responsible (owns) for approving waivers.

- Functional requirements define what functions need to be performed to accomplish the objectives.
- Performance requirements define how well the system needs to perform the functions.

With an overall understanding of the constraints, physical/functional interfaces, and functional/behavioral expectations, the requirements can be further defined by establishing performance and other technical criteria. The expected performance is expressed as a quantitative measure to indicate how well each product function needs to be accomplished.

EXAMPLE OF FUNCTIONAL AND PERFORMANCE REQUIREMENTS

Initial Function Statement

The Thrust Vector Controller (TVC) shall provide vehicle control about the pitch and yaw axes.

This statement describes a high-level function that the TVC must perform. The technical team needs to transform this statement into a set of design-to functional and performance requirements.

Functional Requirements with Associated Performance Requirements

- The TVC shall gimbal the engine a maximum of 9 degrees, ± 0.1 degree.
- The TVC shall gimbal the engine at a maximum rate of 5 degrees/second ± 0.3 degrees/second.
- The TVC shall provide a force of 40,000 pounds, ± 500 pounds.
- The TVC shall have a frequency response of 20 Hz, ± 0.1 Hz.

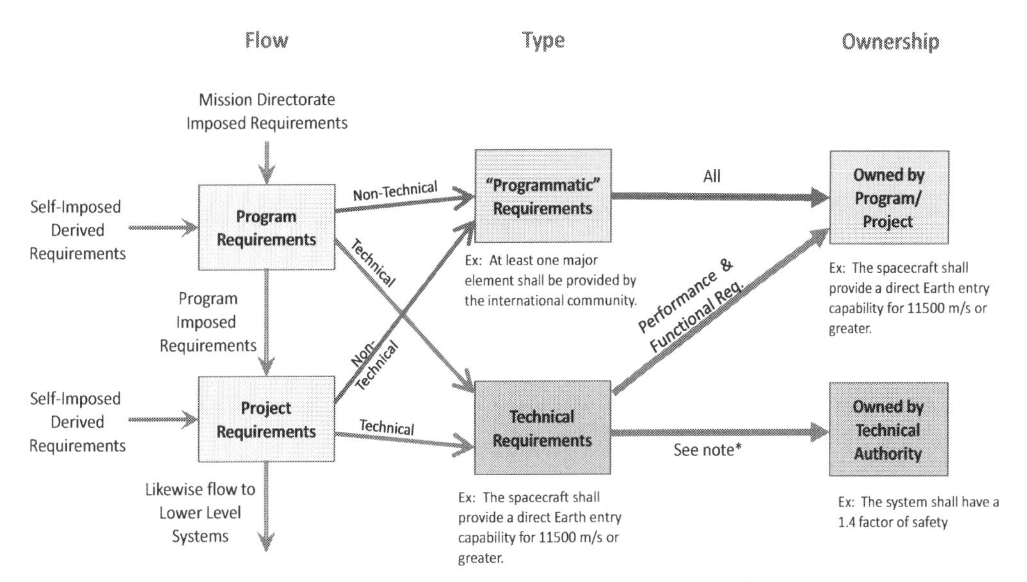

FIGURE 4.2-2 Flow, Type and Ownership of Requirements

NOTE: Requirements can be generated from non-obvious stakeholders and may not directly support the current mission and its objectives, but instead provide an opportunity to gain additional benefits or information that can support the Agency or the Nation. Early in the process, the systems engineer can help identify potential areas where the system can be used to collect unique information that is not directly related to the primary mission. Often outside groups are not aware of the system goals and capabilities until it is almost too late in the process.

Technical requirements come from a number of sources including functional, performance, interface, environmental, safety, human interfaces, standards and in support of the "'ilities" such as reliability, sustainability, producibility and others. Consideration and inclusion of all types of requirements is needed in order to form a complete and consistent set of technical requirements from which the system will be architected and designed. FIGURE 4.2-3 shows an example of parent and child requirement flowdown.

4.2.1.2.3 *Define Requirements in Acceptable Statements*

Finally, the requirements should be defined in acceptable "shall" statements, which are complete sentences with a single "shall" per statement. Rationale for the requirement should also be captured to ensure the reason and context of the requirement is understood. The Key Driving Requirements (KDRs) should be identified. These are requirements that can have a large impact on cost or schedule when implemented. A KDR can have any priority or criticality. Knowing the impact that a KDR has on the design allows better management of requirements.

See *Appendix C* for guidance and a checklist on how to write good requirements and *Appendix E* for validating requirements. A well-written requirements

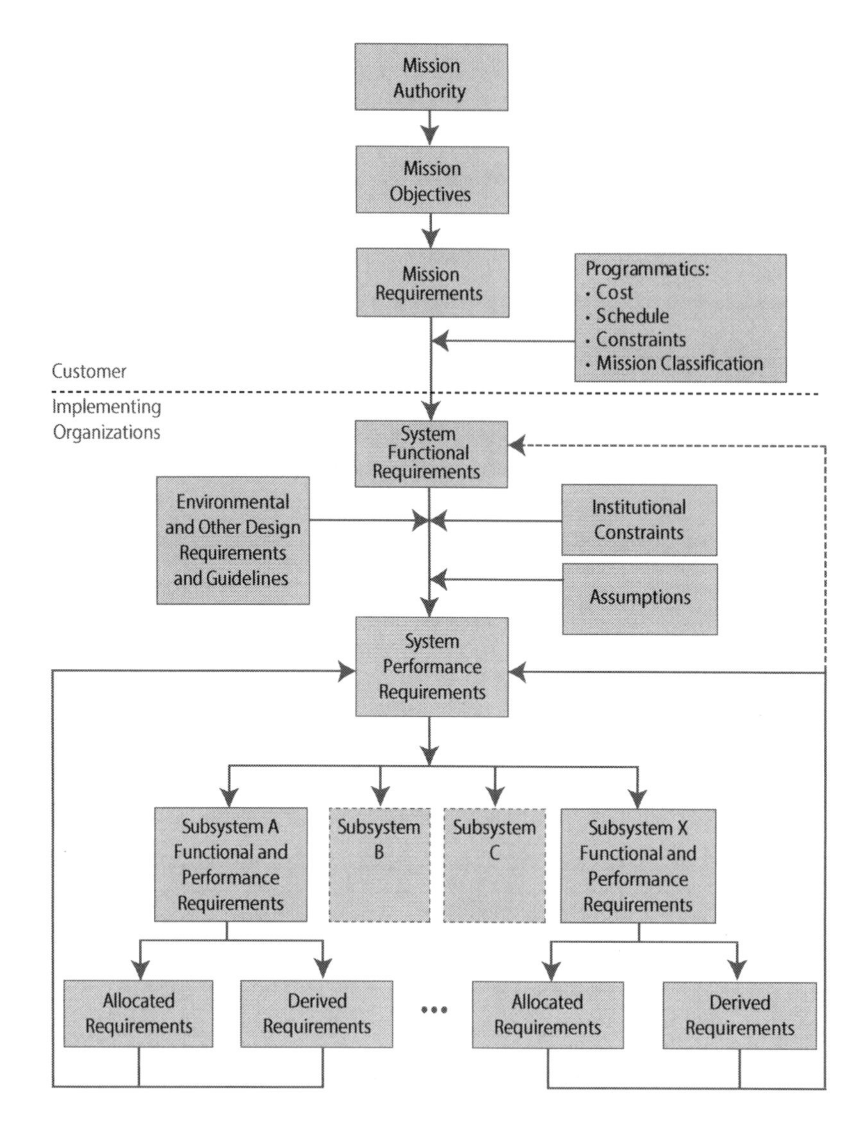

FIGURE 4.2-3 The Flowdown of Requirements

document provides several specific benefits to both the stakeholders and the technical team as shown in TABLE 4.2-1.

It is useful to capture information about each of the requirements, called metadata, for future reference and use. Many requirements management tools will request or have options for storing this type of information. TABLE 4.2-2 provides examples of the types of metadata that might be useful.

4.2.1.2.4 *Validate Technical Requirements*

An important part of requirements definition is the validation of the requirements against the stakeholder

TABLE 4.2-1 Benefits of Well-Written Requirements

Benefit	Rationale
Establish the basis for agreement between the stakeholders and the developers on what the product is to do	The complete description of the functions to be performed by the product specified in the requirements will assist the potential users in determining if the product specified meets their needs or how the product should be modified to meet their needs. During system design, requirements are allocated to subsystems (e.g., hardware, software, and other major components of the system), people, or processes.
Reduce the development effort because less rework is required to address poorly written, missing, and misunderstood requirements	The Technical Requirements Definition Process activities force the relevant stakeholders to rigorously consider all of the requirements before design begins. Careful review of the requirements can reveal omissions, misunderstandings, and inconsistencies early in the development cycle when these problems are easier to correct thereby reducing costly redesign, remanufacture, recoding, and retesting in later life cycle phases.
Provide a basis for estimating costs and schedules	The description of the product to be developed as given in the requirements is a realistic basis for estimating project costs and can be used to evaluate bids or price estimates.
Provide a baseline for verification and validation	Organizations can develop their verification and validation plans much more productively from a good requirements document. Both system and subsystem test plans and procedures are generated from the requirements. As part of the development, the requirements document provides a baseline against which compliance can be measured. The requirements are also used to provide the stakeholders with a basis for acceptance of the system.
Facilitate transfer	The requirements make it easier to transfer the product. Stakeholders thus find it easier to transfer the product to other parts of their organization, and developers find it easier to transfer it to new stakeholders or reuse it.
Serve as a basis for enhancement	The requirements serve as a basis for later enhancement or alteration of the finished product.

TABLE 4.2-2 Requirements Metadata

Item	Function
Requirement ID	Provides a unique numbering system for sorting and tracking.
Rationale	Provides additional information to help clarify the intent of the requirements at the time they were written. (See "Rationale" box below on what should be captured.)
Traced from	Captures the bidirectional traceability between parent requirements and lower level (derived) requirements and the relationships between requirements.
Owner	Person or group responsible for writing, managing, and/or approving changes to this requirement.
Verification method	Captures the method of verification (test, inspection, analysis, demonstration) and should be determined as the requirements are developed.
Verification lead	Person or group assigned responsibility for verifying the requirement.
Verification level	Specifies the level in the hierarchy at which the requirements will be verified (e.g., system, subsystem, element).

RATIONALE

The rationale should be kept up to date and include the following information:

- **Reason for the Requirement:** Often the reason for the requirement is not obvious, and it may be lost if not recorded as the requirement is being documented. The reason may point to a constraint or concept of operations. If there is a clear parent requirement or trade study that explains the reason, then it should be referenced.
- **Document Assumptions:** If a requirement was written assuming the completion of a technology development program or a successful technology mission, the assumption should be documented.
- **Document Relationships:** The relationships with the product's expected operations (e.g., expectations about how stakeholders will use a product) should be documented. This may be done with a link to the ConOps.
- **Document Design Constraints:** Constraints imposed by the results from decisions made as the design evolves should be documented. If the requirement states a method of implementation, the rationale should state why the decision was made to limit the solution to this one method of implementation.

expectations, the mission objectives and constraints, the concept of operations, and the mission success criteria. Validating requirements can be broken into six steps:

1. **Are the Requirements Written Correctly?** Identify and correct requirements "shall" statement format errors and editorial errors.
2. **Are the Requirements Technically Correct?** A few trained reviewers from the technical team identify and remove as many technical errors as possible before having all the relevant stakeholders review the requirements. The reviewers should check that the requirement statements (a) have bidirectional traceability to the baselined stakeholder expectations; (b) were formed using valid assumptions; and (c) are essential to and consistent with designing and realizing the appropriate product solution form that will satisfy the applicable product life cycle phase success criteria.
3. **Do the Requirements Satisfy Stakeholders?** All relevant stakeholder groups identify and remove defects.
4. **Are the Requirements Feasible?** All requirements should make technical sense and be possible to achieve.
5. **Are the Requirements Verifiable?** All requirements should be stated in a fashion and with enough information that it will be possible to verify the requirement after the end product is implemented.
6. **Are the Requirements Redundant or Over-specified?** All requirements should be unique (not redundant to other requirements) and necessary to meet the required functions, performance, or behaviors.

Requirements validation results are often a deciding factor in whether to proceed with the next process of Logical Decomposition or Design Solution Definition. The project team should be prepared to: (1) demonstrate that the project requirements are complete and understandable; (2) demonstrate that evaluation criteria are consistent with requirements and the operations and logistics concepts; (3) confirm that requirements and MOEs are consistent with stakeholder needs; (4) demonstrate that operations and architecture concepts support mission needs, goals, objectives, assumptions, guidelines, and constraints; and (5) demonstrate that the process for managing change in requirements is established, documented in the project information repository, and communicated to stakeholders.

4.2.1.2.5 *Define MOPs and TPMs*

Measures of Performance (MOPs) define the performance characteristics that the system should exhibit when fielded and operated in its intended environment. MOPs are derived from the MOEs but are stated in more technical terms from the supplier's point of view. Typically, multiple MOPs, which are quantitative and measurable, are needed to satisfy a MOE, which can be qualitative. From a verification and acceptance point of view, MOPs reflect the system characteristics deemed necessary to achieve the MOEs.

Technical Performance Measures (TPMs) are physical or functional characteristics of the system associated with or established from the MOPs that are deemed critical or key to mission success. The TPMs are monitored during implementation by comparing the current actual achievement or best estimate of the parameters with the values that were anticipated for the current time and projected for future dates. They are used to confirm progress and identify deficiencies that might jeopardize meeting a critical system requirement or put the project at cost or schedule risk.

For additional information on MOPs and TPMs, their relationship to each other and MOEs, and examples of each, see Section 6.7.2.6.2 of the NASA Expanded Guidance for SE document at *https://nen.nasa.gov/web/se/doc-repository.*

4.2.1.2.6 *Establish Technical Requirement Baseline*

Once the technical requirements are identified and validated to be good (clear, correct, complete, and achievable) requirements, and agreement has been gained by the customer and key stakeholders, they are baselined and placed under configuration control. Typically, a System Requirements Review (SRR) is held to allow comments on any needed changes and to gain agreement on the set of requirements so that it may be subsequently baselined. For additional information on the SRR, see Section 6.7.

4.2.1.2.7 *Capture Work Products*

The work products generated during the above activities should be captured along with key decisions that were made, any supporting decision rationale and assumptions, and lessons learned in performing these activities.

4.2.1.3 *Outputs*

- **Validated Technical Requirements:** This is the approved set of requirements that represents a complete description of the problem to be solved and requirements that have been validated and approved by the customer and stakeholders. Examples of documents that capture the requirements are a System Requirements Document (SRD), Project Requirements Document (PRD), Interface Requirements Document (IRD), and a Software Requirements Specification (SRS).

- **Measures of Performance:** These are the identified quantitative measures that, when met by the design solution, help ensure that one or more MOEs will be satisfied. There may be two or more MOPs for each MOE. See Section 6.7.2.6.2in the NASA Expanded Guidance for Systems Engineering at *https://nen.nasa.gov/web/se/doc-repository* for further details.

- **Technical Performance Measures:** These are the set of performance measures that are monitored and trended by comparing the current actual achievement of the parameters with that expected or required at the time. TPMs are used to confirm progress and identify deficiencies. See Section 6.7.2.6.2 in the NASA Expanded Guidance for Systems Engineering at *https://nen.nasa.gov/web/se/doc-repository* for further details.

4.2.2 Technical Requirements Definition Guidance

Refer to Section 4.2.2 of the NASA Expanded Guidance for SE document at *https://nen.nasa.gov/web/se/doc-repository* for additional information on:

- types of requirements,
- requirements databases, and
- the use of technical standards.

4.3 Logical Decomposition

Logical decomposition is the process for creating the detailed functional requirements that enable NASA programs and projects to meet the stakeholder expectations. This process identifies the "what" that should be achieved by the system at each level to enable a successful project. Logical decomposition utilizes functional analysis to create a system architecture and to decompose top-level (or parent) requirements and allocate them down to the lowest desired levels of the project.

The Logical Decomposition Process is used to:

- Improve understanding of the defined technical requirements and the relationships among the requirements (e.g., functional, performance, behavioral, and temporal etc.), and
- Decompose the parent requirements into a set of logical decomposition models and their associated sets of derived technical requirements for input to the Design Solution Definition Process.

4.3.1 Process Description

FIGURE 4.3-1 provides a typical flow diagram for the Logical Decomposition Process and identifies typical inputs, outputs, and activities to consider in addressing logical decomposition.

4.3.1.1 *Inputs*

Typical inputs needed for the Logical Decomposition Process include the following:

- **Technical Requirements:** A validated set of requirements that represent a description of the problem to be solved, have been established by functional and performance analysis, and have been approved by the customer and other stakeholders. Examples of documents that capture the requirements are an SRD, PRD, and IRD.
- **Technical Measures:** An established set of measures based on the expectations and requirements that will be tracked and assessed to determine overall system or product effectiveness and customer satisfaction. These measures are MOEs, MOPs, and a special subset of these called TPMs. See Section 6.7.2.6.2 in the NASA Expanded Guidance for Systems Engineering at *https://nen.nasa.gov/web/se/doc-repository* for further details.

4.3.1.2 *Process Activities*

4.3.1.2.1 *Define One or More Logical Decomposition Models*

The key first step in the Logical Decomposition Process is establishing the system architecture model. The system architecture activity defines the

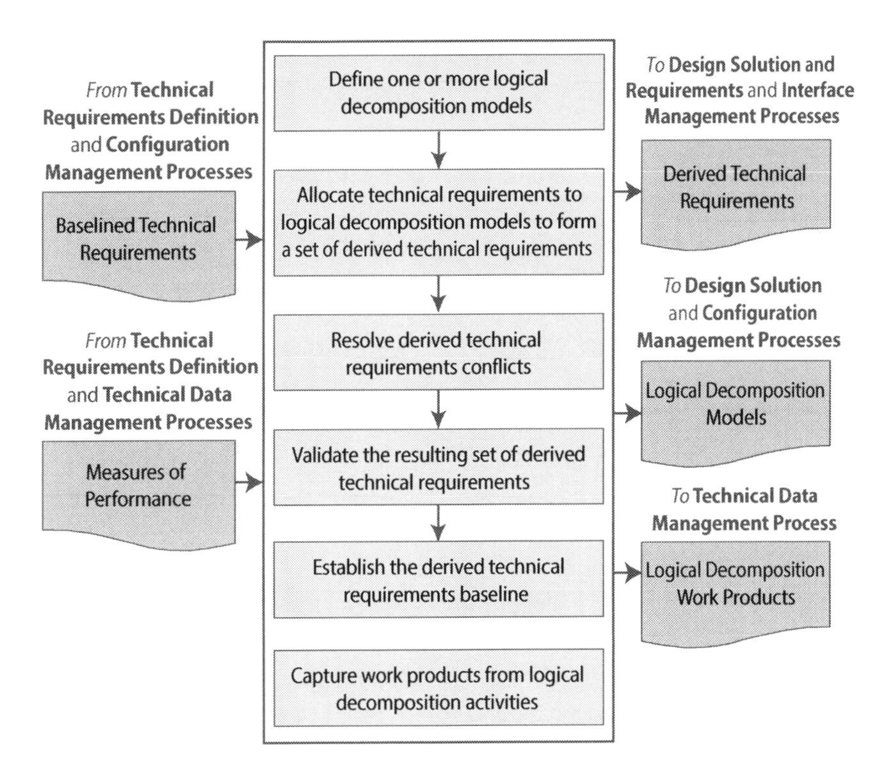

FIGURE 4.3-1 Logical Decomposition Process

underlying structure and relationships of hardware, software, humans-in-the-loop, support personnel, communications, operations, etc., that provide for the implementation of Agency, mission directorate, program, project, and subsequent levels of the requirements. System architecture activities drive the partitioning of system elements and requirements to lower level functions and requirements to the point that design work can be accomplished. Interfaces and relationships between partitioned subsystems and elements are defined as well.

Once the top-level (or parent) functional requirements and constraints have been established, the system designer uses functional analysis to begin to formulate a conceptual system architecture. The system architecture can be seen as the strategic organization of the functional elements of the system, laid out to enable the roles, relationships, dependencies, and interfaces between elements to be clearly defined and understood. It is strategic in its focus on the overarching structure of the system and how its elements fit together to contribute to the whole, instead of on the particular workings of the elements themselves. It enables the elements to be developed separately from each other while ensuring that they work together effectively to achieve the top-level (or parent) requirements.

Much like the other elements of functional decomposition, the development of a good system-level architecture is a creative, recursive, collaborative, and iterative process that combines an excellent understanding of the project's end objectives and constraints with an equally good knowledge of various potential technical means of delivering the end products.

Focusing on the project's ends, top-level (or parent) requirements, and constraints, the system architect should develop at least one, but preferably multiple, concept architectures capable of achieving program objectives. Each architecture concept involves specification of the functional elements (what the pieces do), their relationships to each other (interface definition), and the ConOps, i.e., how the various segments, subsystems, elements, personnel, units, etc., will operate as a system when distributed by location and environment from the start of operations to the end of the mission.

The development process for the architectural concepts should be recursive and iterative with feedback from stakeholders and external reviewers, as well as from subsystem designers and operators, provided as often as possible to increase the likelihood of effectively achieving the program's desired ends while reducing the likelihood of cost and schedule overruns.

In the early stages of development, multiple concepts are generated. Cost and schedule constraints will ultimately limit how long a program or project can maintain multiple architectural concepts. For all NASA programs, architecture design is completed during the Formulation Phase. For most NASA projects (and tightly coupled programs), the baselining of a single architecture happens during Phase A. Architectural changes at higher levels occasionally occur as decomposition to lower levels produces complexity in design, cost, or schedule that necessitates such changes. However, as noted in FIGURE 2.5-1, the later in the development process that changes occur, the more expensive they become.

Aside from the creative minds of the architects, there are multiple tools that can be utilized to develop a system's architecture. These are primarily modeling and simulation tools, functional analysis tools, architecture frameworks, and trade studies. (For example, one way of doing architecture is the Department of Defense (DOD) Architecture Framework (DODAF). A search concept is developed, and analytical models of the architecture, its elements, and their operations are developed with increased fidelity as the project evolves. Functional decomposition, requirements development, and trade studies are subsequently undertaken. Multiple iterations of these activities feed back to the evolving architectural concept as the requirements flow down and the design matures.

4.3.1.2.2 *Allocate Technical Requirements, Resolve Conflicts, and Baseline*

Functional analysis is the primary method used in system architecture development and functional requirement decomposition. It is the systematic process of identifying, describing, and relating the functions a system should perform to fulfill its goals and objectives. Functional analysis identifies and links system functions, trade studies, interface characteristics, and rationales to requirements. It is usually based on the ConOps for the system of interest.

Three key steps in performing functional analysis are:

1. Translate top-level requirements into functions that should be performed to accomplish the requirements.
2. Decompose and allocate the functions to lower levels of the product breakdown structure.
3. Identify and describe functional and subsystem interfaces.

The process involves analyzing each system requirement to identify all of the functions that need to be performed to meet the requirement. Each function identified is described in terms of inputs, outputs, failure modes, consequence of failure, and interface requirements. The process is repeated from the top down so that sub-functions are recognized as part of larger functional areas. Functions are arranged in a logical sequence so that any specified operational usage of the system can be traced in an end-to-end path.

The process is recursive and iterative and continues until all desired levels of the architecture/system have been analyzed, defined, and baselined. There will almost certainly be alternative ways to decompose functions. For example, there may be several ways to communicate with the crew: Radio Frequency (RF), laser, Internet, etc. Therefore, the outcome is highly dependent on the creativity, skills, and experience of the engineers doing the analysis. As the analysis proceeds to lower levels of the architecture and system, and the system is better understood, the systems engineer should keep an open mind and a willingness to go back and change previously established architecture and system requirements. These changes will then have to be decomposed down through the architecture and sub-functions again with the recursive process continuing until the system is fully defined with all of the requirements understood and known to be viable, verifiable, and internally consistent. Only at that point should the system architecture and requirements be baselined.

4.3.1.2.3 *Capture Work Products*

The other work products generated during the Logical Decomposition Process should be captured along with key decisions made, supporting decision rationale and assumptions, and lessons learned in performing the activities.

4.3.1.3 *Outputs*

Typical outputs of the Logical Decomposition Process include the following:

- **Logical Decomposition Models:** These models define the relationship of the requirements and functions and their behaviors. They include the system architecture models that define the underlying structure and relationship of the elements of the system (e.g., hardware, software, humans-in-the-loop, support personnel, communications, operations, etc.) and the basis for the partitioning of requirements into lower levels to the point that design work can be accomplished.
- **Derived Technical Requirements:** These are requirements that arise from the definitions of the selected architecture that were not explicitly stated in the baselined requirements that served as an input to this process. Both the baselined and derived requirements are allocated to the system architecture and functions.
- **Logical Decomposition Work Products:** These are the other products generated by the activities of this process.

4.3.2 Logical Decomposition Guidance

Refer to Section 4.3.2 and Appendix F in the NASA Expanded Guidance for Systems Engineering at *https://nen.nasa.gov/web/se/doc-repository* for additional guidance on:

- Product Breakdown Structures and
- Functional Analysis Techniques.

4.4 Design Solution Definition

The Design Solution Definition Process is used to translate the high-level requirements derived from the stakeholder expectations and the outputs of the Logical Decomposition Process into a design solution. This involves transforming the defined logical decomposition models and their associated sets of derived technical requirements into alternative solutions. These alternative solutions are then analyzed through detailed trade studies that result in the selection of a preferred alternative. This preferred alternative is then fully defined into a final design solution that satisfies the technical requirements. This design solution definition is used to generate the end product specifications that are used to produce the product and to conduct product verification. This process may be further refined depending on whether there are additional subsystems of the end product that need to be defined.

4.4.1 Process Description

FIGURE 4.4-1 provides a typical flow diagram for the Design Solution Definition Process and identifies typical inputs, outputs, and activities to consider in addressing design solution definition.

4.4.1.1 *Inputs*

There are several fundamental inputs needed to initiate the Design Solution Definition Process:

- **Technical Requirements:** These are the customer and stakeholder needs that have been translated into a complete set of validated requirements for the system, including all interface requirements.
- **Logical Decomposition Models:** Requirements are analyzed and decomposed by one or more different methods (e.g., function, time, behavior, data flow, states, modes, system architecture, etc.) in order to gain a more comprehensive understanding of their interaction and behaviors. (See the definition of a model in *Appendix B*.)

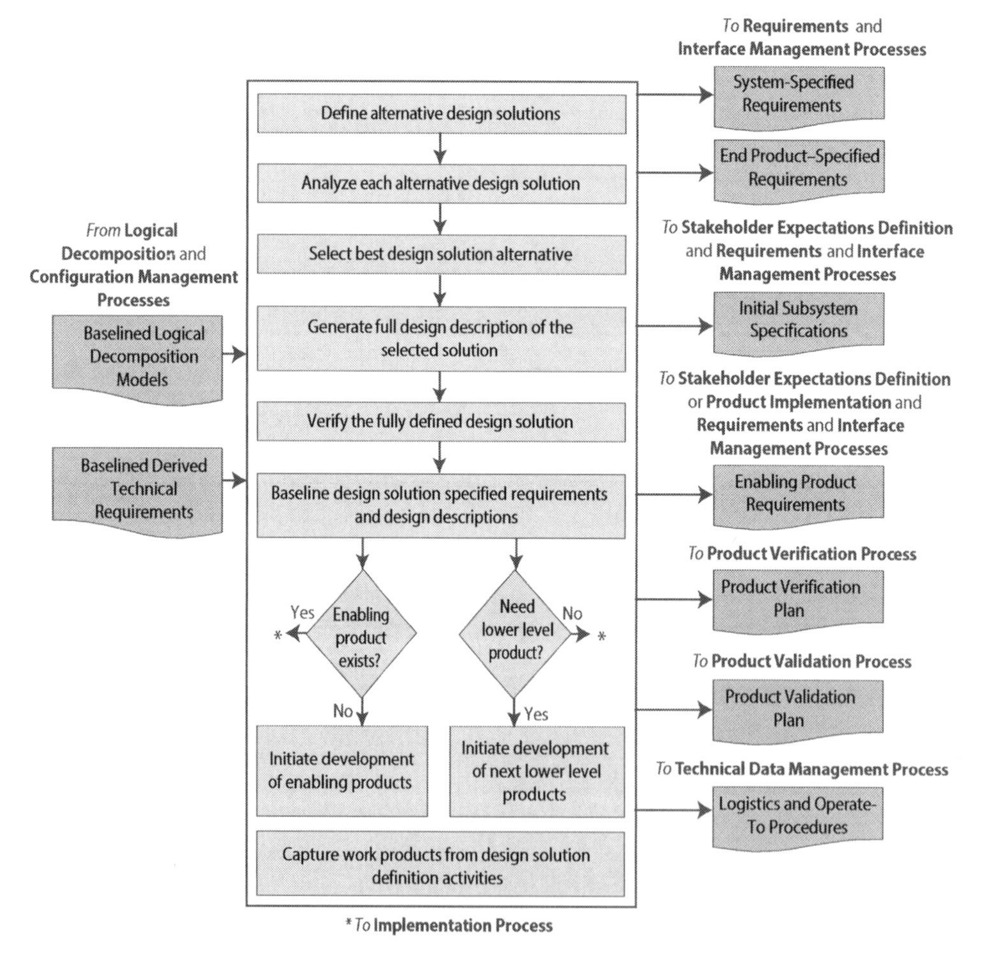

FIGURE 4.4-1 Design Solution Definition Process

4.4.1.2 *Process Activities*

4.4.1.2.1 *Define Alternative Design Solutions*

The realization of a system over its life cycle involves a succession of decisions among alternative courses of action. If the alternatives are precisely defined and thoroughly understood to be well differentiated in the cost-effectiveness space, then the systems engineer can make choices among them with confidence.

To obtain assessments that are crisp enough to facilitate good decisions, it is often necessary to delve more deeply into the space of possible designs than has yet been done, as illustrated in FIGURE 4.4-2. It should be realized, however, that this illustration represents neither the project life cycle, which encompasses the system development process from inception through disposal, nor the product development process by which the system design is developed and implemented.

Each "create concepts" step in FIGURE 4.4-2 involves a recursive and iterative design loop driven by the set of stakeholder expectations where a straw man architecture/design, the associated ConOps, and the derived requirements are developed and programmatic constraints such as cost and schedule are considered. These three products should be consistent with each other and will require iterations and design decisions to achieve this consistency. This recursive and iterative design loop is illustrated in FIGURE 4.0-1.

Each "create concepts" step in FIGURE 4.4-2 also involves an assessment of potential capabilities offered by the continually changing state of technology and potential pitfalls captured through experience-based review of prior program/project lessons learned data. It is imperative that there be a continual interaction between the technology development process, crosscutting processes such as human systems integration, and the design process to ensure that the design reflects the realities of the available technology and that overreliance on immature technology is avoided.

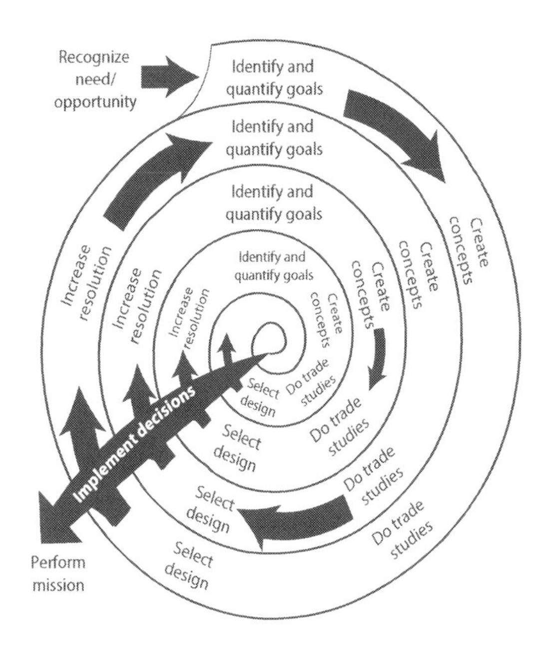

FIGURE 4.4-2 The Doctrine of Successive Refinement

Additionally, the state of any technology that is considered enabling should be properly monitored, and care should be taken when assessing the impact of this technology on the concept performance. This interaction is facilitated through a periodic assessment of the design with respect to the maturity of the technology required to implement the design. (See Section 4.4.2.1 in the NASA Expanded Guidance for Systems Engineering at ***https://nen.nasa.gov/web/se/doc-repository*** for a more detailed discussion of technology assessment.) These technology elements usually exist at a lower level in the PBS. Although the process of design concept development by the integration of lower level elements is a part of the systems engineering process, there is always a danger that the top-down process cannot keep up with the bottom-up process. Therefore, system architecture issues need to be resolved early so that the system can be modeled with sufficient realism to do reliable trade studies.

As the system is realized, its particulars become clearer—but also harder to change. See the rising

"Cost to Change Design Direction" in FIGURE 2.5-1. The purpose of systems engineering is to make sure that the Design Solution Definition Process happens in a way that leads to the most functional, safe, and cost-effective final system while working within any given schedule boundaries. The basic idea is that before those decisions that are hard to undo are made, the alternatives should be carefully and iteratively assessed, particularly with respect both to the maturity of the required technology and to stakeholder expectations for efficient, effective operations.

4.4.1.2.2 *Create Alternative Design Concepts*

Once it is understood what the system is to accomplish, it is possible to devise a variety of ways that those goals can be met. Sometimes, that comes about as a consequence of considering alternative functional allocations and integrating available subsystem design options, all of which can have technologies at varying degrees of maturity. Ideally, as wide a range of plausible alternatives as is consistent with the design organization's charter should be defined, keeping in mind the current stage in the process of successive refinement. When the bottom-up process is operating, a problem for the systems engineer is that the designers tend to become fond of the designs they create, so they lose their objectivity; the systems engineer should stay an "outsider" so that there is more objectivity. This is particularly true in the assessment of the technological maturity of the subsystems and components required for implementation. There is a tendency on the part of technology developers and project management to overestimate the maturity and applicability of a technology that is required to implement a design. This is especially true of "heritage" equipment. The result is that critical aspects of systems engineering are often overlooked.

The creation of alternative design solutions involves assessment of potential capabilities offered by the continually changing state of technology. A continual interaction between the technology development process and the design process ensures that the design reflects the realities of the available technology. This interaction is facilitated through periodic assessment of the design with respect to the maturity of the technology required to implement the design.

After identifying the technology gaps existing in a given design concept, it is frequently necessary to undertake technology development in order to ascertain viability. Given that resources will always be limited, it is necessary to pursue only the most promising technologies that are required to enable a given concept.

If requirements are defined without fully understanding the resources required to accomplish needed technology developments, then the program/project is at risk. Technology assessment should be done iteratively until requirements and available resources are aligned within an acceptable risk posture. Technology development plays a far greater role in the life cycle of a program/project than has been traditionally considered, and it is the role of the systems engineer to develop an understanding of the extent of program/project impacts—maximizing benefits and minimizing adverse effects. Traditionally, from a program/project perspective, technology development has been associated with the development and incorporation of any "new" technology necessary to meet requirements. However, a frequently overlooked area is that associated with the modification of "heritage" systems incorporated into different architectures and operating in different environments from the ones for which they were designed. If the required modifications and/or operating environments fall outside the realm of experience, then these too should be considered technology development.

To understand whether or not technology development is required—and to subsequently quantify the associated cost, schedule, and risk—it is necessary to systematically assess the maturity of each system, subsystem, or component in terms of the architecture and operational environment. It is then necessary to assess what is required in the way of development to advance the maturity to a point where it can successfully be incorporated within cost, schedule, and performance constraints. A process for accomplishing this assessment is described in *Appendix G*. Because technology development has the potential for such significant impacts on a program/project, technology assessment needs to play a role throughout the design and development process from concept development through Preliminary Design Review (PDR). Lessons learned from a technology development point of view should then be captured in the final phase of the program.

On the first turn of the successive refinement in FIGURE 4.4-2, the subject is often general approaches or strategies, sometimes architectural concepts. On the next, it is likely to be functional design, then detailed design, and so on. The reason for avoiding a premature focus on a single design is to permit discovery of the truly best design. Part of the systems engineer's job is to ensure that the design concepts to be compared take into account all interface requirements. Characteristic questions include: "Did you include the cabling?" or "Did you consider how the maintainers can repair the system?" When possible, each design concept should be described in terms of controllable design parameters so that each represents as wide a class of designs as is reasonable. In doing so, the systems engineer should keep in mind that the potentials for change may include organizational structure, personnel constraints, schedules, procedures, and any of the other things that make up a system. When possible, constraints should also be described by parameters.

4.4.1.2.3 *Analyze Each Alternative Design Solution*

The technical team analyzes how well each of the design alternatives meets the system objectives (technology gaps, effectiveness, technical achievability, performance, cost, schedule, and risk, both quantified and otherwise). This assessment is accomplished through the use of trade studies. The purpose of the trade study process is to ensure that the system architecture, intended operations (i.e., the ConOps) and design decisions move toward the best solution that can be achieved with the available resources. The basic steps in that process are:

- Devise some alternative means to meet the functional requirements. In the early phases of the project life cycle, this means focusing on system architectures; in later phases, emphasis is given to system designs.
- Evaluate these alternatives in terms of the MOPs and system life cycle cost. Mathematical models are useful in this step not only for forcing recognition of the relationships among the outcome variables, but also for helping to determine what the MOPs should be quantitatively.
- Rank the alternatives according to appropriate selection criteria.
- Drop less promising alternatives and proceed to the next level of resolution, if needed.

The trade study process should be done openly and inclusively. While quantitative techniques and rules are used, subjectivity also plays a significant role. To make the process work effectively, participants should have open minds, and individuals with different skills—systems engineers, design engineers, crosscutting specialty discipline and domain engineers, program analysts, system end users, decision scientists, maintainers, operators, and project managers—should cooperate. The right quantitative methods and selection criteria should be used. Trade study assumptions, models, and results should be documented as part of the project archives. The participants should remain focused on the functional requirements, including those for enabling products. For an in-depth discussion of the trade study process, see *Section 6.8*. The ability to perform these studies is enhanced by the development of system models that relate the design parameters to those assessments, but it does not depend upon them.

The technical team should consider a broad range of concepts when developing the system model. The model should define the roles of crew, operators, maintainers, logistics, hardware, and software in the system. It should identify the critical technologies required to implement the mission and should consider the entire life cycle from fabrication to disposal. Evaluation criteria for selecting concepts should be established. Cost is always a limiting factor. However, other criteria, such as time to develop and certify a unit, risk, and reliability, also are critical. This stage cannot be accomplished without addressing the roles of operators and maintainers. These contribute significantly to life cycle costs and to the system reliability. Reliability analysis should be performed based upon estimates of component failure rates for hardware and an understanding of the consequences of these failures. If probabilistic risk assessment models are applied, it may be necessary to include occurrence rates or probabilities for software faults or human error events. These models should include hazard analyses and controls implemented through fault management. Assessments of the maturity of the required technology should be done and a technology development plan developed.

Controlled modification and development of design concepts, together with such system models, often permits the use of formal optimization techniques to find regions of the design space that warrant further investigation.

Whether system models are used or not, the design concepts are developed, modified, reassessed, and compared against competing alternatives in a closed-loop process that seeks the best choices for further development. System and subsystem sizes are often determined during the trade studies. The end result is the determination of bounds on the relative cost-effectiveness of the design alternatives, measured in terms of the quantified system goals. (Only bounds, rather than final values, are possible because determination of the final details of the design is intentionally deferred.) Increasing detail associated with the continually improving resolution reduces the spread between upper and lower bounds as the process proceeds.

4.4.1.2.4 *Select the Best Design Solution Alternative*

The technical team selects the best design solution from among the alternative design concepts, taking into account subjective factors that the team was unable to quantify, such as robustness, as well as estimates of how well the alternatives meet the quantitative requirements; the maturity of the available technology; and any effectiveness, cost, schedule, risk, or other constraints.

The Decision Analysis Process, as described in *Section 6.8*, should be used to make an evaluation of the alternative design concepts and to recommend the "best" design solution.

When it is possible, it is usually well worth the trouble to develop a mathematical expression, called an "objective function," that expresses the values of combinations of possible outcomes as a single measure of cost-effectiveness, as illustrated in FIGURE 4.4-3, even if both cost and effectiveness should be described by more than one measure.

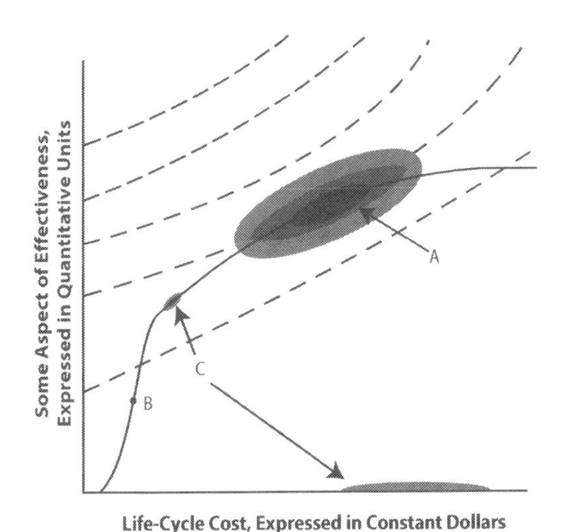

FIGURE 4.4-3 A Quantitative Objective Function, Dependent on Life Cycle Cost and All Aspects of Effectiveness

Note: The different shaded areas indicate different levels of uncertainty. Dashed lines represent constant values of objective function (cost-effectiveness). Higher values of cost-effectiveness are achieved by moving toward upper left. A, B, and C are design concepts with different risk patterns.

The objective function (or "cost function") assigns a real number to candidate solutions or "feasible solutions" in the alternative space or "search space." A feasible solution that minimizes (or maximizes, if that is the goal) the objective function is called an "optimal solution." When achievement of the goals can be quantitatively expressed by such an objective function, designs can be compared in terms of their value. Risks associated with design concepts can cause these evaluations to be somewhat nebulous because they are uncertain and are best described by probability distributions.

In FIGURE 4.4-3, the risks are relatively high for design concept A. There is little risk in either effectiveness or cost for concept B, while the risk of an expensive failure is high for concept C, as is shown by the cloud of probability near the *x* axis with a high cost and essentially no effectiveness. Schedule factors may affect the effectiveness and cost values and the risk distributions.

The mission success criteria for systems differ significantly. In some cases, effectiveness goals may be much more important than all others. Other projects may demand low costs, have an immutable schedule, or require minimization of some kinds of risks. Rarely (if ever) is it possible to produce a combined quantitative measure that relates all of the important factors, even if it is expressed as a vector with several components. Even when that can be done, it is essential that the underlying actors and relationships be thoroughly revealed to and understood by the systems engineer. The systems engineer should weigh the importance of the unquantifiable factors along with the quantitative data.

Technical reviews of the data and analyses, including technology maturity assessments, are an important part of the decision support packages prepared for the technical team. The decisions that are made are generally entered into the configuration management system as changes to (or elaborations of) the system baseline. The supporting trade studies are archived for future use. An essential feature of the systems engineering process is that trade studies are performed before decisions are made. They can then be baselined with much more confidence.

4.4.1.2.5 *Increase the Resolution of the Design*

The successive refinement process of FIGURE 4.4-2 illustrates a continuing refinement of the system design. At each level of decomposition, the baselined derived (and allocated) requirements become the set of high-level requirements for the decomposed elements, and the process begins again. One might ask, "When do we stop refining the design?" The answer is that the design effort proceeds to a depth that is sufficient to meet several needs: the design should penetrate sufficiently to allow analytical validation of the design to the requirements and ConOps; it should also have sufficient depth to support cost and operations modeling and to convince a review team of a feasible design with performance, cost, and risk margins.

The systems engineering engine is applied again and again as the system is developed. As the system is realized, the issues addressed evolve and the particulars of the activity change. Most of the major system decisions (goals, architecture, acceptable life cycle cost, etc.) are made during the early phases of the project, so the successive refinements do not correspond precisely to the phases of the system life cycle. Much of the system architecture can be seen even at the outset, so the successive refinements do not correspond exactly to development of the architectural hierarchy either. Rather, they correspond to the successively greater resolution by which the system is defined.

It is reasonable to expect the system to be defined with better resolution as time passes. This tendency is formalized at some point (in Phase B) by defining a baseline system definition. Usually, the goals, objectives, and constraints are baselined as the requirements portion of the baseline. The entire baseline is then placed under configuration control in an attempt to ensure that any subsequent changes are indeed justified and affordable.

At this point in the systems engineering process, there is a logical branch point. For those issues for which the process of successive refinement has proceeded far enough, the next step is to implement the decisions at that level of resolution. For those issues that are still insufficiently resolved, the next step is to refine the development further.

4.4.1.2.6 *Fully Describe the Design Solution*

Once the preferred design alternative has been selected and the proper level of refinement has been completed, then the design is fully defined into a final design solution that will satisfy the technical requirements and ConOps. The design solution definition will be used to generate the end product specifications that will be used to produce the product and to conduct product verification. This process may be further refined depending on whether there are additional subsystems of the end product that need to be defined.

The scope and content of the full design description should be appropriate for the product life cycle phase, the phase success criteria, and the product position in the PBS (system structure). Depending on these factors, the form of the design solution definition could be simply a simulation model or a paper study report. The technical data package evolves from phase to phase, starting with conceptual sketches or models and ending with complete drawings, parts list, and other details needed for product implementation or product integration. Typical output definitions from the Design Solution Definition Process are shown in FIGURE 4.4-1 and are described in *Section 4.4.1.3*.

4.4.1.2.7 *Verify the Design Solution*

Once an acceptable design solution has been selected from among the various alternative designs and documented in a technical data package, the design solution should next be verified against the system requirements and constraints. A method to achieve this verification is by means of a peer review to evaluate the resulting design solution definition. Guidelines for conducting a peer review are discussed in Section 6.7.2.4.5.

In addition, peer reviews play a significant role as a detailed technical component of higher level technical and programmatic reviews. For example, the peer review of a component battery design can go into much more technical detail on the battery than the integrated power subsystem review. Peer reviews can cover the components of a subsystem down to the level appropriate for verifying the design against the requirements. Concerns raised at the peer review might have implications on the power subsystem design and verification and therefore should be reported at the next higher level review of the power subsystem.

The verification should show that the design solution definition:

- Is realizable within the constraints imposed on the technical effort;
- Has specified requirements that are stated in acceptable statements and have bidirectional traceability with the technical requirements and stakeholder expectations; and
- Has decisions and assumptions made in forming the solution consistent with its set of technical requirements and identified system product and service constraints.

This design solution verification is in contrast to the verification of the end product described in the end product verification plan which is part of the technical data package. That verification occurs in a later life cycle phase and is a result of the Product Verification Process (see Section 5.3) applied to the realization of the design solution as an end product.

4.4.1.2.8 *Validate the Design Solution*

The validation of the design solution is a recursive and iterative process as shown in FIGURE 4.0-1. Each alternative design concept is validated against the set of stakeholder expectations. The stakeholder expectations drive the iterative design loop in which a straw man architecture/design, the ConOps, and the derived requirements are developed. These three products should be consistent with each other and will require iterations and design decisions to achieve this consistency. Once consistency is achieved, functional analyses allow the study team to validate the design against the stakeholder expectations. A simplified validation asks the questions: Does the system work as expected? How does the system respond to failures, faults, and anomalies? Is the system affordable? If the answer to any of these questions is no, then changes to the design or stakeholder expectations will be required, and the process is started over again. This process continues until the system—architecture, ConOps, and requirements—meets the stakeholder expectations.

This design solution validation is in contrast to the validation of the end product described in the end-product validation plan, which is part of the technical data package. That validation occurs in a later life cycle phase and is a result of the Product Validation Process (see Section 5.4) applied to the realization of the design solution as an end product.

4.4.1.2.9 *Identify Enabling Products*

Enabling products are the life cycle support products and services (e.g., production, test, deployment, training, maintenance, and disposal) that facilitate the progression and use of the operational end product through its life cycle. Since the end product and its enabling products are interdependent, they are viewed as a system. Project responsibility thus extends to responsibility for acquiring services from the relevant enabling products in each life cycle phase. When a suitable enabling product does not already exist, the project that is responsible for the end product can also be responsible for creating and using the enabling product.

Therefore, an important activity in the Design Solution Definition Process is the identification of the enabling products and personnel that will be required during the life cycle of the selected design solution and then initiating the acquisition or development of those enabling products and personnel. Need dates for the enabling products should be realistically identified on the project schedules, incorporating appropriate schedule slack. Then firm commitments in the form of contracts, agreements, and/or operational plans should be put in place to ensure that the enabling products will be available when needed to support the product life cycle phase activities. The enabling product requirements are documented as part of the technical data package for the Design Solution Definition Process.

An environmental test chamber is an example of an enabling product whose use would be acquired at an appropriate time during the test phase of a space flight system.

Special test fixtures or special mechanical handling devices are examples of enabling products that would have to be created by the project. Because of long development times as well as oversubscribed facilities, it is important to identify enabling products and secure the commitments for them as early in the design phase as possible.

4.4.1.2.10 *Baseline the Design Solution*

As shown earlier in FIGURE 4.0-1, once the selected system design solution meets the stakeholder expectations, the study team baselines the products and prepares for the next life cycle phase. Because of the recursive nature of successive refinement, intermediate levels of decomposition are often validated and baselined as part of the process. In the next level of decomposition, the baselined requirements become the set of high-level requirements for the decomposed elements, and the process begins again.

Baselining a particular design solution enables the technical team to focus on one design out of all the alternative design concepts. This is a critical point in the design process. It puts a stake in the ground and gets everyone on the design team focused on the same concept. When dealing with complex systems, it is difficult for team members to design their portion of the system if the system design is a moving target. The baselined design is documented and placed under configuration control. This includes the system requirements, specifications, and configuration descriptions.

While baselining a design is beneficial to the design process, there is a danger if it is exercised too early in the Design Solution Definition Process. The early exploration of alternative designs should be free and open to a wide range of ideas, concepts, and implementations. Baselining too early takes the inventive nature out of the concept exploration. Therefore, baselining should be one of the last steps in the Design Solution Definition Process.

4.4.1.3 *Outputs*

Outputs of the Design Solution Definition Process are the specifications and plans that are passed on to the product realization processes. They contain the design-to, build-to, train-to, and code-to documentation that complies with the approved baseline for the system.

As mentioned earlier, the scope and content of the full design description should be appropriate for the product life cycle phase, the phase success criteria, and the product position in the PBS.

Outputs of the Design Solution Definition Process include the following:

- **The System Specification:** The system specification contains the functional baseline for the system that is the result of the Design Solution Definition Process. The system design specification provides sufficient guidance, constraints, and system requirements for the design engineers to begin developing the design.

- **The System External Interface Specifications:** The system external interface specifications describe the functional baseline for the behavior and characteristics of all physical interfaces that the system has with the external world. These include all structural, thermal, electrical, and signal interfaces, as well as the human-system interfaces.
- **The End-Product Specifications:** The end-product specifications contain the detailed build-to and code-to requirements for the end product. They are detailed, exact statements of design particulars, such as statements prescribing materials, dimensions, and quality of work to build, install, or manufacture the end product.
- **The End-Product Interface Specifications:** The end-product interface specifications contain the detailed build-to and code-to requirements for the behavior and characteristics of all logical and physical interfaces that the end product has with external elements, including the human-system interfaces.
- **Initial Subsystem Specifications:** The end-product subsystem initial specifications provide detailed information on subsystems if they are required.
- **Enabling Product Requirements:** The requirements for associated supporting enabling products provide details of all enabling products. Enabling products are the life cycle support products, infrastructures, personnel, logistics, and services that facilitate the progression and use of the operational end product through its life cycle. They are viewed as part of the system since the end product and its enabling products are interdependent.
- **Product Verification Plan:** The end-product verification plan (generated through the Technical Planning Process) provides the content and depth of detail necessary to provide full visibility of all verification activities for the end product. Depending on the scope of the end product, the plan encompasses qualification, acceptance, prelaunch, operational, and disposal verification activities for flight hardware and software.
- **Product Validation Plan:** The end-product validation plan (generated through the Technical Planning Process) provides the content and depth of detail necessary to provide full visibility of all activities to validate the end product against the baselined stakeholder expectations. The plan identifies the type of validation, the validation procedures, and the validation environment that are appropriate to confirm that the realized end product conforms to stakeholder expectations.
- **Logistics and Operate-to Procedures:** The applicable logistics and operate-to procedures for the system describe such things as handling, transportation, maintenance, long-term storage, and operational considerations for the particular design solution.

Other outputs may include:

- **Human Systems Integration Plan:** The system HSI Plan should be updated to indicate the numbers, skills, and development (i.e., training) required for humans throughout the full life cycle deployment and operations of the system.

4.4.2 Design Solution Definition Guidance

Refer to Section 4.4.2 in the NASA Expanded Guidance for Systems Engineering at *https://nen.nasa.gov/web/se/doc-repository* for additional guidance on:

- technology assessment,
- human capability assessment, and
- integrating engineering specialties into the SE process.

5.0 Product Realization

This chapter describes the activities in the product realization processes listed in FIGURE 2.1-1. The chapter is separated into sections corresponding to steps 5 through 9 listed in FIGURE 2.1-1. The processes within each step are discussed in terms of the inputs, the activities, and the outputs. Additional guidance is provided using examples that are relevant to NASA projects.

In the product realization side of the SE engine, five interdependent processes result in systems that meet the design specifications and stakeholder expectations. These products are produced, acquired, reused, or coded; integrated into higher level assemblies; verified against design specifications; validated against stakeholder expectations; and transitioned to the next level of the system. As has been mentioned in previous sections, products can be models and simulations, paper studies or proposals, or hardware and software. The type and level of product depends on the phase of the life cycle and the product's specific objectives. But whatever the product, all should effectively use the processes to ensure the system meets the intended operational concept.

This effort starts with the technical team taking the output from the system design processes and using the appropriate crosscutting functions, such as data and configuration management, and technical assessments to make, buy, or reuse subsystems. Once these subsystems are realized, they should be integrated to the appropriate level as designated by the appropriate interface requirements. These products are then verified through the Technical Assessment Process to ensure that they are consistent with the technical data package and that "the product was built right." Once consistency is achieved, the technical team validates the products against the stakeholder expectations to ensure that "the right product was built." Upon successful completion of validation, the products are transitioned to the next level of the system. FIGURE 5.0-1 illustrates these processes.

This is an iterative and recursive process. Early in the life cycle, paper products, models, and simulations are run through the five realization processes. As the system matures and progresses through the life cycle, hardware and software products are run through these processes. It is important to detect as many errors and failures as possible at the lowest level of integration and early in the life cycle so that changes can be made through the design processes with minimum impact to the project.

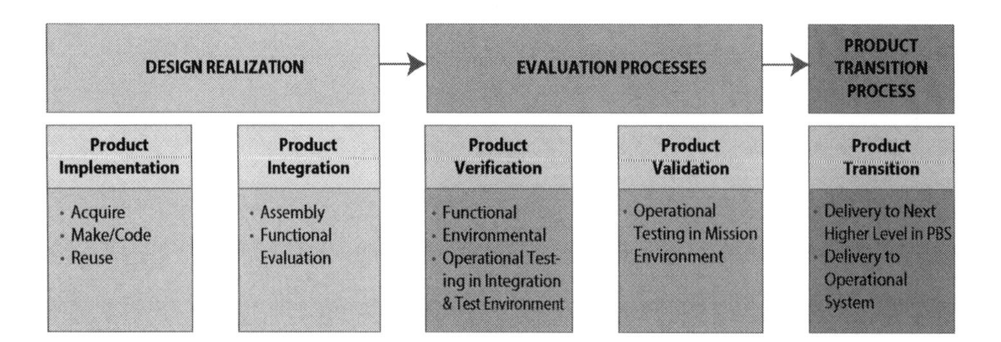

FIGURE 5.0-1 Product Realization

PRODUCT REALIZATION KEYS

- Define and execute production activities.
- Generate and manage requirements for off-the-shelf hardware/software products as for all other products.
- Understand the differences between verification testing and validation testing.
- Consider all customer, stakeholder, technical, programmatic, and safety requirements when evaluating the input necessary to achieve a successful product transition.
- Analyze for any potential incompatibilities with interfaces as early as possible.
- Completely understand and analyze all test data for trends and anomalies.
- Understand the limitations of the testing and any assumptions that are made.
- Ensure that a reused product meets the verification and validation required for the relevant system in which it is to be used, as opposed to relying on the original verification and validation it met for the system of its original use. Then ensure that it meets the same verification and validation as a purchased product or a built product. The "pedigree" of a reused product in its original application should not be relied upon in a different system, subsystem, or application.

The next sections describe each of the five product realization processes and their associated products for a given NASA mission.

5.1 Product Implementation

Product implementation is the first process encountered in the SE engine that begins the movement from the bottom of the product hierarchy up towards the Product Transition Process. This is where the plans, designs, analysis, requirements development, and drawings are realized into actual products.

Product implementation is used to generate a specified product of a project or activity through buying, making/coding, or reusing previously developed

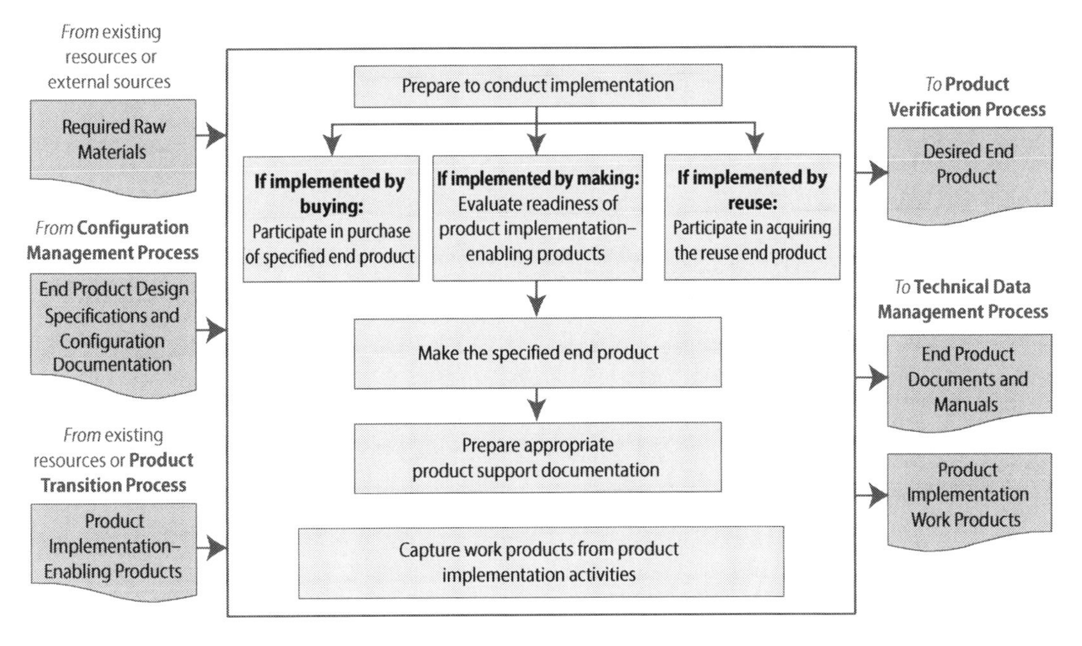

FIGURE 5.1-1 Product Implementation Process

hardware, software, models, or studies to generate a product appropriate for the phase of the life cycle. The product should satisfy the design solution and its specified requirements.

The Product Implementation Process is the key activity that moves the project from plans and designs into realized products. Depending on the project and life cycle phase within the project, the product may be hardware, software, a model, simulations, mock-ups, study reports, or other tangible results. These products may be realized through their purchase from commercial or other vendors, through partial or complete reuse of products from other projects or activities, or they may be generated from scratch. The decision as to which of these realization strategies or combination of strategies will be used for the products of this project will have been made early in the life cycle using the Decision Analysis Process.

5.1.1 Process Description

FIGURE 5.1-1 provides a typical flow diagram for the Product Implementation Process and identifies typical inputs, outputs, and activities to consider in addressing product implementation.

5.1.1.1 *Inputs*

Inputs to the Product Implementation Process depend primarily on the decision about whether the end product will be purchased, developed from scratch, or formed by reusing part or all of products from other projects. Typical inputs are shown in FIGURE 5.1-1.

- **Inputs If Purchasing the End Product:** If the decision was made to purchase part or all of the products for this project, the end product design specifications are obtained from the configuration management system as well as other applicable documents.

- **Inputs If Making/Coding the End Product:** For end products that will be made/coded by the technical team, the inputs will be the configuration-controlled design specifications, manufacturing plans, manufacturing processes, manufacturing procedures, and raw materials as provided to or purchased by the project.

- **Inputs Needed If Reusing an End Product:** For end products that will reuse part or all of products generated by other projects, the inputs may be the documentation associated with the product as well as the product itself. Care should be taken to ensure that these products will indeed meet the specifications and environments for this project. These would have been factors involved in the Decision Analysis Process to determine the make/buy/reuse decision.

- **Enabling Products:** These would be any enabling products necessary to make, code, purchase, or reuse the product (e.g., drilling fixtures, production facilities, production lines, software development facilities, software test facilities, system integration and test facilities).

5.1.1.2 *Process Activities*

Implementing the product can take one of three forms:

1. Purchase/buy
2. Make/code
3. Reuse

These three forms will be discussed in the following subsections. FIGURE 5.1-1 shows what kind of inputs, outputs, and activities are performed during product implementation regardless of where in the product hierarchy or life cycle it is. These activities include preparing to conduct the implementation, purchasing/making/reusing the product, and capturing the product implementation work product. In some cases, implementing a product may have aspects of more than one of these forms (such as a build-to-print). In those cases, the appropriate aspects of the applicable forms are used.

5.1.1.2.1 *Prepare to Conduct Implementation*

Preparing to conduct the product implementation is a key first step regardless of what form of implementation has been selected. For complex projects, implementation strategy and detailed planning or procedures need to be developed and documented. For less complex projects, the implementation strategy and planning need to be discussed, approved, and documented as appropriate for the complexity of the project.

The documentation, specifications, and other inputs also need to be reviewed to ensure they are ready and at an appropriate level of detail to adequately complete the type of implementation form being employed and for the product life cycle phase. For example, if the "make" implementation form is being employed, the design specifications need to be reviewed to ensure they are at a design-to level that allows the product to be developed. If the product is to be bought as a pure Commercial Off-the-Shelf (COTS) item, the specifications need to be checked to make sure they adequately describe the vendor characteristics to narrow to a single make/model of their product line.

Finally, the availability and skills of personnel needed to conduct the implementation as well as the availability of any necessary raw materials, enabling products, or special services should also be reviewed. Any special training necessary for the personnel to perform their tasks needs to be performed by this time. This is a key part of the Acceptance Data Package.

5.1.1.2.2 *Purchase, Make, or Reuse the Product*

Purchase the Product

In the first case, the end product is to be purchased from a commercial or other vendor. Design/purchase specifications will have been generated during requirements development and provided as inputs. The technical team needs to review these specifications and ensure they are in a form adequate for the contract or purchase order. This may include the generation of contracts, Statements of Work (SOWs), requests for proposals, purchase orders, or other purchasing mechanisms. For major end products purchased from a vendor, the responsibilities of the Government and contractor team should be documented in the SEMP and Integration Plan. This will define, for example, whether NASA expects the vendor to provide a fully verified and validated product or whether the NASA technical team will be performing those duties. The team needs to work with the acquisition team to ensure the accuracy of the contract SOW or purchase order and to ensure that adequate documentation, certificates of compliance, or other specific needs are requested from the vendor.

For contracted purchases, as proposals come back from the vendors, the technical team should work with the contracting officer and participate in the review of the technical information and in the selection of the vendor that best meets the design requirements for acceptable cost and schedule.

As the purchased products arrive, the technical team should assist in the inspection of the delivered product and its accompanying documentation. The team should ensure that the requested product was indeed the one delivered, and that all necessary documentation, such as source code, operator manuals, certificates of compliance, safety information, or drawings have been received.

The NASA technical team should also ensure that any enabling products necessary to provide test, operations, maintenance, and disposal support for the product are also ready or provided as defined in the contract.

Depending on the strategy and roles/responsibilities of the vendor, a determination/analysis of the vendor's verification and validation compliance may need to be reviewed. This may be done informally or formally as appropriate for the complexity of the product. For products that were verified and validated by the vendor, after ensuring that all work products from this phase have been captured, the product may be ready to enter the Product Transition Process to be delivered to the next higher level or to its final end user. For products that the technical team will verify and validate, the product will be ready for verification after ensuring that all work products for this phase have been captured.

Make/Code the Product

If the strategy is to make or code the product, the technical team should first ensure that the enabling products are ready. This may include ensuring all piece parts are available, drawings are complete and adequate, software design is complete and reviewed, machines to cut the material are available, interface specifications are approved, operators are trained and available, manufacturing and/or coding procedures/processes are ready, software personnel are trained and available to generate code, test fixtures are developed and ready to hold products while being generated, and software test cases are available and ready to begin model generation.

The product is then made or coded in accordance with the specified requirements, configuration documentation, and applicable standards. Software development must be consistent with NPR 7150.2, NASA Software Engineering Requirements. Throughout this process, the technical team should work with the quality organization to review, inspect, and discuss progress and status within the team and with higher levels of management as appropriate. Progress should be documented within the technical schedules. Peer reviews, audits, unit testing, code inspections, simulation checkout, and other techniques may be used to ensure the made or coded product is ready for the verification process. Some production and coding can also be separately contracted. This is sometimes pursued as a cost control feature providing motivation for the design contractor to keep the operations costs low and not roll costs into the operations phase of a long-term contract. This is also valuable when the design contractor is not well suited for long-term continuing production operations. Small projects and activities often use small manufacturing shops to fabricate the system or major portions and small software companies to code their software. In these cases, the production and software engineers may specify some portion of the hardware production or software coding and request the remaining portions, including as-built documentation, from the manufacturing or software provider. The specified portions are contained as part of the contract statement of work in these cases. The level of process control and information provided to or from the vendor is dependent on the criticality of the systems obtained. As production proceeds and components are produced, there is a need to establish a method (Material Review Boards (MRBs) are typically used for large projects) to review any nonconformance to specifications and disposition whether the components can be accepted, reworked, or scrapped and remade.

Reuse

If the strategy is to reuse a product that already exists, extreme care should be taken to ensure that the product is truly applicable to this project and for the intended uses and the environment in which it will be used. This should have been a major factor used in the decision strategy to make/buy/reuse. If the new environment is more extreme, requalification is needed for the component or system. Design factors of safety, margins, and other required design and construction standards should also be assessed. If the program/project requires higher factor of safety or margins, the component may not be usable or a waiver may have to be approved.

The documentation available (e.g., as-built documentation, user's guides, operations manuals, discrepancy reports, waivers and deviations) from the reuse product should be reviewed by the technical team so that they can become completely familiar with the product and ensure it will meet the requirements in the intended environment. Any supporting manuals, drawings, or other documentation available should also be gathered.

The availability of any supporting or enabling products or infrastructure needed to complete the fabrication, coding, testing, analysis, verification, validation, or shipping of the product needs to be determined. Supporting products may be found in product manufacturing plans, processes, and procedures. If any of these products or services are lacking, they will need to be developed or arranged for before progressing to the next phase.

Special arrangements may need to be made or forms such as nondisclosure agreements may need to be acquired before the reuse product can be received.

A reused product often needs to undergo the same verification and validation as a purchased product or a built product. Relying on prior verification and validation should only be considered if the product's verification and validation documentation meets or exceeds the verification, validation, and documentation requirements of the current project and the documentation demonstrates that the product was verified and validated against equivalent requirements (including environments) and expectations. The savings gained from reuse is not necessarily from reduced acceptance-level testing of the flight products, but possibly elimination of the need to fully requalify the item (if all elements are the same, including the environment and operation), elimination of the need to specify all of the internal requirements such as printed circuit board specifications or material requirements, reduced internal data products, or the confidence that the item will pass acceptance test and will not require rework.

5.1.1.2.3 *Capture Work Products*

Regardless of what implementation form was selected, all work products from the make/buy/reuse process should be captured, including as-built design drawings, design documentation, design models, code listings, model descriptions, procedures used, operator manuals, maintenance manuals, or other documentation as appropriate.

5.1.1.3 *Outputs*

- **End Product for Verification:** Unless the vendor performs verification, the made/coded, purchased, or reused end product in a form appropriate for the life cycle phase is provided for the verification process. The form of the end product is a function of the life cycle phase and the placement within the system structure (the form of the end product could be hardware, software, model, prototype, first article for test, or single operational article or multiple production articles).

- **End Product Documents and Manuals:** Appropriate documentation is also delivered with the end product to the verification process and to the technical data management process. Documentation may include applicable as-built design drawings; close out photos; operation, user, maintenance, or training manuals; applicable baseline documents (configuration information such as as-built specifications or stakeholder expectations); certificates of compliance; or other vendor documentation.

- **Product Implementation Work Products:** Any additional work products providing reports, records, lesson learned, assumptions, updated CM products, and other outcomes of these activities.

The process is complete when the following activities have been accomplished:

- End products are fabricated, purchased, or reuse modules are acquired.

- End products are reviewed, checked, and ready for verification.

- Procedures, decisions, assumptions, anomalies, corrective actions, lessons learned, etc., resulting from the make/buy/reuse are recorded.

5.1.2 Product Implementation Guidance

Refer to Section 5.1.2 in the NASA Expanded Guidance for Systems Engineering at *https://nen.nasa.gov/web/se/doc-repository* for additional guidance on:

- buying off-the-shelf products and
- the need to consider the heritage of products.

5.2 Product Integration

Product integration is a key activity of the systems engineer. Product integration is the engineering of the subsystem interactions and their interactions with the system environments (both natural and induced). Also in this process, lower-level products are assembled into higher-level products and checked to make sure that the integrated product functions properly and that there are no adverse emergent behaviors. This integration begins during concept definition and continues throughout the system life cycle. Integration involves several activities focused on the interactions of the subsystems and environments. These include system analysis to define and understand the interactions, development testing including qualification testing, and integration with external systems (e.g., launch operations centers, space vehicles, mission operations centers, flight control centers, and aircraft) and objects (i.e., planetary bodies or structures). To accomplish this integration, the systems engineer is active in integrating the different discipline and design teams to ensure system and environmental interactions are being properly balanced by the differing design teams. The result of a well-integrated and balanced system is an elegant design and operation.

Integration begins with concept development, ensuring that the system concept has all necessary functions and major elements and that the induced and natural environment domains in which the system is expected to operate are all identified. Integration continues during requirements development, ensuring that all system and environmental requirements are compatible and that the system has a proper balance of functional utility to produce a robust and efficient system. Interfaces are defined in this phase and are the pathway of system interactions. Interfaces include mechanical (i.e., structure, loads), fluids, thermal, electrical, data, logical (i.e., algorithms and software), and human. These interfaces may include support for assembly, maintenance, and testing functions in addition to the system main performance functions. The interactions that occur through all of these interfaces can be subtle and complex, leading to both intended and unintended consequences. All of these interactions need to be engineered to produce an elegant and balanced system.

Integration during the design phase continues the engineering of these interactions and requires constant analysis and management of the subsystem functions and the subsystem interactions between themselves and with their environments. Analysis of the system interactions and managing the balance of the system is the central function of the systems engineer during the design process. The system needs to create and maintain a balance between the subsystems, optimizing the system performance over any one subsystem to achieve an elegant and efficient design. The design phase often involves development testing at the component, assembly, or system level. This is a key source of data on system interactions, and the developmental test program should be structured to include subsystem interactions, human-in-the-loop evaluations, and environmental interaction test data as appropriate.

Integration continues during the operations phase, bringing together the system hardware, software, and human operators to perform the mission. The interactions between these three integrated natures of the system need to be managed throughout development and into operations for mission success. The systems engineer, program manager, and the operations team (including the flight crew from crewed missions) need to work together to perform this management. The systems engineer is not only cognizant of these operations team interactions, but is also involved in the design responses and updates to changes in mission parameters and unintended consequences (through fault management).

Finally, integration or de-integration occurs during system closeout (i.e., decommissioning and disposal). The system capabilities to support de-integration and/or disposal need to be engineered into the system from the concept definition phase. The closeout phase involves the safe disposal of flight assets consistent with U.S. policy and law and international treaties. This disposal can involve the safe reentry and recovery or impact in the ocean, impact on the moon, or solar trajectory. This can also involve the disassembly or repurposing of terrestrial equipment used in manufacturing, assembly, launch, and flight operations. Dispositioning of recovered flight assets also occurs during this phase. Capture of system data and archiving for use in future analysis also occurs. In all of these activities, the systems engineer is involved in ensuring a smooth and logical disassembly of the system and associated program assets.

The Product Integration Process applies not only to hardware and software systems but also to service-oriented solutions, requirements, specifications, plans, and concepts. The ultimate purpose of product integration is to ensure that the system elements function as a whole.

Product integration involves many activities that need to be planned early in the program or project in order to effectively and timely accomplish the integration. Some integration activities (such as system tests) can require many years of work and costs that need to be identified and approved through the budget cycles. An integration plan should be developed and documented to capture this planning. Small projects and activities may be able to include this as part of their SEMP. Some activities may have their integration plans captured under the integration plan of the sponsoring flight program or R&T program. Larger programs and projects need to have a separate integration plan to clearly lay out the complex analysis and tests that need to occur. An example outline for a separate integration plan is provided in *Appendix H*.

During project closeout, a separate closeout plan should be produced describing the decommissioning and disposal of program assets. (For example, see National Space Transportation System (NSTS) 60576, *Space Shuttle Program, Transition Management Plan*). For smaller projects and activities, particularly with short life cycles (i.e., short mission durations), the closeout plans may be contained in the SEMP.

5.2.1 Process Description

FIGURE 5.2-1 provides a typical flow diagram for the Product Integration Process and identifies typical inputs, outputs, and activities to consider in addressing product integration. The activities of the Product Integration Process are truncated to indicate the action and object of the action.

5.2.1.1 *Inputs*

- **Lower-level products to be integrated:** These are the products developed in the previous lower-level tier in the product hierarchy. These products will be integrated/assembled to generate the product for this product layer.

- **End product design specifications and configuration documentation:** These are the specifications, Interface Control Documents (ICDs), drawings, integration plan, procedures, or other documentation or models needed to perform the integration including documentation for each of the lower-level products to be integrated.

- **Product integration-enabling products:** These would include any enabling products, such as holding fixtures, necessary to successfully integrate the lower-level products to create the end product for this product layer.

5.2.1.2 *Process Activities*

This subsection addresses the approach to the implementation of the Product Integration Process, including the activities required to support the process. The basic tasks that need to be established involve the management of internal and external interactions of the various levels of products and operator tasks to support product integration and are as follows:

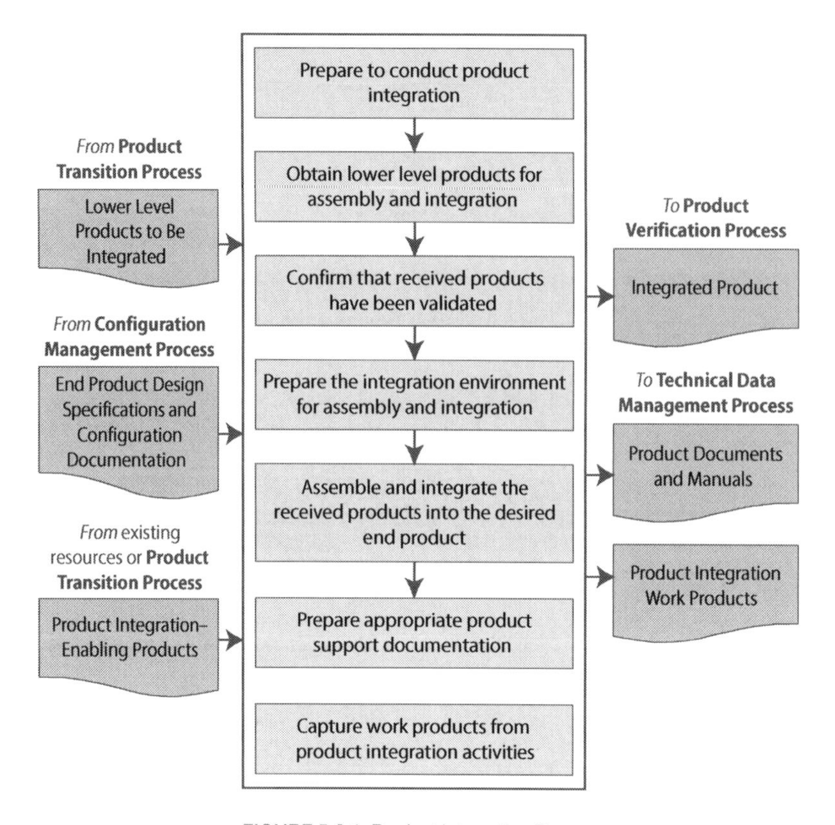

FIGURE 5.2-1 Product Integration Process

5.2.1.2.1 *Prepare to Conduct Product Integration*

Prepare to conduct product integration by (1) reviewing the product integration strategy/plan (see Section 6.1.2.4.4), generating detailed planning for the integration, and developing integration sequences and procedures; and (2) determining whether the product configuration documentation is adequate to conduct the type of product integration applicable for the product life cycle phase, location of the product in the system structure, and management phase success criteria.

An integration strategy is developed and documented in an integration plan. This plan, as well as supporting documentation, identifies the optimal sequence of receipt, assembly, and activation of the various components that make up the system. This strategy should use technical, cost, and schedule factors to ensure an assembly, activation, and loading sequence that minimizes cost and assembly difficulties. The larger or more complex the system or the more delicate the element, the more critical the proper sequence becomes, as small changes can cause large impacts on project results.

The optimal sequence of assembly is built from the bottom up as components become sub-elements, elements, and subsystems, each of which should be checked prior to fitting it into the next higher assembly. The sequence will encompass any effort needed to establish and equip the assembly facilities; e.g., raised floor, hoists, jigs, test equipment, input/output, and power connections. Once established, the sequence should be periodically reviewed to ensure that variations in production and delivery schedules have not had an adverse impact on the sequence or compromised the factors on which earlier decisions were made.

5.2.1.2.2 *Obtain Lower-Level Products for Assembly and Integration*

Each of the lower-level products that is needed for assembly and integration is obtained from the transitioning lower-level product owners or a storage facility as appropriate. Received products should be inspected to ensure no damages occurred during the transitioning process.

5.2.1.2.3 *Confirm That Received Products Have Been Validated*

Confirm that the received products that are to be assembled and integrated have been validated to demonstrate that the individual products satisfy the agreed-to set of stakeholder expectations, including interface requirements. This validation can be conducted by the receiving organization or by the providing organization if fully documented or witnessed by the receiving representative.

5.2.1.2.4 *Prepare the Integration Environment for Assembly and Integration*

Prepare the integration environment in which assembly and integration will take place, including evaluating the readiness of the product integration-enabling products and the assigned workforce. These enabling products may include facilities, equipment jigs, tooling, and assembly/production lines. The integration environment includes test equipment, simulators, models, storage areas, and recording devices.

5.2.1.2.5 *Assemble and Integrate the Received Products into the Desired End Product*

Assemble and integrate the received products into the desired end product in accordance with the specified requirements, configuration documentation, interface requirements, applicable standards, and integration sequencing and procedures. This activity includes managing, evaluating, and controlling physical, functional, and data interfaces among the products being integrated.

Functional testing of the assembled or integrated unit is conducted to ensure that assembly is ready to enter verification testing and ready to be integrated into the next level. Typically, all or key representative functions are checked to ensure that the assembled system is functioning as expected. Formal product verification and validation will be performed in the next process.

5.2.1.2.6 *Prepare Appropriate Product Support Documentation*

Prepare appropriate product support documentation, such as special procedures for performing product verification and product validation. Drawings or accurate models of the assembled system are developed and confirmed to be representative of the assembled system.

5.2.1.2.7 *Capture Product Integration Work Products*

Capture work products and related information generated while performing the Product Integration Process activities. These work products include system models, system analysis data and assessment reports, derived requirements, the procedures that were used in the assembly, decisions made and supporting rationale, assumptions that were made, identified anomalies and associated corrective actions, lessons learned in performing the assembly, and updated product configuration and support documentation.

5.2.1.3 *Outputs*

The following are typical outputs from this process and destinations for the products from this process:

- **Integrated product(s)** with all system interactions identified and properly balanced.
- **Documentation and manuals**, including system analysis models, data, and reports supporting flight-readiness rationale and available for future analysis during the operation of the system in the mission-execution phase.
- **Work products**, including reports, records, and non-deliverable outcomes of product integration activities (to support the Technical Data Management Process); integration strategy document; assembly/check area drawings; system/component documentation sequences and rationale for selected assemblies; interface management documentation; personnel requirements; special handling requirements; system documentation; shipping schedules; test equipment and drivers' requirements; emulator requirements; and identification of limitations for both hardware and software.

5.2.2 Product Integration Guidance

Refer to Section 5.2.2 in the NASA Expanded Guidance for Systems Engineering at *https://nen.nasa.gov/web/se/doc-repository* for additional guidance on:

- product integration strategies,
- the relationship to product implementation,
- product integration support,
- product integration of the design solution,
- system analysis, and
- interface system integration.

5.3 Product Verification

The Product Verification Process is the first of the verification and validation processes conducted on an end product. As used in the context of the systems engineering common technical processes, a product is one provided by either the Product Implementation Process or the Product Integration Process in a form suitable for meeting applicable life cycle phase success criteria. Realization is the act of implementing, integrating, verifying, validating, and transitioning the end product for use at the next level up of the system structure or to the customer. At this point, the end product can be referred to as a "realized product" or "realized end product."

Product verification proves that an end product (whether built, coded, bought, or reused) for any element within the system structure conforms to its requirements or specifications. Such specifications and other design description documentation establish the configuration baseline of that product, which may have to be modified at a later time. Without a verified baseline and appropriate configuration controls, such later modifications could be costly or cause major performance problems.

From a process perspective, product verification and validation may be similar in nature, but the objectives are fundamentally different. A customer is interested in whether the end product provided will do what the customer intended within the environment of use. Examination of this condition is validation. Simply put, the Product Verification Process answers the critical question, "Was the end product realized right?" The Product Validation Process addresses the equally critical question, "Was the right end product realized?" When cost effective and warranted by analysis, the expense of validation testing alone can be mitigated by combining tests to perform verification and validation simultaneously.

DIFFERENCES BETWEEN VERIFICATION AND VALIDATION TESTING

Testing is a detailed evaluation method of both verification and validation

Verification Testing: Verification testing relates back to the approved requirements set (such as an SRD) and can be performed at different stages in the product life cycle. Verification tests are the official "for the record" testing performed on a system or element to show that it meets its allocated requirements or specifications including physical and functional interfaces. Verification tests use instrumentation and measurements and are generally accomplished by engineers, technicians, or operator-maintainer test personnel in a controlled environment to facilitate failure analysis.

Validation Testing: Validation relates back to the ConOps document. Validation testing is conducted under realistic conditions (or simulated conditions) on any end product to determine the effectiveness and suitability of the product for use in mission operations by typical users and to evaluate the results of such tests. It ensures that the system is operating as expected when placed in a realistic environment.

The outcome of the Product Verification Process is confirmation that the end product, whether achieved by implementation or integration, conforms to its specified requirements, i.e., verification of the end product. This subsection discusses the process activities, inputs, outcomes, and potential product deficiencies.

5.3.1 Process Description

FIGURE 5.3-1, taken from NPR 7123.1, provides a typical flow diagram for the Product Verification Process and identifies typical inputs, outputs, and activities to consider in addressing product verification.

5.3.1.1 *Inputs*

Key inputs to the process are:

- **The product to be verified:** This product will have been transitioned from either the Product Implementation Process or the Product Integration Process. The product will likely have been through at least a functional test to ensure it was assembled correctly. Any supporting documentation should be supplied with the product.
- **Verification plan:** This plan will have been developed under the Technical Planning Process and baselined before entering this verification.
- **Specified requirements baseline:** These are the requirements that have been identified to be verified for this product. Acceptance criteria should have been identified for each requirement to be verified.
- **Enabling products:** Any other products needed to perform the Product Verification Process. This may include test fixtures and support equipment.

Additional work products such as the ConOps, mission needs and goals, interface control drawings, testing standards and policies, and Agency standards and policies may also be needed to put verification activities into context.

DIFFERENCES BETWEEN VERIFICATION, QUALIFICATION, ACCEPTANCE AND CERTIFICATION

Verification: Verification is a formal process, using the method of test, analysis, inspection or demonstration, to confirm that a system and its associated hardware and software components satisfy all specified requirements. The Verification program is performed once regardless of how many flight units may be generated (as long as the design doesn't change).

Qualification: Qualification activities are performed to ensure that the flight unit design will meet functional and performance requirements in anticipated environmental conditions. A subset of the verification program is performed at the extremes of the environmental envelope and will ensure the design will operate properly with the expected margins. Qualification is performed once regardless of how many flight units may be generated (as long as the design doesn't change).

Acceptance: smaller subset of the verification program is selected as criteria for the acceptance program. The selected Acceptance activities are performed on each of the flight units as they are manufactured and readied for flight/use. An Acceptance Data Package is prepared for each of the flight units and shipped with the unit. The acceptance test/analysis criteria are selected to show that the manufacturing/workmanship of the unit conforms to the design that was previously verified/qualified. Acceptance testing is performed for each flight unit produced.

Certification: Certification is the audit process by which the body of evidence that results from the verification activities and other activities are provided to the appropriate certifying authority to indicate the design is certified for flight/use. The Certification activity is performed once regardless of how many flight units may be generated.

5.3.1.2 *Process Activities*

There are five major activities in the Product Verification Process: (1) prepare to conduct product verification; (2) perform verification; (3) analyze verification results; (4) preparing a product verification report; and (5) capture work products generated during the verification activities.

Product Verification is often performed by the developer that produced the end product with participation of the end user and customer. Quality Assurance (QA) personnel are also critical in the verification planning and execution activities.

A verification approach should be adapted (tailored) to the project it supports. The project manager and systems engineer should work with the verification lead engineer to develop a verification approach and plan the activities. Many factors need to be considered in developing this approach and the subsequent verification program. These factors include:

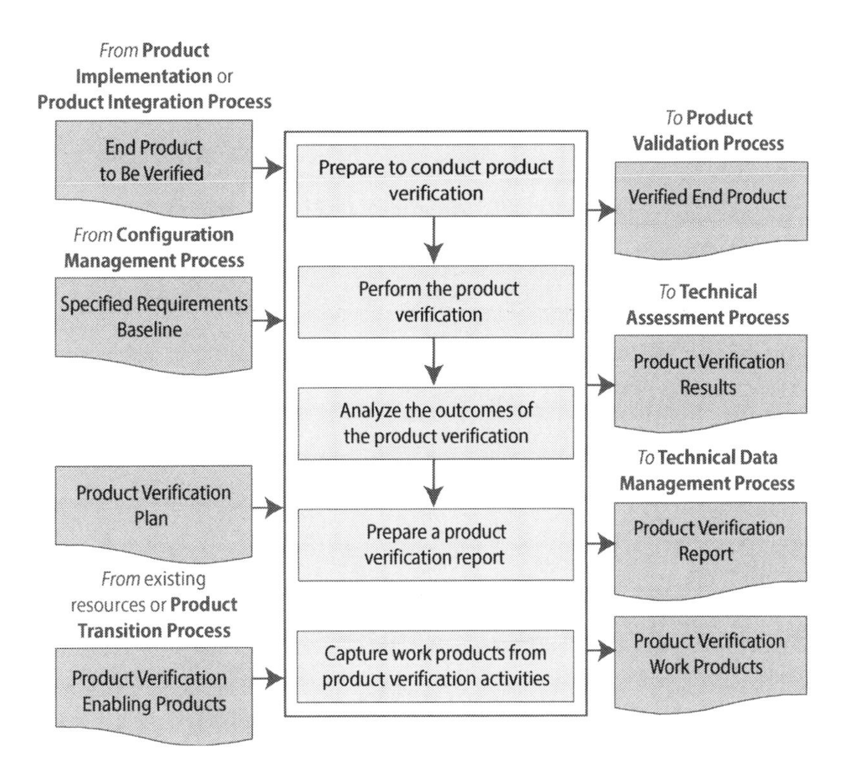

FIGURE 5.3-1 Product Verification Process

- Project type, especially for flight projects. Verification activities and timing depend on the following:
  - » The type of flight article involved (e.g., an experiment, payload, or launch vehicle).
  - » For missions required to follow NPR 7120.5, NASA Space Flight Program and Project Management Requirements, NASA payload classification (NPR 8705.4, Risk Classification for NASA Payloads) guidelines are intended to serve as a starting point for establishing the formality of verification approaches that can be adapted to the needs of a specific project based on the "A–D" payload classification. Further flexibility is imparted to projects following NPR 7120.8, NASA Research and Technology Program and Project Management Requirements.
  - » Project cost and schedule implications. Verification activities can be significant drivers of a project's cost and schedule, and these implications should be considered early in the development of the verification plan. Trade studies should be performed early in the life cycle to support decisions about verification methods and types and the selection of facility capabilities and locations. For example, a trade study might be made to decide between performing a test at a centralized facility or at several decentralized locations.
  - » Risk management should be considered in the development of the verification approach. Qualitative risk assessments and quantitative risk analyses (e.g., a Failure Mode and Effects Analysis (FMEA)) often identify new concerns that can be mitigated by additional verifications, thus increasing the extent of

verification activities. Other risk assessments contribute to trade studies that determine the preferred methods of verification to be used and when those methods should be performed. For example, a trade might be made between performing a model test versus determining model characteristics by a less costly but less revealing analysis. The project manager/systems engineer should determine what risks are acceptable in terms of the project's cost and schedule.

- Availability of verification facilities/sites and transportation assets to move an article from one location to another (when needed). This requires coordination with the Integrated Logistics Support (ILS) engineer.
- Availability of appropriately trained users for interaction with systems having human interfaces.
- Acquisition strategy; i.e., in-house development or system contract. A NASA field Center can often shape a contractor's verification process through the project's SOW.
- Degree of design heritage and hardware/software reuse.

5.3.1.2.1 *Product Verification Preparation*

In preparation for verification, the verification plan and the specified requirements are collected, reviewed, and confirmed. The product to be verified is obtained (output from the Product Implementation Process or the Product Integration Process) along with any enabling products, such as those representing external interfacing products and support resources (including personnel) that are necessary for verification. Procedures capturing detailed step-by-step activities and based on the verification type and methods are finalized and approved. Development of procedures typically begins during the design phase of the project life cycle and matures as the design is matured. The verification environment is considered as part of procedure development. Operational scenarios are assessed to explore all possible verification activities to be performed. The final element is preparation of the verification environment; e.g., facilities, equipment, tools, measuring devices, and climatic conditions.

When operator or other user interaction is involved, it is important to ensure that humans are properly represented in the verification activities. This includes physical size, skills, knowledge, training, clothing, special gear, and tools. Note: Testing that includes representatives of the human in the system is often referred to as "human-in-the-loop" testing.

NOTE: Depending on the nature of the verification effort and the life cycle phase the program is in, some type of review to assess readiness for verification (as well as validation later) is typically held. In earlier phases of the life cycle, these Test Readiness Reviews (TRRs) may be held informally; in later phases of the life cycle, this review may become a more formal event. TRRs and other technical reviews are an activity of the Technical Assessment Process.

On most projects, a number of TRRs with tailored entrance/success criteria are held to assess the readiness and availability of test ranges, test facilities, trained testers, instrumentation, integration labs, support equipment, and other enabling products.

Peer reviews are additional reviews that may be conducted formally or informally to ensure readiness for verification (as well as the results of the verification process). Guidelines for conducting a peer review are discussed in Section 6.7.2.4.5.

TABLE 5.3-1 provides an example of the type of information that may be included in a verification procedure and a verification report.

METHODS OF VERIFICATION

Analysis: The use of mathematical modeling and analytical techniques to predict the suitability of a design to stakeholder expectations based on calculated data or data derived from lower system structure end product verifications. Analysis is generally used when a prototype; engineering model; or fabricated, assembled, and integrated product is not available. Analysis includes the use of modeling and simulation as analytical tools. A model is a mathematical representation of reality. A simulation is the manipulation of a model. Analysis can include verification by similarity of a heritage product.

Demonstration: Showing that the use of an end product achieves the individual specified requirement. It is generally a basic confirmation of performance capability, differentiated from testing by the lack of detailed data gathering. Demonstrations can involve the use of physical models or mock-ups; for example, a requirement that all controls shall be reachable by the pilot could be verified by having a pilot perform flight-related tasks in a cockpit mock-up or simulator. A demonstration could also be the actual operation of the end product by highly qualified personnel, such as test pilots, who perform a one-time event that demonstrates a capability to operate at extreme limits of system performance, an operation not normally expected from a representative operational pilot.

Inspection: The visual examination of a realized end product. Inspection is generally used to verify physical design features or specific manufacturer identification. For example, if there is a requirement that the safety arming pin has a red flag with the words "Remove Before Flight" stenciled on the flag in black letters, a visual inspection of the arming pin flag can be used to determine if this requirement was met. Inspection can include inspection of drawings, documents, or other records.

Test: The use of an end product to obtain detailed data needed to verify performance or provide sufficient information to verify performance through further analysis. Testing can be conducted on final end products, breadboards, brassboards, or prototypes. Testing produces data at discrete points for each specified requirement under controlled conditions and is the most resource-intensive verification technique. As the saying goes, "Test as you fly, and fly as you test." (See Section 5.3.2.5 in the NASA Expanded Guidance for Systems Engineering at ***https://nen.nasa.gov/web/se/doc-repository***)

Outcomes of verification preparation include the following:

- The verification plan, approved procedures, and an appropriate baseline set of specified requirements and supporting configuration documentation is available and on hand;
- Articles/models to be verified and verification-enabling products are on hand, assembled, and integrated with the verification environment according to verification plans and schedules;
- The resources (funding, facilities, and people including appropriately skilled operators) needed

TABLE 5.3-1 Example information in Verification Procedures and Reports

Verification Procedure	Verification Report
Nomenclature and identification of the test article or material;	Verification objectives and the degree to which they were met;
Identification of test configuration and any differences from flight operational configuration;	Description of verification activity including deviations from nominal results (discrepancies);
Identification of objectives and criteria established for the verification by the applicable requirements specification;	Test configuration and differences from the flight operational configuration;
Characteristics and design criteria to be inspected, demonstrated, or tested, including values with tolerances for acceptance or rejection;	Specific result of each activity and each procedure, including the location or link to verification data/artifacts;
Description, in sequence, of steps, operations, and observations to be taken;	Specific result of each analysis including those associated with test-data analysis;
Identification of computer software required;	Test performance data tables, graphs, illustrations, and pictures;
Identification of measuring, test, and recording equipment to be used, specifying range, accuracy, and type;	Summary of nonconformance/discrepancy reports, including dispositions with approved corrective actions and planned retest activity if available;
Provision for recording equipment calibration or software version data;	Conclusions and recommendations relative to the success of verification activity;
Credentials showing that required computer test programs/support equipment and software have been verified prior to use with flight operational hardware;	Status of Government-Supplied Equipment (GSE) and other enabling support equipment as affected by test;
Any special instructions for operating data recording equipment or other automated test equipment as applicable;	Copy of the as-run procedure (may include redlines); and
Layouts, schematics, or diagrams showing identification, location, and interconnection of test equipment, test articles, and measuring points and any other associated design or configuration work products;	Authentication of test results and authorization of acceptability.
Identification of hazardous situations or operations;	
Precautions and safety instructions to ensure safety of personnel and prevent degradation of test articles and measuring equipment;	
Environmental and/or other conditions to be maintained with tolerances;	
Constraints on inspection or testing;	
Provision or instructions for the recording of verification results and other artifacts;	
Special instructions for instances of nonconformance and anomalous occurrences or results; and	
Specifications for facility, equipment maintenance, housekeeping, quality inspection, and safety and handling requirements before, during, and after the total verification activity.	

to conduct the verification are available according to the verification plans and schedules; and

- The verification environment is evaluated for adequacy, completeness, readiness, and integration.

5.3.1.2.2 *Perform Product Verification*

The actual act of verifying the end product is performed as spelled out in the plans and procedures, and conformance is established with each specified product requirement. The verification lead should ensure that the procedures were followed and performed as planned, the verification-enabling products and instrumentation were calibrated correctly, and the data were collected and recorded for required verification measures.

A verification program may include verifications at several layers in the product hierarchy. Some verifications need to be performed at the lowest component level if the ability to verify a requirement at a higher assembly is not possible. Likewise, there may be verifications at assemblies, sub-systems and system levels. If practicable, a final set of testing with as much of the end-to-end configuration as possible is important.

The purpose of end-to-end testing is to demonstrate interface compatibility and desired total functionality among different elements of a mission system, between systems (the system of interest and external enabling systems), and within a system as a whole. It can involve real or representative input and operational scenarios. End-to-end tests performed on the integrated ground and flight assets include all elements of the flight article (payload or vehicle), its control, stimulation, communications, and data processing to demonstrate that the entire integrated mission system is operating in a manner to fulfill all mission requirements and objectives. End-to-end tests may be performed as part of investigative engineering tests, verification testing, or validation testing. These are some of the most important tests for the systems engineers to participate in or to lead. They review the overall compatibility of the various systems and demonstrate compliance with system-level requirements and whether the system behaves as expected by the stakeholders.

End-to-end testing includes executing complete threads or operational scenarios across multiple configuration items, ensuring that all mission requirements are verified and validated. Operational scenarios are used extensively to ensure that the mission system (or collections of systems) will successfully execute mission requirements. Operational scenarios are a step-by-step description of how the system should operate and interact with its users and its external interfaces (e.g., other systems). Scenarios should be described in a manner that allows engineers to walk through them and gain an understanding of how all the various parts of the system should function and interact as well as verify that the system will satisfy the user's goals and expectations (MOEs). Operational scenarios should be described for all operational modes, mission phases (e.g., installation, startup, typical examples of normal and contingency operations, shutdown, and maintenance), and critical sequences of activities for all classes of users identified. Each scenario should include events, actions, stimuli, information, and interactions as appropriate to provide a comprehensive understanding of the operational aspects of the system.

FIGURE 5.3-2 presents an example of an end-to-end data flow for a scientific satellite mission. Each arrow in the diagram represents one or more data or control flows between two hardware, software, subsystem, or system configuration items. End-to-end testing verifies that the data flows throughout the multisystem environment are correct, that the system provides the required functionality, and that the outputs at the eventual end points correspond to expected results.

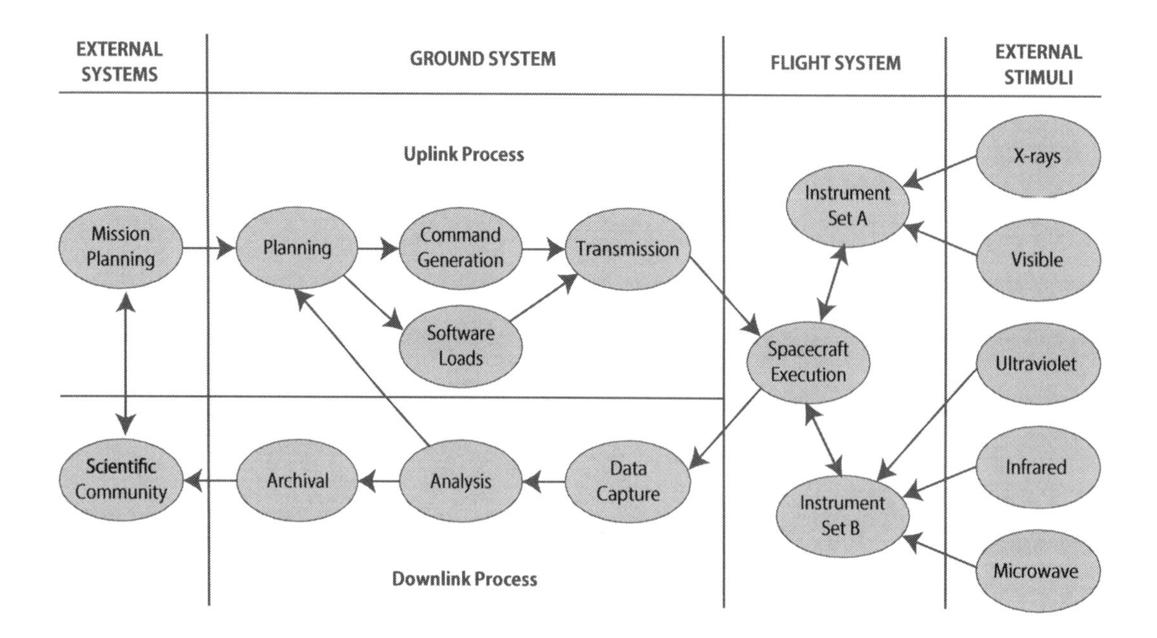

FIGURE 5.3-2 Example of End-to-End Data Flow for a Scientific Satellite Mission

Since the test environment is as close an approximation as possible to the operational environment, system performance requirements testing is also included. This figure is not intended to show the full extent of end-to-end testing. Each system shown would need to be broken down into a further level of granularity for completeness.

End-to-end testing is an integral part of the verification and validation of the total (mission) system. It is a set of activities that can be employed during selected hardware, software, and system phases throughout the life cycle using developmental forms and external simulators. However, final end-to-end testing should be done on the flight articles in the flight configuration if possible and prior to deployment and launch. In comparison with configuration item testing, end-to-end testing addresses each configuration item (end product) only down to the level designated by the verification plan (generally, the segment r element) and focuses on external interfaces, which can be either hardware, software, or human-based. Internal interfaces (e.g., software subroutine calls, analog-to-digital conversion) of a designated configuration item are not within the scope of end-to-end testing.

When a "discrepancy" is observed (i.e., any variance, lack of agreement, or contradiction with the required or expected outcome, configuration, or result), verification activities should stop and a discrepancy report should be generated. The activities and events leading up to the discrepancy should be analyzed to determine if a nonconforming product exists or there is an issue with the verification procedure or conduct. The Decision Analysis Process should be used to make decisions with respect to needed changes in the verification plans, environment, and/or procedures.

Outcomes of performing product verification include the following:

- A verified product is established with supporting confirmation that the product (in the appropriate form for the life cycle phase) complies with its specified requirements, and if it does not, a nonconformance report delineating the variance is available.

- A determination is made as to whether the appropriate results were collected and evaluated to show completion of verification objectives throughout their performance envelope.

- A determination is made that the verification product was appropriately integrated with the enabling products and verification environment.

5.3.1.2.3 *Analyze Product Verification Results and Report*

As the verification activities are completed, the results are collected and analyzed. The data are analyzed for quality, integrity, correctness, consistency, and validity. Any verification discrepancies (anomalies, variations, and out-of-compliance conditions) are identified and reviewed to determine if there is a nonconforming product not resulting from poor verification conduct, procedure, or conditions. If possible, this analysis is performed while the test/analysis configuration is still intact. This allows a quick turnaround in case the data indicates that a correction to the test or analysis run needs to be performed again.

Discrepancies and nonconforming products should be recorded and reported for follow-up action and closure. Verification results should be recorded in a requirements compliance or verification matrix or other method developed during the Technical Requirements Definition Process to trace compliance for each product requirement. Waivers needed as a result of verification to request relief from or modify a requirement are identified.

NOTE: Nonconformance and discrepancy reports may be directly linked with the Technical Risk Management Process. Depending on the nature of the nonconformance, approval through such bodies as a Material Review Board or Configuration Control Board (which typically includes risk management participation) may be required.

System design and product realization process activities may be required to resolve product nonconformance. If the mitigation of the nonconformance results in a change to the product, the verification may need to be planned and performed again.

Outcomes of analyzing the verification results include the following:

- Product nonconformance (not compliant with product requirement) is identified.

- Appropriate replanning, redefinition of requirements, redesign, implementation/integration, modification, and reverification have been accomplished for resolution of the nonconforming product.

- Appropriate facility modifications, procedure corrections, enabling product modification, and reverification have been performed for non-product-related discrepancies.

- Waivers for nonconforming products are accepted.

- Discrepancy and nonconformance reports including corrective actions have been generated as needed.

- The verification report is completed.

Re-engineering

Based on analysis of verification results, it could be necessary to re-realize the end product used for verification or to re-engineer the end products assembled and integrated into the product being verified, based on where and what type of nonconformance was found.

Re-engineering could require the reapplication of the system design processes (Stakeholder Expectations Definition Process, Technical Requirements Definition Process, Logical Decomposition Process, and Design Solution Definition Process).

5.3.1.2.4 *Capture Product Verification Work Products*

Verification work products (inputs to the Technical Data Management Process) take many forms and involve many sources of information. The capture and recording of verification results and related data is a very important, but often underemphasized, step in the Product Verification Process.

Verification results, peer review reports, anomalies, and any corrective action(s) taken should be captured, as should all relevant results from the application of the Product Verification Process (related decisions, rationale for the decisions made, assumptions, and lessons learned).

Outcomes of capturing verification work products include the following:

- Verification of work products is recorded, e.g., method of verification, procedures, environments, outcomes, decisions, assumptions, corrective actions, and lessons learned.
- Variations, anomalies, and out-of-compliance conditions have been identified and documented, including the actions taken to resolve them.
- Proof that the realized end product did or did not satisfy the specified requirements is documented.
- The verification report is developed, including:
  - » recorded test/verification results/data;
  - » version of the set of specified requirements used;
  - » version of the product verified;
  - » version or standard for tools, data, and equipment used;
  - » results of each verification including pass or fail declarations; and
  - » discrepancies.

5.3.1.3 *Outputs*

Key outputs from the process are:

- **Verified product ready for validation:** After the product is verified, it will next pass through the Product Validation Process.
- **Product verification results:** Results from executed procedures are passed to technical assessment.
- **Product verification report(s):** A report shows the results of the verification activities. It includes the requirement that was to be verified and its bidirectional traceability, the verification method used, and reference to any special equipment, conditions, or procedures used. It also includes the results of the verification, any anomalies, variations or out-of-compliance results noted and associated corrective actions taken.
- **Product verification work products:** These include discrepancy and nonconformance reports with identified correction actions; updates to requirements compliance documentation; changes needed to the procedures, equipment or

environment; configuration drawings; calibrations; operator certifications; and other records.

Criteria for completing verification of the product include: (1) documented objective evidence of compliance with requirements or waiver and (2) closure of all discrepancy and nonconformance reports.

5.3.2 Product Verification Guidance

Refer to Section 5.3.2 in the NASA Expanded Guidance for Systems Engineering at *https://nen.nasa.gov/web/se/doc-repository* for additional guidance on:

- the verification approach,
- verification in the life cycle,
- verification procedures,
- verification reports
- end-to-end testing,
- use of modeling and simulations, and
- hardware-in-the-loop testing.

5.4 Product Validation

The Product Validation Process is the second of the verification and validation processes conducted on an implemented or integrated end product. While verification proves whether "the product was done right," validation proves whether "the right product was done." In other words, verification provides objective evidence that every "shall" statement in the requirements document or specification was met, whereas validation is performed for the benefit of the customers and users to ensure that the system functions in the expected manner when placed in the intended environment. This is achieved by examining the products of the system at every level of the product structure and comparing them to the stakeholder expectations for that level. A well-structured validation process can save cost and schedule while meeting the stakeholder expectations.

System validation confirms that the integrated realized end products conform to stakeholder expectations as captured in the MOEs, MOPs, and ConOps. Validation also ensures that any anomalies discovered are appropriately resolved prior to product delivery. This section discusses the process activities, methods of validation, inputs and outputs, and potential deficiencies.

See *Section 2.4* for a discussion about the distinctions between Product Verification and Product Validation.

5.4.1 Process Description

FIGURE 5.4-1, taken from NPR 7123.1, provides a typical flow diagram for the Product Validation Process and identifies typical inputs, outputs, and activities to consider in addressing product validation.

5.4.1.1 *Inputs*

Key inputs to the process are:

- **End product to be validated:** This is the end product that is to be validated and which has successfully passed through the verification process.
- **Validation plan:** This plan would have been developed under the Technical Planning Process and baselined prior to entering this process. This plan may be a separate document or a section within the Verification and Validation Plan.
- **Baselined stakeholder expectations:** These would have been developed for the product at this level during the Stakeholder Expectations Definition Process. It includes the needs, goals, and objectives as well as the baselined and updated concept of operations and MOEs.
- **Any enabling products:** These are any special equipment, facilities, test fixtures, applications,

or other items needed to perform the Product Validation Process.

5.4.1.2 *Process Activities*

The Product Validation Process demonstrates that the end product satisfies its stakeholder (customer and other interested party) expectations (MOEs) within the intended operational environments, with validation performed by anticipated operators and/or users whenever possible. The method of validation is a function of the life cycle phase and the position of the end product within the system structure.

There are five major steps in the validation process: (1) preparing to conduct validation, (2) conduct planned validation (perform validation), (3) analyze validation results, (4) prepare a validation report, and (5) capture the validation work products.

The objectives of the Product Validation Process are:

- To confirm that the end product fulfills its intended use when operated in its intended environment:
  - » Validation is performed for each implemented or integrated and verified end product from the lowest end product in a system structure branch up to the top level end product (the system).
  - » Evidence is generated as necessary to confirm that products at each layer of the system structure meet the capability and other operational expectations of the customer/user/operator and other interested parties for that product.

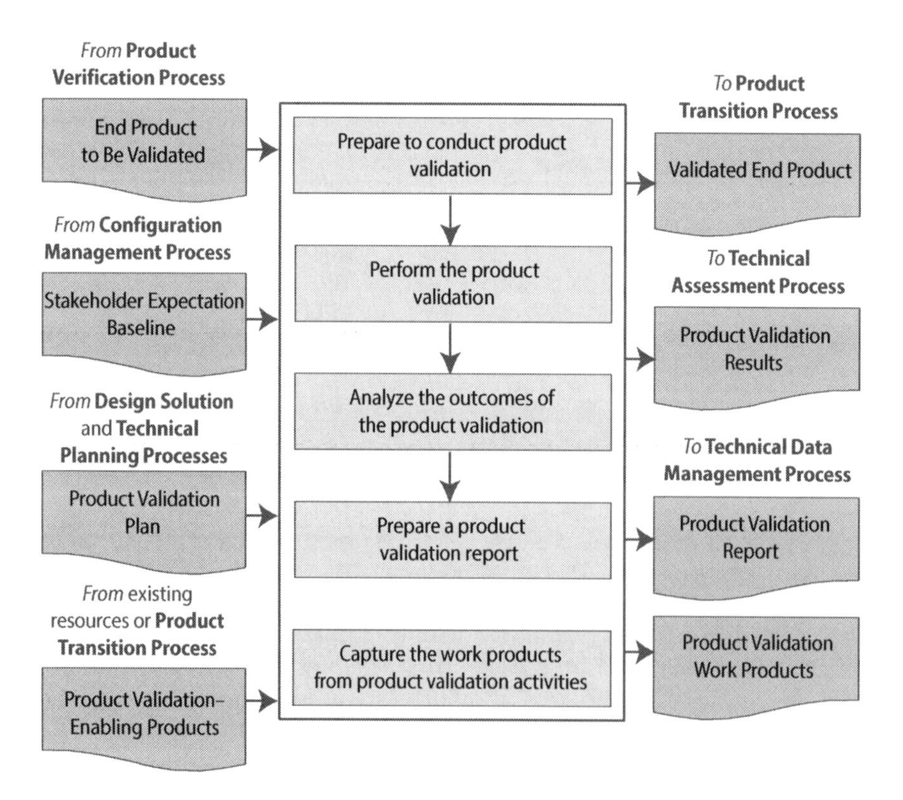

FIGURE 5.4-1 Product Validation Process

- To ensure the human has been properly integrated into the system:
    - » The user interface meets human engineering criteria.
    - » Operators and maintainers have the required skills and abilities.
    - » Instructions are provided and training programs are in place.
    - » The working environment supports crew health and safety.
- To ensure that any problems discovered are appropriately resolved prior to delivery of the end product (if validation is done by the supplier of the product) or prior to integration with other products into a higher level assembled product (if validation is done by the receiver of the product).

5.4.1.2.1 *Product Validation Preparation*

To prepare for performing product validation, the appropriate set of expectations, including MOEs and MOPs, against which the validation is to be made

METHODS OF VALIDATION

Analysis: The use of mathematical modeling and analytical techniques to predict the suitability of a design to stakeholder expectations based on calculated data or data derived from lower system structure end product verifications. Analysis is generally used when a prototype; engineering model; or fabricated, assembled, and integrated product is not available. Analysis includes the use of modeling and simulation as analytical tools. A model is a mathematical representation of reality. A simulation is the manipulation of a model.

Demonstration: Showing that the use of an end product achieves the stakeholder expectations as defined in the NGOs and the ConOps. It is generally a basic confirmation of behavioral capability, differentiated from testing by the lack of detailed data gathering. Demonstrations can involve the use of physical models or mock-ups; for example, an expectation that controls are readable by the pilot in low light conditions could be validated by having a pilot perform flight-related tasks in a cockpit mock-up or simulator under those conditions.

Inspection: The visual examination of a realized end product. Inspection is generally used to validate the presence of a physical design features or specific manufacturer identification. For example, if there is an expectation that the safety arming pin has a red flag with the words "Remove Before Flight" stenciled on the flag in black letters, a visual inspection of the arming pin flag can be used to determine if this expectation has been met.

Test: The use of an end product to obtain detailed data needed to determine a behavior, or provide sufficient information to determine a behavior through further analysis. Testing can be conducted on final end products, breadboards, brassboards, or prototypes. Testing produces information at discrete points for each specified expectation under controlled conditions and is the most resource-intensive validation technique.

should be obtained. In addition to the V&V Plan, other documentation such as the ConOps and HSI Plan may be useful. The product to be validated (output from implementation, or integration and verification), as well as the appropriate enabling products and support resources (requirements identified and acquisition initiated by design solution activities) with which validation will be conducted should be collected. Enabling products includes those representing external interfacing products and special test equipment. Support resources include personnel necessary to support validation and operators. Procedures, capturing detailed step-by-step activities and based on the validation type and methods are finalized and approved. Development of procedures typically begins during the design phase of the project life cycle and matures as the design is matured. The validation environment is considered as part of procedure development. Operational scenarios are assessed to explore all possible validation activities to be performed. The final element is preparation of the validation environment; e.g., facilities, equipment, software, and climatic conditions.

When operator or other user interaction is involved, it is important to ensure that humans are properly represented in the validation activities. This includes physical size, skills, knowledge, training, clothing, special gear, and tools. When possible, actual end users/operators should be used and other stakeholders should participate or observe activities as appropriate and practical.

Outcomes of validation preparation include the following:

- The validation plan, approved procedures, supporting configuration documentation, and an appropriate baseline set of stakeholder expectations are available and on hand;
- Enabling products are integrated within the validation environment according to plans and schedules;
- Users/operators and other resources are available according to validation plans and schedules; and
- The validation environment is evaluated for adequacy, completeness, readiness, and integration.

5.4.1.2.2 *Perform Product Validation*

The act of validating the end product is performed as spelled out in the validation plans and procedures, and the conformance established to each specified stakeholder expectation (MOEs and ConOps) shows that the validation objectives were met. Validation differs from qualification testing. Validation testing is focused on the expected environments and operations of the system where as qualification testing includes the worst case loads and environmental requirements within which the system is expected to perform or survive. The verification lead should ensure that the procedures were followed and performed as planned, the validation-enabling products and instrumentation were calibrated correctly, and the data were collected and recorded for required validation measures.

When a discrepancy is observed, the validation should be stopped and a discrepancy report generated. The activities and events leading up to the discrepancy should be analyzed to determine if a nonconforming product exists or there is an issue with the verification procedure, conduct, or conditions. If there are no product issues, the validation is replanned as necessary, the environment preparation anomalies are corrected, and the validation is conducted again with improved or correct procedures and resources. The Decision Analysis Process should be used to make decisions with respect to needed changes to the validation plans, environment, and/or conduct.

Outcomes of performing validation include the following:

- A validated product is established with supporting confirmation that the appropriate results were collected and evaluated to show completion of validation objectives.
- A determination is made as to whether the fabricated/manufactured or assembled and integrated products (including software or firmware builds and human element allocations) comply with their respective stakeholder expectations.
- A determination is made that the validated product was appropriately integrated with the validation environment and the selected stakeholder expectations set was properly validated.
- A determination is made that the product being validated functions together with interfacing products throughout their operational envelopes.

5.4.1.2.3 *Analyze Product Validation Results*

Once the validation activities have been completed, the results are collected and the data are analyzed to confirm that the end product provided will supply the customer's needed capabilities within the intended environments of use, validation procedures were followed, and enabling products and supporting resources functioned correctly. The data are also analyzed for quality, integrity, correctness, consistency, and validity, and any unsuitable products or product attributes are identified and reported.

It is important to compare the actual validation results to the expected results. If discrepancies are found, it needs to be determined if they are a result of the test configuration or analysis assumptions or whether they are a true characteristic or behavior of the end product. If it is found to be a result of the test configuration, the configuration should be corrected and the validation repeated. If it is found to be a result of the end product being validated, discussions with the customer should be held and any required system design and product realization process activities should be conducted to resolve deficiencies. The deficiencies along with recommended corrective actions and resolution results should be recorded, and validation should be repeated, as required.

Outcomes of analyzing validation results include the following:

- Product anomalies, variations, deficiencies, nonconformance and/or issues are identified.
- Assurances that appropriate replanning, redefinition of requirements, design, and revalidation have been accomplished for resolution of anomalies, variations, deficiencies or out-of-compliance conditions (for problems not caused by poor validation conduct).
- Discrepancy and corrective action reports are generated as needed.
- The validation report is completed.

Re-engineering

Based on the results of the Product Validation Process, it could become necessary to re-engineer a deficient end product. Care should be taken that correcting a deficiency or set of deficiencies does not generate a new issue with a part or performance that had previously operated satisfactorily. Regression testing, a formal process of rerunning previously used acceptance tests (primarily used for software), is one method to ensure a change does not affect function or performance that was previously accepted.

Validation Deficiencies

Validation outcomes can be unsatisfactory for several reasons. One reason is poor conduct of the validation (e.g., enabling products and supporting resources missing or not functioning correctly, untrained operators, procedures not followed, equipment not calibrated, or improper validation environmental conditions) and failure to control other variables not involved in validating a set of stakeholder expectations. A second reason could be a shortfall in the verification process of the end product. This could create the need for:

- Re-engineering end products lower in the system structure that make up the end product that was found to be deficient (i.e., that failed to satisfy validation requirements); and/or
- Re-performing any needed verification and validation processes.

Other reasons for validation deficiencies (particularly when M&S are involved) may be incorrect and/or inappropriate initial or boundary conditions; poor formulation of the modeled equations or behaviors; the impact of approximations within the modeled equations or behaviors; failure to provide the required geometric and physics fidelities needed for credible simulations for the intended purpose; and/or poor spatial, temporal, and perhaps, statistical resolution of physical phenomena used in M&S.

NOTE: Care should be exercised to ensure that the corrective actions identified to remove validation deficiencies do not conflict with the baselined stakeholder expectations without first coordinating such changes with the appropriate stakeholders.

Of course, the ultimate reason for performing validation is to determine if the design itself is the right design for meeting stakeholder expectations. After any and all validation test deficiencies are ruled out, the true value of validation is to identify design changes needed to ensure the program/product's mission. Validation should be performed as early and as iteratively as possible in the SE process since the earlier re-engineering needs are discovered, the less expensive they are to resolve.

Pass Verification but Fail Validation?

Sometimes systems successfully complete verification but then are unsuccessful in some critical phase of the validation process, delaying development and causing extensive rework and possible compromises with the stakeholder. Developing a solid ConOps in early phases of the project (and refining it through the requirements development and design phases) is critical to preventing unsuccessful validation. Similarly, developing clear expectations for user community involvement in the HSI Plan is critical to successful validation. Frequent and iterative communications with stakeholders helps to identify operational scenarios and key needs that should be understood when designing and implementing the end product. Should the product fail validation, redesign may be a necessary reality. Review of the understood requirements set, the existing design, operational scenarios, user population numbers and skills, training, and support material may be necessary, as well as negotiations and compromises with the customer, other stakeholders, and/or end users to determine what, if anything, can be done to correct or resolve the situation. This can add time and cost to the overall project or, in some cases, cause the project to fail or be canceled. However, recall from FIGURE 2.5-1 that the earlier design issues are discovered, the less costly the corrective action.

5.4.1.2.4 *Prepare Report and Capture Product Validation Work Products*

Validation work products (inputs to the Technical Data Management Process) take many forms and involve many sources of information. The capture and recording of validation-related data is a very important, but often underemphasized, step in the Product Validation Process.

Validation results, deficiencies identified, and corrective actions taken should be captured, as should all relevant results from the application of the Product Validation Process (related decisions, rationale for decisions made, assumptions, and lessons learned).

Outcomes of capturing validation work products include the following:

- Work products and related information generated while doing Product Validation Process activities and tasks are recorded; i.e., method of validation conducted, the form of the end product used for validation, validation procedures used, validation environments, outcomes, decisions, assumptions, corrective actions, lessons learned, etc. (often captured in a matrix or other tool—see *Appendix E*).
- Deficiencies (e.g., variations and anomalies and out-of-compliance conditions) are identified and documented, including the actions taken to resolve.
- Proof is provided that the end product is in conformance with the stakeholder expectation set used in the validation.
- Validation report including:
  - » Recorded validation results/data;
  - » Version of the set of stakeholder expectations used;
  - » Version and form of the end product validated;
  - » Version or standard for tools and equipment used, together with applicable calibration data;
  - » Outcome of each validation including pass or fail declarations; and
  - » Discrepancy between expected and actual results.

NOTE: For systems where only a single deliverable item is developed, the Product Validation Process normally completes acceptance testing of the system. However, for systems with several production units, it is important to understand that continuing verification and validation is not an appropriate approach to use for the items following the first deliverable. Instead, acceptance testing is the preferred means to ensure that subsequent deliverables meet stakeholder expectations.

5.4.1.3 *Outputs*

Key outputs of validation are:

- **Validated end product:** This is the end product that has successfully passed validation and is ready to be transitioned to the next product layer or to the customer.
- **Product validation results:** These are the raw results of performing the validations.
- **Product validation report:** This report provides the evidence of product conformance with the stakeholder expectations that were identified as being validated for the product at this layer. It includes any nonconformance, anomalies, or other corrective actions that were taken.
- **Work products:** These include procedures, required personnel training, certifications,

configuration drawings, and other records generated during the validation activities.

Success criteria for this process include: (1) objective evidence of performance and the results of each system-of-interest validation activity are documented, and (2) the validation process should not be considered or designated as complete until all issues and actions are resolved.

5.4.2 Product Validation Guidance

Refer to Section 5.4.2 in the NASA Expanded Guidance for Systems Engineering at *https://nen.nasa.gov/web/se/doc-repository* for additional guidance on:

- use of modeling and simulation,
- software validation, and
- taking credit for validation.

5.5 Product Transition

The Product Transition Process is used to transition a verified and validated end product that has been generated by product implementation or product integration to the customer at the next level in the system structure for integration into an end product or, for the top-level end product, transitioned to the intended end user. The form of the product transitioned will be a function of the product life cycle phase success criteria and the location within the system structure of the WBS model in which the end product exists. The systems engineer involvement in this process includes ensuring the product being transitioned has been properly tested and verified/validated prior to being shipped to the next level stakeholder/customer.

Product transition occurs during all phases of the life cycle. During the early phases, the technical team's products are documents, models, studies, and reports. As the project moves through the life cycle, these paper or soft products are transformed through implementation and integration processes into hardware and software solutions to meet the stakeholder expectations. They are repeated with different degrees of rigor throughout the life cycle. The Product Transition Process includes product transitions from one level of the system architecture upward. The Product Transition Process is the last of the product realization processes, and it is a bridge from one level of the system to the next higher level.

The Product Transition Process is the key to bridge from one activity, subsystem, or element to the overall engineered system. As the system development nears completion, the Product Transition Process is again applied for the end product, but with much more rigor since now the transition objective is delivery of the system-level end product to the actual end user. Depending on the kind or category of system developed, this may involve a Center or the Agency and impact thousands of individuals storing, handling, and transporting multiple end products; preparing user sites; training operators and maintenance personnel; and installing and sustaining, as applicable. Examples are transitioning the external tank, solid rocket boosters, and orbiter to Kennedy Space Center (KSC) for integration and flight. Another example is the transition of a software subsystem for integration into a combined hardware/software system.

5.5.1 Process Description

FIGURE 5.5-1 provides a typical flow diagram for the Product Transition Process and identifies typical inputs, outputs, and activities to consider in addressing product transition.

5.5.1.1 *Inputs*

Inputs to the Product Transition Process depend primarily on the transition requirements, the product that is being transitioned, the form of the product

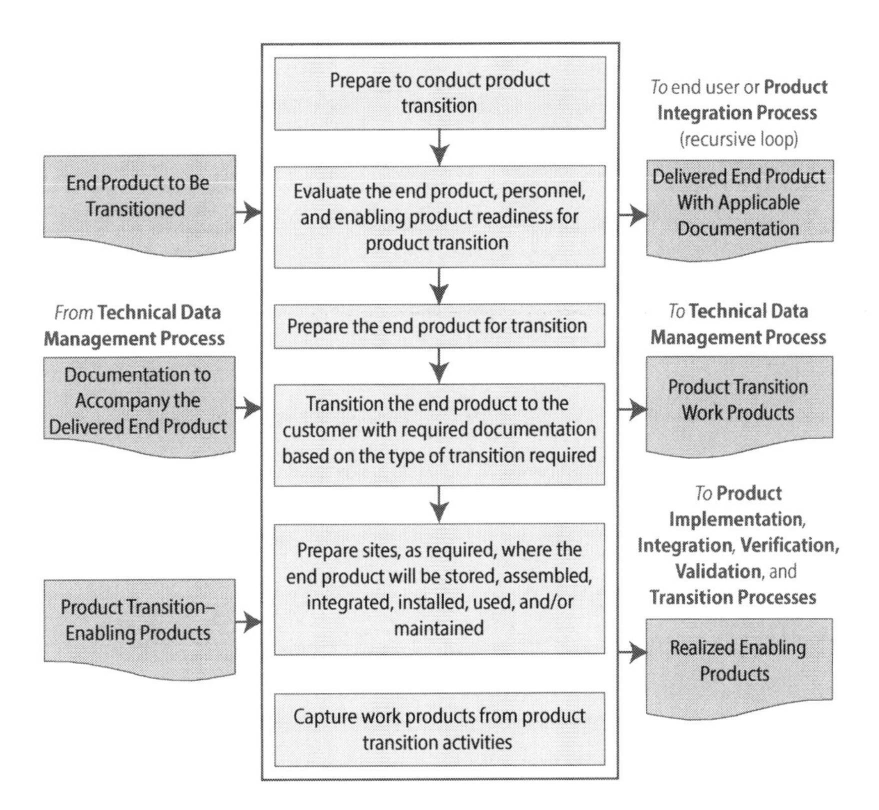

FIGURE 5.5-1 Product Transition Process

transition that is taking place, and the location to which the product is transitioning. Typical inputs are shown in FIGURE 5.5-1 and described below.

- **The end product or products to be transitioned (from the Product Validation Process):** The product to be transitioned can take several forms. It can be a subsystem component, system assembly, or top-level end product. It can be hardware, analytical models, or software. It can be newly built, purchased, or reused. A product can transition from a lower system product to a higher one by being integrated with other transitioned products. This process may be repeated until the final end product is achieved. Each succeeding transition requires unique input considerations when preparing the validated product for transition to the next level.

  Early phase products can take the form of information or data generated from basic or applied research using analytical or physical models and are often in paper or electronic form. In fact, the end product for many NASA research projects or science activities is a report, paper, model, or even an oral presentation. In a sense, the dissemination of information gathered through NASA research and development is an important form of product transition.

- **Documentation including manuals, procedures, and processes that accompany the end product (from the Technical Data Management Process):** The documentation required for the Product Transition Process depends on the specific end product; its current location within the system structure; and the requirements identified in various agreements, plans, or requirements documents. Typically, a product has a unique identification (i.e., serial or version number) and may have a pedigree (documentation) that specifies its heritage and current state. Pertinent information may be controlled using a configuration control process or work order system as well as design drawings and test reports. Documentation often includes proof of verification and validation conformance. A COTS product would typically contain a manufacturer's specification or fact sheet. Documentation may include operations manuals, installation instructions, and other information.

The documentation level of detail is dependent upon where the product is within the product hierarchy and the life cycle. Early in the life cycle, this documentation may be conceptual or preliminary in nature. Later in the life cycle, the documentation may be detailed design documents, user manuals, drawings, or other work products. Documentation that is gathered during the input process for the transition phase may require editing, assembling, or repackaging to ensure it is in the required condition for acceptance by the customer.

Special consideration should be given to safety, including clearly identifiable tags and markings that identify the use of hazardous materials, special handling instructions, and storage requirements.

- **Product transition-enabling products, including packaging materials; containers; handling equipment; and storage, receiving, and shipping facilities (from existing resources or the Product Transition Process for enabling product realization):** Product transition-enabling products may be required to facilitate the implementation, integration, evaluation, transition, training, operations, support, and/or retirement of the transition product at its next higher level or for the transition of the final end product. Some or all of the enabling products may be defined in transition-related agreements, system requirements documents, or project plans. In some cases, product transition-enabling products are developed during the realization of the product itself or may be required to be developed during the transition stage.

As a product is developed, special containers, holders, or other devices may also be developed to aid in the storing and transporting of the product through development and realization. These may be temporary accommodations that do not satisfy all the transition requirements, but allow the product to be initiated into the transition process. In such cases, the temporary accommodations will have to be modified or new accommodations will need to be designed and built or procured to meet specific transportation, handling, storage, and shipping requirements.

Sensitive or hazardous products may require special enabling products such as monitoring equipment, security features, inspection devices, safety devices, and personnel training to ensure adequate safety and environmental requirements are achieved and maintained.

5.5.1.2 *Process Activities*

Transitioning the product can take one of two forms:

- The delivery of lower system end products to higher ones for integration into another end product; or
- The delivery of the final end product to the customer or user that will use it in its operational environment.

In the first case, the end product is one of perhaps several other pieces that will ultimately be integrated together to form the item. In the second case, the end product is for final delivery to the customer. For example, the end product might be one of several circuit cards that will be integrated together to form the final unit that is delivered. Or that unit might also be one of several units that have to be integrated together to form the final product.

The form of the product transitioned is not only a function of the location of that product within the system product hierarchy, but also a function of the life cycle phase. Early life cycle phase products may be in the form of paper, electronic files, physical models, or technology demonstration prototypes. Later phase products may be preproduction prototypes (engineering models), the final study report, or the flight units.

FIGURE 5.5-1 shows what kind of inputs, outputs, and activities are performed during product transition regardless of where in the product hierarchy or life cycle the product is. These activities include preparing to conduct the transition; making sure the end product, all personnel, and any enabling products are ready for transitioning; preparing the site; and performing the transition including capturing and documenting all work products.

How these activities are performed and what form the documentation takes depends on where the end items are in the product hierarchy and the life cycle phase.

5.5.1.2.1 *Prepare to Conduct Transition*

The first task is to identify which of the two forms of transition is needed: (1) the delivery of lower system end products to higher ones for integration into another end product; or (2) the delivery of the final end product to the customer or user that will use the end product in its operational environment. The form of the product being transitioned affects transition planning and the kind of packaging, handling, storage, and transportation that is required. The customer and other stakeholder expectations, as well as the specific design solution, may indicate special transition procedures or enabling product needs for packaging, storage, handling, shipping/transporting, site preparation, installation, and/or sustainability. These requirements need to be reviewed during the preparation stage.

Other tasks in preparing to transition a product involve making sure the end product, personnel, and any enabling products are ready for that transition. This includes the availability of the documentation or models that will be sent with the end product, including proof of verification and validation conformance. The appropriateness of detail for that documentation depends upon where the product is within the product hierarchy and the life cycle. Early in the life cycle, this documentation may be preliminary in nature. Later in the life cycle, the documentation may be detailed design documents, user manuals, drawings, or other work products. Procedures necessary for conducting the transition should be reviewed and approved by this time.

Finally, the availability and skills of personnel needed to conduct the transition as well as the availability of any necessary packaging materials/containers, handling equipment, storage facilities, and shipping/transporter services should also be reviewed. Any special training necessary for the personnel to perform their tasks needs to be performed by this time.

5.5.1.2.2 *Prepare the Site to Receive the Product*

For either of the forms of product transition, the receiving site needs to be prepared to receive the product. Here the end product is stored, assembled, integrated, installed, used, and/or maintained as appropriate for the life cycle phase, position of the end product in the system structure, and customer agreement.

A vast number of key complex activities, many of them outside direct control of the technical team, need to be synchronized to ensure smooth transition to the end user. If transition activities are not carefully controlled, there can be impacts on schedule, cost, and safety of the end product.

A site survey may need to be performed to determine the issues and needs. This should address the adequacy of existing facilities to accept, store, and operate the new end product and identify any logistical-support-enabling products and services required but not planned for. Additionally, any modifications to existing facilities should be planned well in advance of fielding; therefore, the site survey should be made during an early phase in the product life cycle. These may include logistical enabling products and services to provide support for end-product use, operations, maintenance, and disposal. Training for users, operators, maintainers, and other support personnel may need to be conducted. National Environmental Policy Act documentation or approvals may need to be obtained prior to the receipt of the end product.

Prior to shipment or after receipt, the end product may need to be stored in suitable storage conditions to protect and secure the product and prevent damage or the deterioration of it. These conditions should have been identified early in the design life cycle.

5.5.1.2.3 *Prepare the Product for Transition*

Whether transitioning a product to the next room for integration into the next higher assembly, or for final transportation across the country to the customer, care should be taken to ensure the safe transportation of the product. The requirements for packaging, handling, storage, training, and transportation should have been identified during system design. Preparing the packaging for protection, security, and prevention of deterioration is critical for products placed in storage or when it is necessary to transport or ship between and within organizational facilities or between organizations by land, air, and/or water vehicles. Particular emphasis needs to be on protecting surfaces from physical damage, preventing corrosion, eliminating damage to electronic wiring or cabling, shock or stress damage, heat warping or cold fractures, moisture, and other particulate intrusion that could damage moving parts.

The design requirements should have already addressed the ease of handling or transporting the product such as component staking, addition of transportation hooks, crating, etc. The ease and safety of packing and unpacking the product should also have been addressed. Additional measures may also need to be implemented to show accountability and to securely track the product during transportation. In cases where hazardous materials are involved, special labeling or handling needs, including transportation routes, need to be in place.

5.5.1.2.4 *Transition the Product*

The end product is then transitioned (i.e., moved, transported, or shipped) with required documentation to the customer based on the type of transition required, e.g., to the next higher level item in the product hierarchy (often called the Product Breakdown Structure (PBS)) for product integration or to the end user. Documentation may include operations manuals, installation instructions, and other information.

The end product is finally installed into the next higher assembly or into the customer/user site using the preapproved installation procedures.

Confirm Ready to Support

After installation, whether into the next higher assembly or into the final customer site, functional and acceptance testing of the end product should be conducted. This ensures no damage from the shipping/handling process has occurred and that the product is ready for support. Any final transitional work products should be captured as well as documentation of product acceptance.

5.5.1.2.5 *Capture Product Transition Work Products*
Other work products generated during the transition process are captured and archived as appropriate. These may include site plans, special handling procedures, training, certifications, videos, inspections, or other products from these activities.

5.5.1.3 *Outputs*

- **Delivered end product with applicable documentation:** This may take one of two forms:

    1. **Delivered end product for integration to next level up in system structure:** This includes the appropriate documentation. The form of the end product and applicable documentation are a function of the life cycle phase and the placement within the system structure. (The form of the end product could be hardware, software, model, prototype, first article for test, or single operational article or multiple production articles.) Documentation includes applicable draft installation, operation, user, maintenance, or training manuals; applicable baseline documents (configuration baseline, specifications, and stakeholder expectations); and test results that reflect completion of verification and validation of the end product.

    2. **Delivered operational end product for end users:** The appropriate documentation is to accompany the delivered end product as well as the operational end product appropriately packaged. Documentation includes applicable final installation, operation, user, maintenance, or training manuals; applicable baseline documents (configuration baseline, specifications, stakeholder expectations); and test results that reflect completion of verification and validation of the end product. If the end user will perform end product validation, sufficient documentation to support end user validation activities is delivered with the end product.

- **Work products from transition activities to technical data management:** Work products could include the transition plan, site surveys, measures, training modules, procedures, decisions, lessons learned, corrective actions, etc.

- **Realized enabling end products to appropriate life cycle support organization:** Some of the enabling products that were developed during the various phases could include fabrication or integration specialized machines; tools; jigs; fabrication processes and manuals; integration processes and manuals; specialized inspection, analysis, demonstration, or test equipment; tools; test stands; specialized packaging materials and containers; handling equipment; storage-site environments; shipping or transportation vehicles or equipment; specialized courseware; instructional site environments; and delivery of the training instruction. For the later life cycle phases, enabling products that are to be delivered may include specialized mission control equipment; data collection equipment; data analysis equipment; operations manuals; specialized maintenance equipment, tools, manuals, and spare parts; specialized recovery equipment; disposal equipment; and readying recovery or disposal site environments.

The process is complete when the following activities have been accomplished:

- For deliveries to the integration path, the end product is delivered to intended usage sites in a condition suitable for integration with other end products or composites of end products.

Procedures, decisions, assumptions, anomalies, corrective actions, lessons learned, etc., resulting from transition for integration are recorded.

- For delivery to the end user path, the end products are installed at the appropriate sites; appropriate acceptance and certification activities are completed; training of users, operators, maintainers, and other necessary personnel is completed; and delivery is closed out with appropriate acceptance documentation.
- Any realized enabling end products are also delivered as appropriate including procedures, decisions, assumptions, anomalies, corrective actions, lessons learned, etc., resulting from transition-enabling products.

5.5.2 Product Transition Guidance

Refer to Section 5.5.2 in the NASA Expanded Guidance for Systems Engineering at *https://nen.nasa.gov/web/se/doc-repository* for additional guidance on:

- additional product transition considerations and
- what's next after product transition to the end user.

6.0 Crosscutting Technical Management

This chapter describes the activities in the technical management processes listed in the systems engineering engine (FIGURE 2.1-1). The processes described in *Chapters 4* and *5* are performed through the design and realization phases of a product. These processes can occur throughout the product life cycle, from concept through disposal. They may occur simultaneously with any of the other processes. The chapter is separated into sections corresponding to the technical management processes 10 through 17 listed in FIGURE 2.1-1. Each technical management process is discussed in terms of the inputs, the activities, and the outputs. Additional guidance is provided using examples that are relevant to NASA projects.

The technical management processes are the bridges between project management and the technical team. In this portion of the engine, eight processes provide the crosscutting functions that allow the design solution to be developed, realized, and to operate. Even though every technical team member might not be directly involved with these eight processes, they are indirectly affected by these key functions. Every member of the technical team relies on technical planning; management of requirements, interfaces, technical risk, configuration, and technical data; technical assessment; and decision analysis to meet the project's objectives. Without these crosscutting processes, individual members and tasks cannot be integrated into a functioning system that meets the ConOps within cost and schedule. These technical processes also support the project management team in executing project control.

The next sections describe each of the eight technical management processes and their associated products for a given NASA mission.

6.1 Technical Planning

The Technical Planning Process, the first of the eight technical management processes contained in the systems engineering engine, establishes a plan for applying and managing each of the common technical processes that will be used to drive the development of system products and associated work products. This process also establishes a plan for identifying and defining the technical effort required to satisfy the project objectives and life cycle phase success criteria within the cost, schedule, and risk constraints of the project.

This effort starts with the technical team conducting extensive planning early in Pre-Phase A. With

CROSSCUTTING TECHNICAL MANAGEMENT KEYS

- Thoroughly understand and plan the scope of the technical effort by investing time upfront to develop the technical product breakdown structure, the technical schedule and workflow diagrams, and the technical resource requirements and constraints (funding, budget, facilities, and long-lead items) that will be the technical planning infrastructure. The systems engineer also needs to be familiar with the non-technical aspects of the project.
- Define all interfaces and assign interface authorities and responsibilities to each, both intra-and inter-organizational. This includes understanding potential incompatibilities and defining the transition processes.
- Control of the configuration is critical to understanding how changes will impact the system. For example, changes in design and environment could invalidate previous analysis results.
- Conduct milestone reviews to enable a critical and valuable assessment to be performed. These reviews are not to be solely used to meet contractual or scheduling incentives. These reviews have specific entrance criteria and should be conducted when these are met.
- Understand any biases, assumptions, and constraints that impact the analysis results.
- Place all analysis under configuration control to be able to track the impact of changes and understand when the analysis needs to be reevaluated.

this early planning, technical team members will understand the roles and responsibilities of each team member, and can establish cost and schedule goals and objectives. From this effort, the Systems Engineering Management Plan (SEMP) and other technical plans are developed and baselined. Once the SEMP and technical plans have been established, they should be synchronized with the project master plans and schedule. In addition, the plans for establishing and executing all technical contracting efforts are identified.

This is a recursive and iterative process. Early in the life cycle, the technical plans are established and synchronized to run the design and realization processes. As the system matures and progresses through the life cycle, these plans should be updated as necessary to reflect the current environment and resources and to control the project's performance, cost, and schedule. At a minimum, these updates will occur at every Key Decision Point (KDP). However, if there is a significant change in the project, such as new stakeholder expectations, resource adjustments, or other constraints, all plans should be analyzed for the impact of these changes on the baselined project.

6.1.1 Process Description

FIGURE 6.1-1 provides a typical flow diagram for the Technical Planning Process and identifies typical inputs, outputs, and activities to consider in addressing technical planning.

6.1.1.1 *Inputs*

Input to the Technical Planning Process comes from both the project management and technical teams as outputs from the other common technical processes. Initial planning utilizing external inputs from the project to determine the general scope and framework

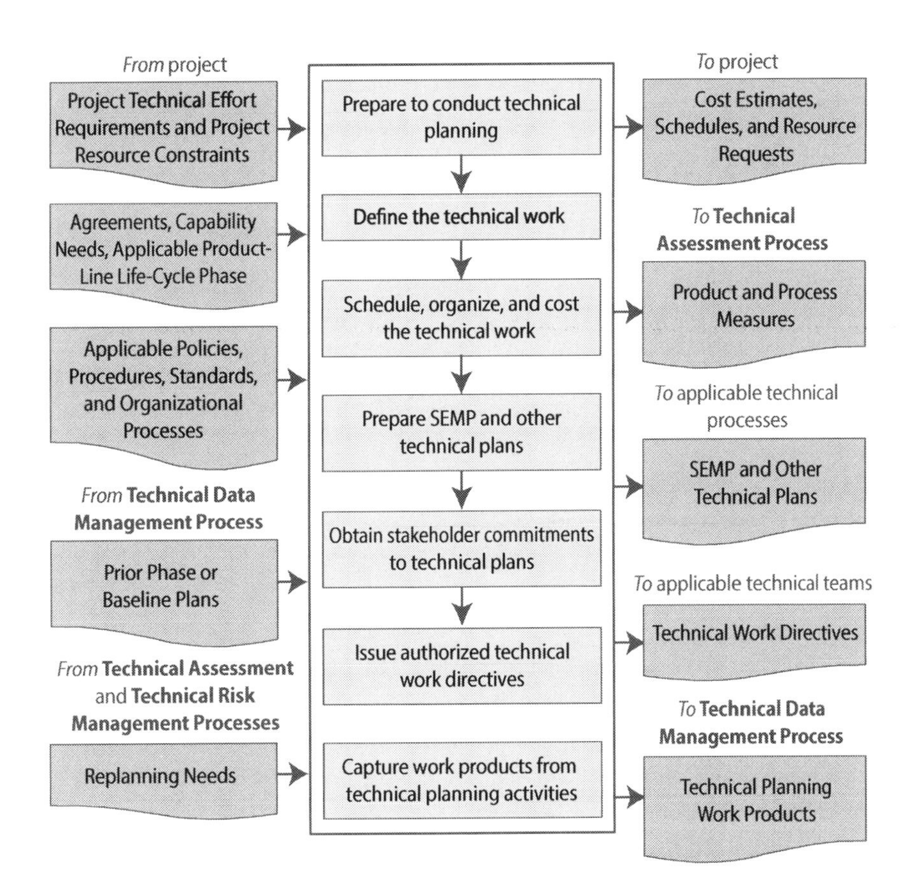

FIGURE 6.1-1 Technical Planning Process

of the technical effort will be based on known technical and programmatic requirements, constraints, policies, and processes. Throughout the project's life cycle, the technical team continually incorporates results into the technical planning strategy and documentation and any internal changes based on decisions and assessments generated by the other processes of the SE engine or from requirements and constraints mandated by the project.

- **Project Technical Effort Requirements and Project Resource Constraints:** The program/project plan provides the project's top-level technical requirements, the available budget allocated to the program/project from the program, and the desired schedule to support overall program needs. Although the budget and schedule allocated to the program/project serve as constraints, the technical team generates a technical cost estimate and schedule based on the actual work required to satisfy the technical requirements. Discrepancies between the allocated budget and schedule and the technical team's actual cost estimate and schedule should be reconciled continuously throughout the life cycle.

- **Agreements, Capability Needs, Applicable Product Life Cycle Phase:** The program/project plan also defines the applicable life cycle phases and milestones, as well as any internal and external agreements or capability needs required for successful execution. The life cycle phases and programmatic milestones provide the general framework for establishing the technical planning effort and for generating the detailed technical activities and products required to meet the overall milestones in each of the life cycle phases.

- **Applicable Policies, Procedures, Standards, and Organizational Processes:** The program/project plan includes all programmatic policies, procedures, standards, and organizational processes that should be adhered to during execution of the technical effort. The technical team should develop a technical approach that ensures the program/project requirements are satisfied and that any technical procedures, processes, and standards to be used in developing the intermediate and final products comply with the policies and processes mandated in the program/project plan.

- **Prior Phase or Baseline Plans:** The latest technical plans (either baselined or from the previous life cycle phase) from the Data Management or Configuration Management Processes should be used in updating the technical planning for the upcoming life cycle phase.

- **Replanning Needs:** Technical planning updates may be required based on results from technical reviews conducted in the Technical Assessment Process, issues identified during the Technical Risk Management Process, or from decisions made during the Decision Analysis Process.

6.1.1.2 *Process Activities*

Technical planning as it relates to systems engineering at NASA is intended to define how the project will be organized, structured, and conducted and to identify, define, and plan how the 17 common technical processes in NPR 7123.1, NASA Systems Engineering Processes and Requirements will be applied in each life cycle phase for all levels of the product hierarchy (see Section 6.1.2.1 in the NASA Expanded Guidance for Systems Engineering at *https://nen.nasa.gov/web/se/doc-repository.*) within the system structure to meet product life cycle phase success criteria. A key document capturing and updating the details from the technical planning process is the SEMP.

The SEMP is a subordinate document to the project plan. The project plan defines how the project will be managed to achieve its goals and objectives within defined programmatic constraints. The SEMP defines for all project participants how the project will be technically managed within the constraints established by the project. The SEMP also communicates how the systems engineering management techniques will be applied throughout all phases of the project life cycle.

Technical planning should be tightly integrated with the Technical Risk Management Process (see *Section 6.4*) and the Technical Assessment Process (see *Section 6.7*) to ensure corrective action for future activities will be incorporated based on current issues identified within the project.

Technical planning, as opposed to program or project planning, addresses the scope of the technical effort required to develop the system products. While the project manager concentrates on managing the overall project life cycle, the technical team, led by the systems engineer, concentrates on managing the technical aspects of the project. The technical team identifies, defines, and develops plans for performing decomposition, definition, integration, verification, and validation of the system while orchestrating and incorporating the appropriate concurrent and crosscutting engineering. Additional planning includes defining and planning for the appropriate technical reviews, audits, assessments, and status reports and determining crosscutting engineering discipline and/or design verification requirements.

This section describes how to perform the activities contained in the Technical Planning Process shown in FIGURE 6.1-1. The initial technical planning at the beginning of the project establishes the technical team members; their roles and responsibilities; and the tools, processes, and resources that will be utilized in executing the technical effort. In addition, the expected activities that the technical team will perform and the products it will produce are identified, defined, and scheduled. Technical planning continues to evolve as actual data from completed tasks are received and details of near-term and future activities are known.

6.1.1.2.1 *Technical Planning Preparation*

For technical planning to be conducted properly, the processes and procedures that are needed to conduct technical planning should be identified, defined, and communicated. As participants are identified, their roles and responsibilities and any training and/or certification activities should be clearly defined and communicated.

Team Selection

Teams engaged in the early part of the technical planning process need to identify the required skill mix for technical teams that will develop and produce a product. Typically, a technical team consists of a mix of both subsystem and discipline engineers. Considering a spacecraft example, subsystem engineers normally have cognizance over development of a particular subsystem (e.g., mechanical, power, etc.), whereas discipline engineers normally provide specific analyses (e.g., flight dynamics, radiation, etc.). The availability of appropriately skilled personnel also needs to be considered.

To an extent, determining the skill mix required for developing any particular product is a subjective process. Due to this, the skill mix is normally determined in consultation with people experienced in leading design teams for a particular mission or technical application. Some of the subjective considerations involved include the product and its requirements, the mission class, and the project phase.

Continuing with a spacecraft example, most teams typically share a common core of required skills, such as subsystem engineering for mechanical, thermal, power, etc. However, the particular requirements of a spacecraft and mission can cause the skill mix to vary. For example, as opposed to robotic space missions, human-rated systems typically add the need for human systems discipline engineering and environmental control and life support subsystem engineering. As opposed to near Earth space missions, deep space missions may add the need for safety and planetary protection discipline engineering specific to contamination of the Earth or remote solar system bodies. And, as opposed to teams designing spacecraft instruments that operate at moderate temperatures, teams designing spacecraft instruments that operate at cryogenic temperatures will need cryogenics subsystem support.

Mission class and project phase may also influence the required team skill mix. For example, with respect to mission class, certain discipline analyses needed for Class A and B missions may not be required for Class D (or lower) missions. And with respect to project phase, some design and analyses may be performed by a single general discipline in Pre-Phase A and Phase A, whereas the need to conduct design and analyses in more detail in Phases B and C may indicate the need for multiple specialized subsystem design and discipline engineering skills.

An example skill mix for a Pre-Phase A technical team tasked to design a cryogenic interferometer space observatory is shown in TABLE 6.1-1 for purposes of illustration. For simplicity, analysis and technology development is assumed to be included

TABLE 6.1-1 Example Engineering Team Disciplines in Pre-Phase A for Robotic Infrared Observatory

Systems Engineering
Mission Systems Engineer
Instrument Systems Engineer
Spacecraft Bus, Flight Dynamics, Launch Vehicle Interface, Ground System Interface Subteam
Flight Dynamics Analysis
Mission Operations (includes ConOps, & interfaces with ground station, mission ops center, science ops center)
Bus Mechanical Subsystem (includes mechanisms)
Bus Power Subsystem (includes electrical harness)
Bus Thermal Subsystem
Bus Propulsion Subsystem
Bus Attitude Control and Determination Subsystem
Bus Avionics Subsystem
Bus Communications Subsystem
Bus Flight Software Subsystem
Integration and Test (bus, observatory)
Launch Vehicle Integration
Radiation Analysis
Orbital Debris/End of Mission Planning Analysis
System Reliability/Fault Tolerance Analysis (includes analysis of instrument)
Instrument Subteam
Mechanical Subsystem
Mechanisms Subsystem
Thermal Subsystem
Cryogenics Subsystem
Avionics Subsystem (incl. Electrical Harness)
Mechanism Drive Electronics Subsystem
Detector Subsystem
Optics Subsystem
Control Subsystem
Metrology Subsystem
Flight Software Subsystem
Integration and Test
Stray Light/Radiometry Analysis
Other Specialty Disciplines (e.g., Contamination Analysis) as needed

in the subsystem or discipline shown. For example, this means "mechanical subsystem" includes both loads and dynamics analysis and mechanical technology development.

Once the processes, people, and roles and responsibilities are in place, a planning strategy may be formulated for the technical effort. A basic technical planning strategy should address the following:

- The communication strategy within the technical team and for up and out communications;
- Identification and tailoring of NASA procedural requirements that apply to each level of the PBS structure;
- The level of planning documentation required for the SEMP and all other technical planning documents;
- Identifying and collecting input documentation;
- The sequence of technical work to be conducted, including inputs and outputs;
- The deliverable products from the technical work;
- How to capture the work products of technical activities;
- How technical risks will be identified and managed;
- The tools, methods, and training needed to conduct the technical effort;
- The involvement of stakeholders in each facet of the technical effort;
- How the NASA technical team will be involved with the technical efforts of external contractors;
- The entry and success criteria for milestones, such as technical reviews and life cycle phases;
- The identification, definition, and control of internal and external interfaces;
- The identification and incorporation of relevant lessons learned into the technical planning;
- The team's approach to capturing lessons learned during the project and how those lessons will be recorded;
- The approach for technology development and how the resulting technology will be incorporated into the project;
- The identification and definition of the technical metrics for measuring and tracking progress to the realized product;
- The criteria for make, buy, or reuse decisions and incorporation criteria for Commercial Off-the-Shelf (COTS) software and hardware;
- The plan to identify and mitigate off-nominal performance;
- The "how-tos" for contingency planning and replanning;
- The plan for status assessment and reporting;
- The approach to decision analysis, including materials needed, skills required, and expectations in terms of accuracy; and
- The plan for managing the human element in the technical activities and product.

By addressing these items and others unique to the project, the technical team will have a basis for understanding and defining the scope of the technical effort, including the deliverable products that the overall technical effort will produce, the schedule and key milestones for the project that the technical team should support, and the resources required by the technical team to perform the work.

A key element in defining the technical planning effort is understanding the amount of work associated with performing the identified activities. Once the scope of the technical effort begins to coalesce, the technical team may begin to define specific planning activities and to estimate the amount of effort and resources required to perform each task. Historically, many projects have underestimated the resources required to perform proper planning activities and have been forced into a position of continuous crisis management in order to keep up with changes in the project.

TABLE 6.1-2 Examples of Types of Facilities to Consider During Planning

Communications and Tracking Labs	Models and Simulation Labs	Thermal Chambers
Power Systems Labs	Prototype Development Shops	Vibration Labs
Propulsion Test Stands	Calibration Labs	Radiation Labs
Mechanical/Structures Labs	Biological Labs	Animal Care Labs
Instrumentation Labs	Space Materials Curation Labs	Flight Hardware Storage Areas
Human Systems Labs	Electromagnetic Effects Labs	Design Visualization
Guidance and Navigation Labs	Materials Labs	Wiring Shops
Robotics Labs	Vacuum Chambers	NDE Labs
Software Development Environment	Mission Control Center	Logistics Warehouse
Meeting Rooms	Training Facilities	Conference Facilities
Education/Outreach Centers	Server Farms	Project Documentation Centers

Identifying Facilities

The planning process also includes identifying the required facilities, laboratories, test beds, and instrumentation needed to build, test, launch, and operate a variety of commercial and Government products. A sample list of the kinds of facilities that might be considered when planning is illustrated in TABLE 6.1-2.

6.1.1.2.2 *Define the Technical Work*

The technical effort should be defined commensurate with the level of detail needed for the life cycle phase. When performing the technical planning, realistic values for cost, schedule, and labor resources should be used. Whether extrapolated from historical databases or from interactive planning sessions with the project and stakeholders, realistic values should be calculated and provided to the project team. Contingency should be included in any estimate and should be based on the complexity and criticality of the effort. Contingency planning should be conducted. The following are examples of contingency planning:

- Additional, unplanned-for software engineering resources are typically needed during hardware and systems development and testing to aid in troubleshooting errors/anomalies. Frequently, software engineers are called upon to help troubleshoot problems and pinpoint the source of errors in hardware and systems development and testing (e.g., for writing additional test drivers to debug hardware problems). Additional software staff should be planned into the project contingencies to accommodate inevitable component and system debugging and avoid cost and schedule overruns.
- Hardware-In-the-Loop (HWIL) should be accounted for in the technical planning contingencies. HWIL testing is typically accomplished as a debugging exercise where the hardware and software are brought together for the first time in the costly environment of HWIL. If upfront work is not done to understand the messages and errors arising during this test, additional time in the HWIL facility may result in significant cost and schedule impacts. Impacts may be mitigated through upfront planning, such as making appropriate debugging software available to the technical team prior to the test, etc.

- Similarly, Human-In-The-Loop (HITL) evaluations identify contingency operational issues. HITL investigations are particularly critical early in the design process to expose, identify, and cost-effectively correct operational issues—nominal, maintenance, repair, off-nominal, training, etc.—in the required human interactions with the planned design. HITL testing should also be approached as a debugging exercise where hardware, software, and human elements interact and their performance is evaluated. If operational design and/or performance issues are not identified early, the cost of late design changes will be significant.

6.1.1.2.3 *Schedule, Organize, and Budget the Technical Effort*

Once the technical team has defined the technical work to be done, efforts can focus on producing a schedule and cost estimate for the technical portion of the project. The technical team should organize the technical tasks according to the project WBS in a logical sequence of events, taking into consideration the major project milestones, phasing of available funding, and timing of the availability of supporting resources.

Scheduling

Products described in the WBS are the result of activities that take time to complete. These activities have time precedence relationships among them that may be used to create a network schedule explicitly defining the dependencies of each activity on other activities, the availability of resources, and the receipt of receivables from outside sources. Use of a scheduling tool may facilitate the development and maintenance of the schedule.

Scheduling is an essential component of planning and managing the activities of a project. The process of creating a network schedule provides a standard method for defining and communicating what needs to be done, how long it will take, and how each element of the project WBS might affect other elements. A complete network schedule may be used to calculate how long it will take to complete a project; which activities determine that duration (i.e., critical path activities); and how much spare time (i.e., float) exists for all the other activities of the project.

"Critical path" is the sequence of dependent tasks that determines the longest duration of time needed to complete the project. These tasks drive the schedule and continually change, so they should be updated. The critical path may encompass only one task or a series of interrelated tasks. It is important to identify the critical path and the resources needed to complete the critical tasks along the path if the project is to be completed on time and within its resources. As the project progresses, the critical path will change as the critical tasks are completed or as other tasks are delayed. This evolving critical path with its identified tasks needs to be carefully monitored during the progression of the project.

Network scheduling systems help managers accurately assess the impact of both technical and resource changes on the cost and schedule of a project. Cost and technical problems often show up first as schedule problems. Understanding the project's schedule is a prerequisite for determining an accurate project budget and for tracking performance and progress. Because network schedules show how each activity affects other activities, they assist in assessing and predicting the consequences of schedule slips or accelerations of an activity on the entire project.

For additional information on scheduling, refer to *NASA/SP-2010-3403, NASA Schedule Management Handbook.*

Budgeting

Budgeting and resource planning involve establishing a reasonable project baseline budget and the capability to analyze changes to that baseline resulting from technical and/or schedule changes. The project's WBS, baseline schedule, and budget should be viewed as mutually dependent, reflecting the technical content, time, and cost of meeting the project's goals and objectives. The budgeting process needs to take into account whether a fixed cost cap or fixed cost profile exists. When no such cap or profile exists, a baseline budget is developed from the WBS and network schedule. This specifically involves combining the project's workforce and other resource needs with the appropriate workforce rates and other financial and programmatic factors to obtain cost element estimates. These elements of cost include

- direct labor costs,
- overhead costs,
- other direct costs (travel, data processing, etc.),
- subcontract costs,
- material costs,
- equipment costs,
- general and administrative costs,
- cost of money (i.e., interest payments, if applicable),
- fee (if applicable), and
- contingency (Unallocated Future Expenses (UFE)).

For additional information on cost estimating, refer to the *NASA Cost Estimating Handbook and NPR 7120.5, NASA Space Flight Program and Project Management Requirements.*

6.1.1.2.4 *Prepare the SEMP and Other Technical Plans*

Systems Engineering Management Plan

The SEMP is the primary, top-level technical management document for the project and is developed early in the Formulation Phase and updated throughout the project life cycle. The SEMP is driven by the type of project, the phase in the project life cycle, and the technical development risks and is written specifically for each project or project element. While the specific content of the SEMP is tailored to the project, the recommended content is discussed in *Appendix J*. It is important to remember that the main value of the SEMP is in the work that goes into the planning.

The technical team, working under the overall project plan, develops and updates the SEMP as necessary. The technical team works with the project manager to review the content and obtain concurrence. This allows for thorough discussion and coordination of how the proposed technical activities would impact the programmatic, cost, and schedule aspects of the project. The SEMP provides the specifics of the technical effort and describes the technical processes that will be used, how the processes will be applied using appropriate activities, how the project will be organized to accomplish the activities, and the cost and schedule associated with accomplishing the activities.

The physical length of a SEMP is not what is important. This will vary from project to project. The plan needs to be adequate to address the specific technical needs of the project. It is a *living* document that is updated as often as necessary to incorporate new information as it becomes available and as the project develops through the Implementation Phase. The SEMP should not duplicate other project documents; however, the SEMP should reference and summarize the content of other technical plans.

The systems engineer and project manager should identify additional required technical plans based on the project scope and type. If plans are not included in the SEMP, they should be referenced and coordinated in the development of the SEMP. Other plans, such as system safety, probabilistic risk assessment, and an HSI Plan also need to be planned for and coordinated with the SEMP. If a technical plan is a stand-alone, it should be referenced in the SEMP. Depending on the size and complexity of the project, these may be separate plans or they may be included within the SEMP. Once identified, the plans can be developed, training on these plans established, and the plans implemented. Examples of technical plans in addition to the SEMP are listed in *Appendix K*.

The SEMP should be developed during pre-formulation. In developing the SEMP, the technical approach to the project's life cycle is developed. This determines the project's length and cost. The development of the programmatic and technical management approaches requires that the key project personnel develop an understanding of the work to be performed and the relationships among the various parts of that work. Refer to Sections 6.1.2.1 and 6.1.1.2 on WBSs and network scheduling, respectively. The SEMP then flows into the project plan to ensure the proper allocation of resources including cost, schedule, and personnel.

The SEMP's development requires contributions from knowledgeable programmatic and technical experts from all areas of the project that can significantly influence the project's outcome. The involvement of recognized experts is needed to establish a SEMP that is credible to the project manager and to secure the full commitment of the project team.

Role of the SEMP

The SEMP is the rule book that describes to all participants how the project will be technically managed. The NASA technical team on the project should have a SEMP to describe how it will conduct its technical management, and each contractor should have a SEMP to describe how it will manage in accordance with both its contract and NASA's technical management practices. Since the SEMP is unique to a project and contract, it should be updated for each significant programmatic change or it will become outmoded and unused and the project could slide into an uncontrolled state. The lead NASA field Center should have its SEMP developed before attempting to prepare an initial cost estimate since activities that incur cost, such as technical risk reduction and human element accounting, need to be identified and described beforehand. The contractor should have its SEMP developed during the proposal process (prior to costing and pricing) because the SEMP describes the technical content of the project, the potentially costly risk management activities, and the verification and validation techniques to be used, all of which should be included in the preparation of project cost estimates. The SEMPs from the supporting Centers should be developed along with the primary project SEMP. The project SEMP is the senior technical management document for the project; all other technical plans should comply with it. The SEMP should be comprehensive and describe how a fully integrated engineering effort will be managed and conducted.

Verification Plan

The verification plan is developed as part of the Technical Planning Process and is baselined at PDR. As the design matures throughout the life cycle, the plan is updated and refined as needed. The task of preparing the verification plan includes establishing the method of verification to be performed, dependent on the life cycle phase; the position of the product in the system structure; the form of the product used; and the related costs of verification of individual specified requirements. The verification methods include analyses, inspection, demonstration, and test. In some cases, the complete verification of a given requirement might require more than one method. For example, to verify the performance of a product may require looking at many use cases. This might be accomplished by running a Monte Carlo simulation (analysis) and also running actual tests on a few of the key cases. The verification plan, typically written at a detailed technical level, plays a pivotal role in bottom-up product realization.

A phase product can be verified recursively throughout the project life cycle and on a wide variety of product forms. For example:

- simulated (algorithmic models, virtual reality simulator);
- mock-up (plywood, brassboard, breadboard);
- concept description (paper report);
- engineering unit (fully functional but may not be same form/fit);
- prototype (form, fit, and function);

TYPES OF HARDWARE

Breadboard: A low fidelity unit that demonstrates function only without considering form or fit in the case of hardware or platform in the case of software. It often uses commercial and/or ad hoc components and is not intended to provide definitive information regarding operational performance.

Brassboard: A medium fidelity functional unit that typically tries to make use of as much operational hardware/software as possible and begins to address scaling issues associated with the operational system. It does not have the engineering pedigree in all aspects, but is structured to be able to operate in simulated operational environments in order to assess performance of critical functions.

Engineering Unit: A high fidelity unit that demonstrates critical aspects of the engineering processes involved in the development of the operational unit. Engineering test units are intended to closely resemble the final product (hardware/software) to the maximum extent possible and are built and tested so as to establish confidence that the design will function in the expected environments. In some cases, the engineering unit will become the final product, assuming proper traceability has been exercised over the components and hardware handling.

Prototype Unit: The prototype unit demonstrates form, fit, and function at a scale deemed to be representative of the final product operating in its operational environment. A subscale test article provides fidelity sufficient to permit validation of analytical models capable of predicting the behavior of full-scale systems in an operational environment.

Qualification Unit: A unit that is the same as the flight unit (form, fit, function, components, etc.) that will be exposed to the extremes of the environmental criteria (thermal, vibration, etc.). The unit will typically not be flown due to these off-nominal stresses.

Protoflight Unit: In projects that will not develop a qualification unit, the flight unit may be designated as a protoflight unit and a limited version of qualification test ranges will be applied. This unit will be flown.

Flight Unit: The end product that will be flown and will typically undergo acceptance level testing.

- design verification test units (form, fit, and function is the same, but they may not have flight parts);
- qualification units (identical to flight units but may be subjected to extreme environments); and
- flight units (end product that is flown, including protoflight units).

Verification of the end product—that is, the official "run for the record" verification where the program/project takes credit for meeting a requirement—is usually performed on a qualification, protoflight, or flight unit to ensure its applicability to the flight system. However, with discussion and approval from the program/project and systems engineering teams, verification credit may be taken on lower fidelity units if they can be shown to be sufficiently like the flight units in the areas to be verified.

Any of these types of product forms may be in any of these states:

- produced (built, fabricated, manufactured, or coded);
- reused (modified internal non-developmental products or OTS product); or
- assembled and integrated (a composite of lower-level products).

The conditions and environment under which the product is to be verified should be established and the verification should be planned based on the associated entrance/exit criteria that are identified. The Decision Analysis Process should be used to help finalize the planning details.

NOTE: The final, official verification of the end product should be on a controlled unit. Typically, attempting to "buy off" a "shall" on a prototype is not acceptable; it is usually completed on a qualification, flight, or other more final, controlled unit.

Procedures should be prepared to conduct verification based on the method (e.g., analysis, inspection, demonstration, or test) planned. These procedures are typically developed during the design phase of the project life cycle and matured as the design is matured. Operational use scenarios are thought through in order to explore all possible verification activities to be performed.

NOTE: Verification planning begins early in the project life cycle during the requirements development phase. (See *Section 4.2*.) The verification approach to use should be included as part of requirements development to plan for future activities, to establish special requirements derived from identified verification-enabling products, and to ensure that the requirements are verifiable. Updates to verification planning continue throughout logical decomposition and design development, especially as design reviews and simulations shed light on items under consideration. (See *Section 6.1*.)

As appropriate, project risk items are updated based on approved verification strategies that cannot duplicate fully integrated test systems, configurations, and/or target operating environments. Rationales, trade space, optimization results, and implications of the approaches are documented in the new or revised risk statements as well as references to accommodate future design, test, and operational changes to the project baseline.

Validation Plan

The validation plan is one of the work products of the Technical Planning Process and is generated during the Design Solution Process to validate the end product against the baselined stakeholder expectations. This plan can take many forms. The plan describes the total Test and Evaluation (T&E) planning from development of lower-end through higher-end products in the system structure and through operational T&E into production and acceptance. It may combine the verification and validation plans into a single document. (See *Appendix I* for a sample Verification and Validation Plan outline.)

The methods of validation include test, demonstration, inspection, and analysis. While the name of each method is the same as the name of the methods for verification, the purpose and intent as described above are quite different.

Planning to conduct the product validation is a key first step. The method of validation to be used (e.g., analysis, demonstration, inspection, or test) should be established based on the form of the realized end product, the applicable life cycle phase, cost, schedule, resources available, and location of the system product within the system structure.

An established set or subset of expectations or behaviors to be validated should be identified and the validation plan reviewed (an output of the Technical Planning Process, based on design solution outputs) for any specific procedures, constraints, success criteria, or other validation requirements. The conditions and environment under which the product is to be validated should be established and the validation should be planned based on the relevant life cycle phase and associated success criteria identified. The Decision Analysis Process should be used to help finalize the planning details.

It is important to review the validation plans with relevant stakeholders and to understand the relationship between the context of the validation and the context of use (human involvement). As part of the planning process, validation-enabling products should be identified and scheduling and/or acquisition should be initiated.

Procedures should be prepared to conduct validation based on the method planned; e.g., analysis, inspection, demonstration, or test). These procedures are typically developed during the design phase of the project life cycle and matured as the design is matured. Operational and use-case scenarios are thought through in order to explore all possible validation activities to be performed.

Validation is conducted by the user/operator or by the developer as determined by NASA Center directives or the contract with the developers. Systems-level validation (e.g., customer Test and Evaluation (T&E) and some other types of validation) may be performed by an acquirer testing organization. For those portions of validation performed by the developer, appropriate agreements should be negotiated to ensure that validation proof-of-documentation is delivered with the product.

Regardless of the source (buy, make, reuse, assemble and integrate) and the position in the system structure, all realized end products should be validated to demonstrate/confirm satisfaction of stakeholder expectations. Variations, anomalies, and out-of-compliance conditions, where such have been detected, are documented along with the actions taken to resolve the discrepancies. Validation is typically carried out in the intended operational environment or a relevant environment under simulated or actual operational conditions, not necessarily under the tightly controlled conditions usually employed for the Product Verification Process.

Validation of phase products can be performed recursively throughout the project life cycle and on a wide variety of product forms. For example:

- simulated (algorithmic models, virtual reality simulator);
- mock-up (plywood, brassboard, breadboard);

ENVIRONMENTS

Relevant Environment: Not all systems, subsystems, and/or components need to be operated in the operational environment in order to satisfactorily address performance margin requirements or stakeholder expectations. Consequently, the relevant environment is the specific subset of the operational environment that is required to demonstrate critical "at risk" aspects of the final product performance in an operational environment.

Operational Environment: The environment in which the final product will be operated. In the case of space flight hardware/software, it is space. In the case of ground-based or airborne systems that are not directed toward space flight, it is the environments defined by the scope of operations. For software, the environment is defined by the operational platform.

- concept description (paper report);
- engineering unit (functional but may not be same form/fit);
- prototype (product with form, fit, and function);
- design validation test units (form, fit, and function may be the same, but they may not have flight parts);
- qualification unit (identical to flight unit but may be subjected to extreme environments); and
- flight unit (end product that is flown).

Any of these types of product forms may be in any of these states:

- produced (built, fabricated, manufactured, or coded);
- reused (modified internal non-developmental products or off-the-shelf product); or
- assembled and integrated (a composite of lower level products).

NOTE: The final, official validation of the end product should be for a controlled unit. Typically, attempting final validation against the ConOps on a prototype is not acceptable: it is usually completed on a qualification, flight, or other more final, controlled unit.

NOTE: In planning for validation, consideration should be given to the extent to which validation testing will be done. In many instances, off-nominal operational scenarios and nominal operational scenarios should be utilized. Off-nominal testing offers insight into a system's total performance characteristics and often assists in identifying the design issues and human-machine interface, training, and procedural changes required to meet the mission goals and objectives. Off-nominal testing as well as nominal testing should be included when planning for validation.

For additional information on technical plans, refer to the following appendices of this document and to Section 6.1.1.2.4 of the NASA Expanded Guidance for Systems Engineering at ***https://nen.nasa.gov/web/se/doc-repository***:

- *Appendix H* Integration Plan Outline
- *Appendix I* Verification and Validation Plan Outline
- *Appendix J* SEMP Content Outline
- *Appendix K* Technical Plans
- *Appendix L* Interface Requirements Document Outline
- *Appendix M* CM Plan Outline
- *Appendix R* HSI Plan Content Outline
- *Appendix S* Concept of Operations Annotated Outline

6.1.1.2.5 *Obtain Stakeholder Commitments to Technical Plans*

Stakeholder Roles in Project Planning

To obtain commitments to the technical plans from the stakeholders, the technical team should ensure that the appropriate stakeholders, including subject domain experts, have a method to provide inputs and to review the project planning for implementation of stakeholder interests.

During the Formulation Phase, the roles of the stakeholders should be defined in the project plan and the SEMP. Review of these plans and the agreements from the stakeholders to the content of these plans constitutes buy-in from the stakeholders to the technical approach. It is essential to identify the stakeholders and get their concurrence on the technical approach.

Later in the project life cycle, stakeholders may be responsible for delivering products to the project. Initial agreements regarding the responsibilities of the stakeholders are key to ensuring that the project technical team obtains the appropriate deliveries from stakeholders.

Stakeholder Involvement in Defining Requirements

The identification of stakeholders is one of the early steps in the systems engineering process. As the project progresses, stakeholder expectations are flowed down through the Logical Decomposition Process, and specific stakeholders are identified for all of the primary and derived requirements. A critical part of the stakeholders' involvement is in the definition of the technical requirements. As requirements and the ConOps are developed, the stakeholders will be required to agree to these products. Inadequate stakeholder involvement leads to inadequate requirements and a resultant product that does not meet the stakeholder expectations. Status on relevant stakeholder involvement should be tracked and corrective action taken if stakeholders are not participating as planned.

Stakeholder Agreements

Throughout the project life cycle, communication with the stakeholders and commitments from the stakeholders may be accomplished through the use of agreements. Organizations may use an Internal Task Agreement (ITA), a Memorandum Of Understanding (MOU), or other similar documentation to establish the relationship between the project and the stakeholder. These agreements are also used to document the customer and provider responsibilities for defining products to be delivered. These agreements should establish the Measures of Effectiveness (MOEs) or Measures of Performance (MOPs) that will be used to monitor the progress of activities. Reporting requirements and schedule requirements should be established in these agreements. Preparation of these agreements will ensure that the stakeholders' roles and responsibilities support the project goals and that the project has a method to address risks and issues as they are identified.

Stakeholder Support for Forums

During development of the project plan and the SEMP, forums are established to facilitate communication and document decisions during the life cycle of the project. These forums include meetings, working groups, decision panels, and control boards. Each of these forums should establish a charter to define the scope and authority of the forum and identify necessary voting or nonvoting participants. Ad hoc members may be identified when the expertise or input of specific stakeholders is needed when specific topics are addressed. It is important to ensure that stakeholders have been identified to support the forum.

6.1.1.2.6 *Issue Technical Work Directives*

The technical team provides technical work directives to Cost Account Managers (CAMs). This enables the CAMs to prepare detailed plans that are mutually consistent and collectively address all of the work to be performed. These plans include the detailed schedules and budgets for cost accounts that are needed for cost management and EVM.

Issuing technical work directives is an essential activity during Phase B of a project when a detailed planning baseline is required. If this activity is not implemented, then the CAMs are often left with insufficient guidance for detailed planning. The schedules and budgets that are needed for EVM will then be based on assumptions and local interpretations of project-level information. If this is the case, it is highly likely that substantial variances will occur between the baseline plan and the work performed. Providing technical work directives to CAMs produces a more organized technical team. This activity may be repeated when replanning occurs.

This "technical work directives" step produces: (1) planning directives to CAMs that result in (2) a consistent set of cost account plans. Where EVM is called for, it produces (3) an EVM planning baseline, including a Budgeted Cost of Work Scheduled (BCWS).

This activity is not limited to systems engineering. This is a normal part of project planning wherever there is a need for an accurate planning baseline. For additional information on Technical Work Directives, refer to Section 6.1.1.2.6 in the NASA Expanded Guidance for Systems Engineering at *https://nen.nasa.gov/web/se/doc-repository*.

6.1.1.2.7 *Capture Technical Planning Work Products*

The work products from the Technical Planning Process should be managed using either the Technical Data Management Process or the Configuration Management Process as required. Some of the more important products of technical planning (i.e., the WBS, the SEMP, and the schedule, etc.) are kept under configuration control and captured using the CM process. The Technical Data Management Process is used to capture trade studies, cost estimates, technical analyses, reports, and other important documents not under formal configuration control. Work products, such as meeting minutes and correspondence (including e-mail) containing decisions or agreements with stakeholders also should be retained and stored in project files for later reference.

6.1.1.3 *Outputs*

Typical outputs from technical planning activities are:

- **Technical work cost estimates, schedules, and resource needs:** e.g., funds, workforce, facilities, and equipment (to the project) within the project resources;
- **Product and process measures:** Those needed to assess progress of the technical effort and the effectiveness of processes (to the Technical Assessment Process);
- **SEMP and other technical plans:** Technical planning strategy, WBS, SEMP, HSI Plan, V&V Plan, and other technical plans that support implementation of the technical effort (to all processes; applicable plans to technical processes);

- **Technical work directives:** e.g., work packages or task orders with work authorization (to applicable technical teams); and

- **Technical Planning Process work products:** Includes products needed to provide reports, records, and non-deliverable outcomes of process activities (to the Technical Data Management Process).

The resulting technical planning strategy constitutes an outline, or rough draft, of the SEMP. This serves as a starting point for the overall Technical Planning Process after initial preparation is complete. When preparations for technical planning are complete, the technical team should have a cost estimate and schedule for the technical planning effort. The budget and schedule to support the defined technical planning effort can then be negotiated with the project manager to resolve any discrepancies between what is needed and what is available. The SEMP baseline needs to be completed. Planning for the update of the SEMP based on programmatic changes needs to be developed and implemented. The SEMP needs to be approved by the appropriate level of authority.

6.1.2 Technical Planning Guidance

Refer to Section 6.1.2 in the NASA Expanded Guidance for Systems Engineering at *https://nen.nasa.gov/web/se/doc-repository* for additional guidance on:

- Work Breakdown Structure (WBS),
- cost definition and modeling, and
- lessons learned.

Additional information on the WBS can also be found in *NASA/SP-2010-3404, NASA Work Breakdown Structure Handbook* and on costing in the *NASA Cost Estimating Handbook.*

6.2 Requirements Management

Requirements management activities apply to the management of all stakeholder expectations, customer requirements, and technical product requirements down to the lowest level product component requirements (hereafter referred to as expectations and requirements). This includes physical functional and operational requirements, including those that result from interfaces between the systems in question and other external entities and environments. The Requirements Management Process is used to:

- Identify, control, decompose, and allocate requirements across all levels of the WBS.

- Provide bidirectional traceability.

DEFINITIONS

Traceability: A discernible association between two or more logical entities such as requirements, system elements, verifications, or tasks.

Bidirectional traceability: The ability to trace any given requirement/expectation to its parent requirement/expectation and to its allocated children requirements/expectations.

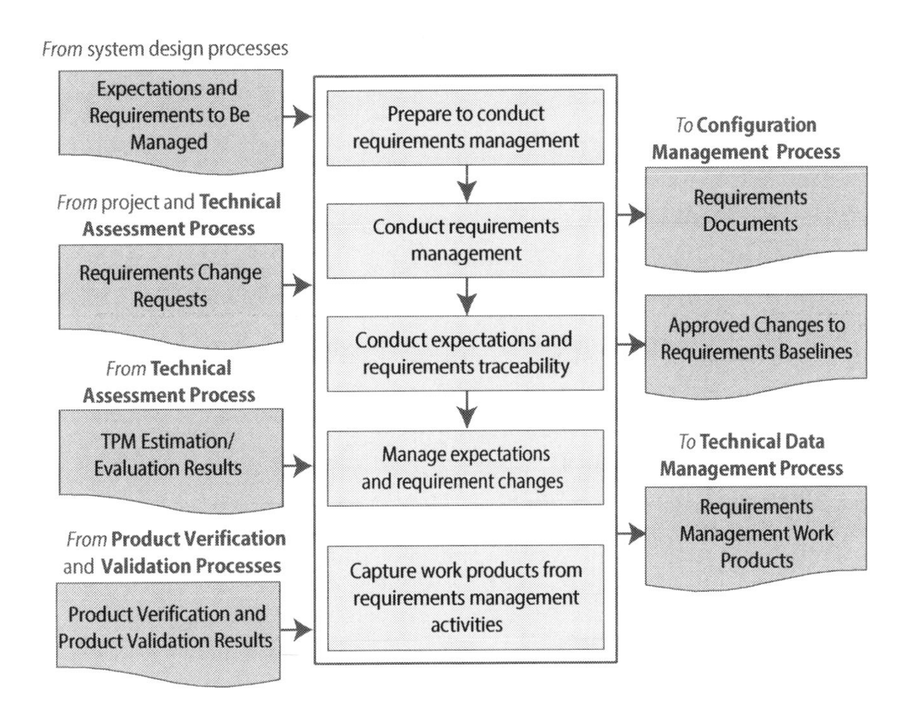

FIGURE 6.2-1 Requirements Management Process

- Manage the changes to established requirement baselines over the life cycle of the system products.

6.2.1 Process Description

FIGURE 6.2-1 provides a typical flow diagram for the Requirements Management Process and identifies typical inputs, outputs, and activities to consider in addressing requirements management.

6.2.1.1 *Inputs*

There are several fundamental inputs to the Requirements Management Process.

- **Expectations and requirements to be managed:** Requirements and stakeholder expectations are identified during the system design processes, primarily from the Stakeholder Expectations Definition Process and the Technical Requirements Definition Process.
- **Requirement change requests:** The Requirements Management Process should be prepared to deal with requirement change requests that can be generated at any time during the project life cycle or as a result of reviews and assessments as part of the Technical Assessment Process.
- **TPM estimation/evaluation results:** TPM estimation/evaluation results from the Technical Assessment Process provide an early warning of the adequacy of a design in satisfying selected critical technical parameter requirements. Variances from expected values of product performance may trigger changes to requirements.
- **Product verification and validation results:** Product verification and product validation results from the Product Verification and Product Validation Processes are mapped into the

requirements database with the goal of verifying and validating all requirements.

6.2.1.2 *Process Activities*

6.2.1.2.1 *Prepare to Conduct Requirements Management*

Preparing to conduct requirements management includes gathering the requirements that were defined and baselined during the Requirements Definition Process. Identification of the sources/owners of each requirement should be checked for currency. The organization (e.g., change board) and procedures to perform requirements management are established.

6.2.1.2.2 *Conduct Requirements Management*

The Requirements Management Process involves managing all changes to expectations and requirements baselines over the life of the product and maintaining bidirectional traceability between stakeholder expectations, customer requirements, technical product requirements, product component requirements, design documents, and test plans and procedures. The successful management of requirements involves several key activities:

- Establish a plan for executing requirements management.
- Receive requirements from the system design processes and organize them in a hierarchical tree structure.
- Maintain bidirectional traceability between requirements.
- Evaluate all change requests to the requirements baseline over the life of the project and make changes if approved by change board.
- Maintain consistency between the requirements, the ConOps, and the architecture/design, and initiate corrective actions to eliminate inconsistencies.

6.2.1.2.3 *Conduct Expectations and Requirements Traceability*

As each requirement is documented, its bidirectional traceability should be recorded. Each requirement should be traced back to a parent/source requirement or expectation in a baselined document or identified as *self-derived* and concurrence on it sought from the next higher level requirements sources. Examples of self-derived requirements are requirements that are locally adopted as good practices or are the result of design decisions made while performing the activities of the Logical Decomposition and Design Solution Processes.

The requirements should be evaluated, independently if possible, to ensure that the requirements trace is correct and that it fully addresses its parent requirements. If it does not, some other requirement(s) should complete fulfillment of the parent requirement and be included in the traceability matrix. In addition, ensure that all top-level parent document requirements have been allocated to the lower level requirements. If there is no parent for a particular requirement and it is not an acceptable self-derived requirement, it should be assumed either that the traceability process is flawed and should be redone or that the requirement is "gold plating" and should be eliminated. Duplication between levels should be resolved. If a requirement is simply repeated at a lower level and it is not an externally imposed constraint, it may not belong at the higher level. Requirements traceability is usually recorded in a requirements matrix or through the use of a requirements modeling application.

6.2.1.2.4 *Managing Expectations and Requirement Changes*

Throughout early Phase A, changes in requirements and constraints will occur as they are initially defined and matured. It is imperative that all changes be thoroughly evaluated to determine the impacts on the cost, schedule, architecture, design, interfaces, ConOps, and higher and lower level requirements. Performing functional and sensitivity analyses will ensure that the requirements are realistic and evenly allocated. Rigorous requirements verification and validation will ensure that the requirements can be satisfied and conform to mission objectives. All changes should be subjected to a review and approval cycle to maintain traceability and to ensure that the impacts are fully assessed for all parts of the system.

Once the requirements have been validated and reviewed in the System Requirements Review (SRR) in late Phase A, they are placed under formal configuration control. Thereafter, any changes to the requirements should be approved by a Configuration Control Board (CCB) or equivalent authority. The systems engineer, project manager, and other key engineers usually participate in the CCB approval processes to assess the impact of the change including cost, performance, programmatic, and safety.

Requirement changes during Phases B and C are more likely to cause significant adverse impacts to the project cost and schedule. It is even more important that these late changes are carefully evaluated to fully understand their impact on cost, schedule, and technical designs.

The technical team should also ensure that the approved requirements are communicated in a timely manner to all relevant people. Each project should have already established the mechanism to track and disseminate the latest project information. Further information on Configuration Management (CM) can be found in *Section 6.5*.

6.2.1.2.5 *Key Issues for Requirements Management*

Requirements Changes

Effective management of requirements changes requires a process that assesses the impact of the proposed changes prior to approval and implementation of the change. This is normally accomplished through the use of the Configuration Management Process. In order for CM to perform this function, a baseline configuration should be documented and tools used to assess impacts to the baseline. Typical tools used to analyze the change impacts are as follows:

- **Performance Margins:** This tool is a list of key performance margins for the system and the current status of the margin. For example, the propellant performance margin will provide the necessary propellant available versus the propellant necessary to complete the mission. Changes should be assessed for their impact on performance margins.

- **CM Topic Evaluators List:** This list is developed by the project office to ensure that the appropriate persons are evaluating the changes and providing impacts to the change. All changes need to be routed to the appropriate individuals to ensure that the change has had all impacts identified. This list will need to be updated periodically.

- **Risk System and Threats List:** The risk system can be used to identify risks to the project and the cost, schedule, and technical aspects of the risk. Changes to the baseline can affect the consequences and likelihood of identified risk or can introduce new risk to the project. A threats list is normally used to identify the costs associated with all the risks for the project. Project reserves are used to mitigate the appropriate risk. Analyses of the reserves available versus the needs identified by the threats list assist in the prioritization for reserve use.

The process for managing requirements changes needs to take into account the distribution of information related to the decisions made during the change process. The Configuration Management Process needs to communicate the requirements change decisions to the affected organizations. During a board meeting to approve a change, actions to update documentation need to be included as part of the change package. These actions should be tracked to ensure that affected documentation is updated in a timely manner.

Requirements Creep

"Requirements creep" is the term used to describe the subtle way that requirements grow imperceptibly during the course of a project. The tendency for the set of requirements is to relentlessly increase in size during the course of development, resulting in a system that is more expensive and complex than originally intended. Often the changes are quite innocent and what appear to be changes to a system are really enhancements in disguise.

However, some of the requirements creep involves truly new requirements that did not exist, and could not have been anticipated, during the Technical Requirements Definition Process. These new requirements are the result of evolution, and if we are to build a relevant system, we cannot ignore them.

There are several techniques for avoiding or at least minimizing requirements creep:

- The first line of defense is a good ConOps that has been thoroughly discussed and agreed-to by the customer and relevant stakeholders.
- In the early requirements definition phase, flush out the conscious, unconscious, and undreamed-of requirements that might otherwise not be stated.
- Establish a strict process for assessing requirement changes as part of the Configuration Management Process.
- Establish official channels for submitting change requests. This will determine who has the authority to generate requirement changes and submit them formally to the CCB (e.g., a contractor-designated representative, project technical leads, customer/science team lead, or user).
- Measure the functionality of each requirement change request and assess its impact on the rest of the system. Compare this impact with the consequences of not approving the change. What is the risk if the change is not approved?
- Determine if the proposed change can be accommodated within the fiscal and technical resource budgets. If it cannot be accommodated within the established resource margins, then the change most likely should be denied.

6.2.1.2.6 *Capture Work Products*

These products include maintaining and reporting information on the rationale for and disposition and implementation of change actions, current requirement compliance status and expectation, and requirement baselines.

6.2.1.3 *Outputs*

Typical outputs from the requirements management activities are:

- **Requirements Documents:** Requirements documents are submitted to the Configuration Management Process when the requirements are baselined. The official controlled versions of these documents are generally maintained in electronic format within the requirements management tool that has been selected by the project. In this way, they are linked to the requirements matrix with all of its traceable relationships.

- **Approved Changes to the Requirements Baselines:** Approved changes to the requirements baselines are issued as an output of the Requirements Management Process after careful assessment of all the impacts of the requirements change across the entire product or system. A single change can have a far-reaching ripple effect, which may result in several requirement changes in a number of documents.

- **Various Requirements Management Work Products:** Requirements management work products are any reports, records, and undeliverable outcomes of the Requirements Management Process. For example, the bidirectional traceability status would be one of the work products that would be used in the verification and validation reports.

6.2.2 Requirements Management Guidance

Refer to Section 6.2.2 in the NASA Expanded Guidance for Systems Engineering at *https://nen.nasa.gov/web/se/doc-repository* for additional guidance on:

- the Requirements Management Plan and
- requirements management tools.

6.3 Interface Management

The definition, management, and control of interfaces are crucial to successful programs or projects. Interface management is a process to assist in controlling product development when efforts are divided among parties (e.g., Government, contractors, geographically diverse technical teams, etc.) and/or to define and maintain compliance among the products that should interoperate.

The basic tasks that need to be established involve the management of internal and external interfaces of the various levels of products and operator tasks to support product integration. These basic tasks are as follows:

- Define interfaces;

- Identify the characteristics of the interfaces (physical, electrical, mechanical, human, etc.);

- Ensure interface compatibility at all defined interfaces by using a process documented and approved by the project;

- Strictly control all of the interface processes during design, construction, operation, etc.;

- Identify lower level products to be assembled and integrated (from the Product Transition Process);

- Identify assembly drawings or other documentation that show the complete configuration of the product being integrated, a parts list, and any assembly instructions (e.g., torque requirements for fasteners);

- Identify end-product, design-definition-specified requirements (specifications), and configuration documentation for the applicable work breakdown structure model, including interface specifications, in the form appropriate to satisfy the product life cycle phase success criteria (from the Configuration Management Process); and

- Identify product integration-enabling products (from existing resources or the Product Transition Process for enabling product realization).

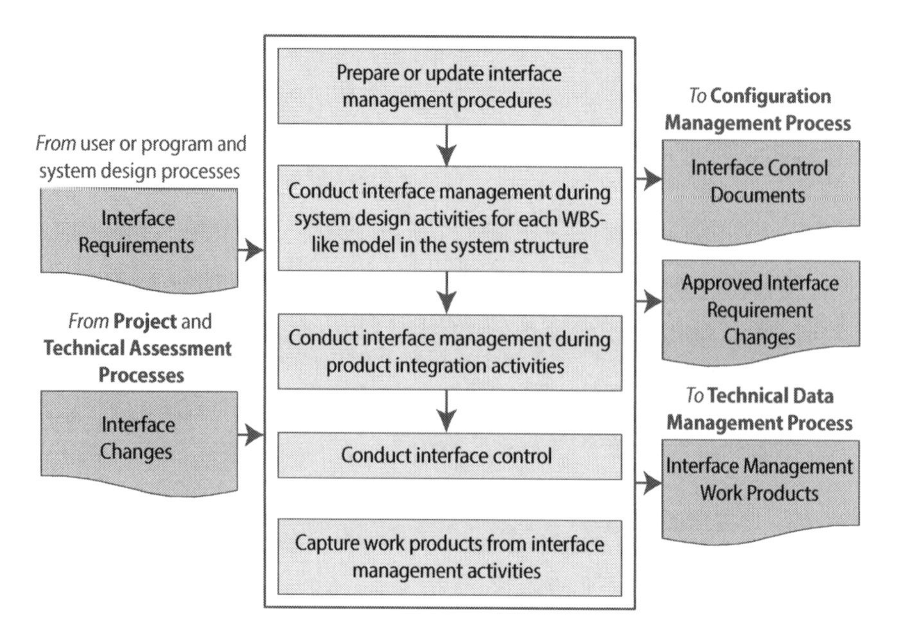

FIGURE 6.3-1 Interface Management Process

6.3.1 Process Description

FIGURE 6.3-1 provides a typical flow diagram for the Interface Management Process and identifies typical inputs, outputs, and activities to consider in addressing interface management.

6.3.1.1 *Inputs*

Typical inputs needed to understand and address interface management would include the following:

- **Interface Requirements:** These include the internal and external functional, physical, and performance interface requirements developed as part of the Technical Requirements Definition Process for the product(s).
- **Interface Change Requests:** These include changes resulting from program or project agreements or changes on the part of the technical team as part of the Technical Assessment Process.

Other inputs that might be useful are:

- **System Description:** This allows the design of the system to be explored and examined to determine where system interfaces exist. Contractor arrangements will also dictate where interfaces are needed.
- **System Boundaries:** Documented physical boundaries, components, and/or subsystems, which are all drivers for determining where interfaces exist.
- **Organizational Structure:** Decisions on which organization will dictate interfaces, particularly when there is the need to jointly agree on shared interface parameters of a system. The program and project WBS will also provide organizational interface boundaries.
- **Boards Structure:** Defined board structure that identifies organizational interface responsibilities.

6.3.1.2 *Process Activities*

6.3.1.2.1 *Prepare or Update Interface Management Procedures*

These procedures establish the interface management responsibilities, what process will be used to maintain and control the internal and external functional and physical interfaces (including human), and how the change process will be conducted. Training of the technical teams or other support may also be required and planned.

6.3.1.2.2 *Conduct Interface Management during System Design Activities*

During project Formulation, the ConOps of the product is analyzed to identify both external and internal interfaces. This analysis will establish the origin, destination, stimuli, and special characteristics of the interfaces that need to be documented and maintained. As the system structure and architecture emerges, interfaces will be added and existing interfaces will be changed and should be maintained. Thus, the Interface Management Process has a close relationship to other areas, such as requirements definition and configuration management, during this period.

6.3.1.2.3 *Conduct Interface Management during Product Integration*

During product integration, interface management activities would support the review of integration and assembly procedures to ensure interfaces are properly marked and compatible with specifications and interface control documents. The interface management process has a close relationship to verification and validation. Interface control documentation and approved interface requirement changes are used as inputs to the Product Verification Process and the Product Validation Process, particularly where verification test constraints and interface parameters are needed to set the test objectives and test plans. Interface requirements verification is a critical aspect of the overall system verification.

6.3.1.2.4 *Conduct Interface Control*

Typically, an Interface Working Group (IWG) establishes communication links between those responsible for interfacing systems, end products, enabling products, and subsystems. The IWG has the responsibility to ensure accomplishment of the planning, scheduling, and execution of all interface activities. An IWG is typically a technical team with appropriate technical membership from the interfacing parties (e.g., the project, the contractor, etc.). The IWG may work independently or as a part of a larger change control board.

6.3.1.2.5 *Capture Work Products*

Work products include the strategy and procedures for conducting interface management, rationale for interface decisions made, assumptions made in approving or denying an interface change, actions taken to correct identified interface anomalies, lessons learned and updated support and interface agreement documentation.

6.3.1.3 *Outputs*

Typical outputs needed to capture interface management would include:

- **Interface control documentation.** This is the documentation that identifies and captures the interface information and the approved interface change requests. Types of interface documentation include the Interface Requirements Document (IRD), Interface Control Document/Drawing (ICD), Interface Definition Document (IDD), and Interface Control Plan (ICP). These outputs will then be maintained and approved using the Configuration Management Process and become a part of the overall technical data package for the project.

- **Approved interface requirement changes.** After the interface requirements have been baselined, the Requirements Management Process should be used to identify the need for changes, evaluate the impact of the proposed change, document the final approval/disapproval, and update the requirements documentation/tool/database. For interfaces that require approval from all sides, unanimous approval is required. Changing interface requirements late in the design or implementation life cycle is more likely to have a significant impact on the cost, schedule, or technical design/operations.

- **Other work products.** These work products include the strategy and procedures for conducting interface management, the rationale for interface decisions made, the assumption made in approving or denying an interface change, the actions taken to correct identified interface anomalies, the lessons learned in performing the interface management activities, and the updated support and interface agreement documentation.

6.3.2 Interface Management Guidance

Refer to Section 6.3.2 in the NASA Expanded Guidance for Systems Engineering at *https://nen.nasa.gov/web/se/doc-repository* for additional guidance on:

- interface requirements documents,
- interface control documents,
- interface control drawings,
- interface definition documents,
- the interface control plans, and
- interface management tasks.

6.4 Technical Risk Management

The Technical Risk Management Process is one of the crosscutting technical management processes. Risk is the potential for performance shortfalls, which may be realized in the future, with respect to achieving explicitly established and stated performance requirements. The performance shortfalls may be related to institutional support for mission execution or related to any one or more of the following mission execution domains:

- safety
- technical
- cost
- schedule

Systems engineers are involved in this process to help identify potential technical risks, develop mitigation plans, monitor progress of the technical effort to determine if new risks arise or old risks can be retired, and to be available to answer questions and resolve issues. The following is guidance in implementation of risk management in general. Thus, when implementing risk management on any given program/project, the responsible systems engineer should direct the effort accordingly. This may involve more or less rigor and formality than that specified in governing documents such as NPRs. Of course, if deviating from NPR "requirements," the responsible engineer must follow the deviation approval process. The idea is to tailor the risk management process so that it meets the needs of the individual program/project being executed while working within the bounds of the governing documentation (e.g., NPRs). For detailed information on the Risk Management Process, refer to the *NASA Risk Management Handbook (NASA/SP-2011-3422).*

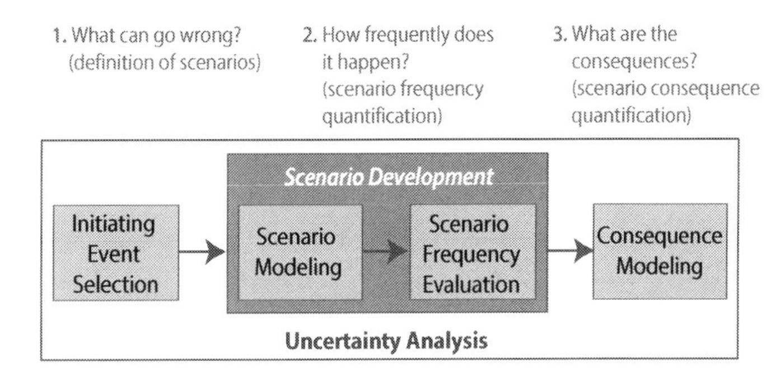

FIGURE 6.4-1 Risk Scenario Development (Source: NASA/SP-2011-3421)

Risk is characterized by three basic components:

1. The scenario(s) leading to degraded performance with respect to one or more performance measures (e.g., scenarios leading to injury, fatality, destruction of key assets; scenarios leading to exceedance of mass limits; scenarios leading to cost overruns; scenarios leading to schedule slippage);

2. The likelihood(s) (qualitative or quantitative) of those scenario(s); and

3. The consequence(s) (qualitative or quantitative severity of the performance degradation) that would result if the scenario(s) was (were) to occur.

Uncertainties are included in the evaluation of likelihoods and consequences.

Scenarios begin with a set of initiating events that cause the activity to depart from its intended state. For each initiating event, other events that are relevant to the evolution of the scenario may (or may not) occur and may have either a mitigating or exacerbating effect on the scenario progression. The frequencies of scenarios with undesired consequences are determined. Finally, the multitude of such scenarios is put together, with an understanding of the uncertainties, to create the risk profile of the system.

This "risk triplet" conceptualization of risk is illustrated in FIGURES 6.4-1 and 6.4-2.

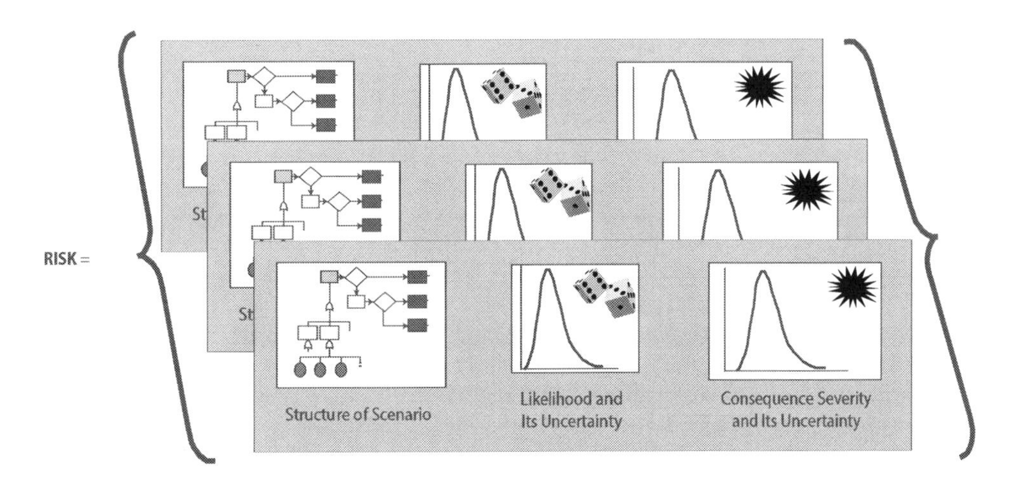

FIGURE 6.4-2 Risk as an Aggregate Set of Risk Triplets

Key Concepts in Risk Management Risk: Risk is the potential for shortfalls, which may be realized in the future with respect to achieving explicitly-stated requirements. The performance shortfalls may be related to institutional support for mission execution, or related to any one or more of the following mission execution domains: safety, technical, cost, schedule. Risk is characterized as a set of triplets:

> The scenario(s) leading to degraded performance in one or more performance measures.
> The likelihood(s) of those scenarios.
> The consequence(s), impact, or severity of the impact on performance that would result if those scenarios were to occur.

Uncertainties are included in the evaluation of likelihoods and consequences.

Cost Risk: This is the risk associated with the ability of the program/project to achieve its life-cycle cost objectives and secure appropriate funding. Two risk areas bearing on cost are (1) the risk that the cost estimates and objectives are not accurate and reasonable; and (2) the risk that program execution will not meet the cost objectives as a result of a failure to handle cost, schedule, and performance risks.

Schedule Risk: Schedule risks are those associated with the adequacy of the time estimated and allocated for the development, production, implementation, and operation of the system. Two risk areas bearing on schedule risk are (1) the risk that the schedule estimates and objectives are not realistic and reasonable; and (2) the risk that program execution will fall short of the schedule objectives as a result of failure to handle cost, schedule, or performance risks.

Technical Risk: This is the risk associated with the evolution of the design and the production of the system of interest affecting the level of performance necessary to meet the stakeholder expectations and technical requirements. The design, test, and production processes (process risk) influence the technical risk and the nature of the product as depicted in the various levels of the PBS (product risk).

Programmatic Risk: This is the risk associated with action or inaction from outside the project, over which the project manager has no control, but which may have significant impact on the project. These impacts may manifest themselves in terms of technical, cost, and/or schedule. This includes such activities as: International Traffic in Arms Regulations (ITAR), import/export control, partner agreements with other domestic or foreign organizations, congressional direction or earmarks, Office of Management and Budget (OMB) direction, industrial contractor restructuring, external organizational changes, etc.

Scenario: A sequence of credible events that specifies the evolution of a system or process from a given state to a future state. In the context of risk management, scenarios are used to identify the ways in which a system or process in its current state can evolve to an undesirable state.

Undesired scenario(s) might come from technical or programmatic sources (e.g., a cost overrun, schedule slippage, safety mishap, health problem, malicious activities, environmental impact, or failure to achieve a needed scientific or technological objective or success criterion). Both the likelihood and consequences may have associated uncertainties.

6.4.1 Risk Management Process Description

FIGURE 6.4-3 provides a typical flow diagram for the Risk Management Process and identifies typical inputs, activities, and outputs to consider in addressing risk management.

6.4.1.1 *Inputs*

The following are typical inputs to risk management:

- **Project Risk Management Plan:** The Risk Management Plan is developed under the Technical Planning Process and defines how risk will be identified, mitigated, monitored, and controlled within the project.
- **Technical Risk Issues:** These will be the technical issues identified as the project progresses that pose a risk to the successful accomplishment of the project mission/goals.
- **Technical Risk Status Measurements:** These are any measures that are established that help to monitor and report the status of project technical risks.

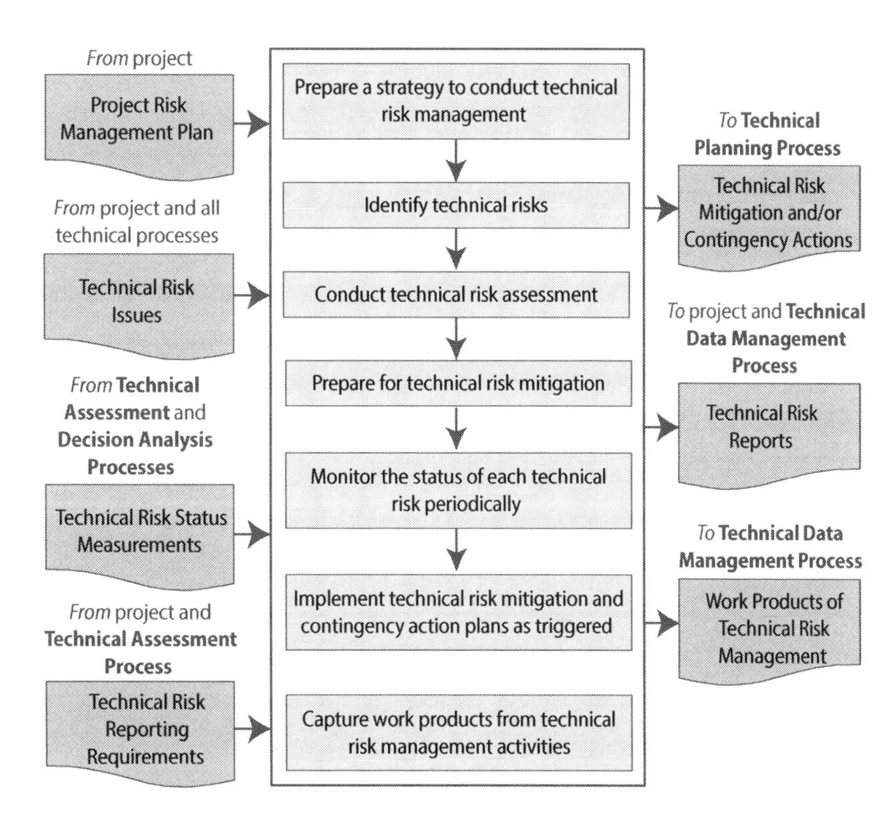

FIGURE 6.4-3 Risk Management Process

- **Technical Risk Reporting Requirements:** Includes requirements of how technical risks will be reported, how often, and to whom.

Additional inputs that may be useful:

- **Other Plans and Policies:** Systems Engineering Management Plan, form of technical data products, and policy input to metrics and thresholds.

- **Technical Inputs:** Stakeholder expectations, concept of operations, imposed constraints, tracked observables, current program baseline, performance requirements, and relevant experience data.

6.4.1.2 *Activities*

6.4.1.2.1 *Prepare a Strategy to Conduct Technical Risk Management*

This strategy would include documenting how the program/project risk management plan (as developed during the Technical Planning Process) will be implemented, identifying any additional technical risk sources and categories not captured in the plan, identifying what will trigger actions and how these activities will be communicated to the internal and external teams.

6.4.1.2.2 *Identify Technical Risks*

On a continuing basis, the technical team will identify technical risks including their source, analyze the potential consequence and likelihood of the risks occurring, and prepare clear risk statements for entry into the program/project risk management system. Coordination with the relevant stakeholders for the identified risks is included. For more information on identifying technical risks, see Section 6.4.2.1.

6.4.1.2.3 *Conduct Technical Risk Assessment*

Until recently, NASA's Risk Management (RM) approach was based almost exclusively on Continuous Risk Management (CRM), which stresses the management of individual risk issues during implementation. In December of 2008, NASA revised its RM approach in order to more effectively foster proactive risk management. The new approach, which is outlined in NPR 8000.4, Agency Risk Management Procedural Requirements and further developed in *NASA/SP-2011-3422, NASA Risk Management Handbook*, evolves NASA's risk management to entail two complementary processes: Risk-Informed Decision Making (RIDM) and CRM. RIDM is intended to inform direction-setting systems engineering (SE) decisions (e.g., design decisions) through better use of risk and uncertainty information in selecting alternatives and establishing baseline performance requirements (for additional RIDM technical information, guidance, and process description, see *NASA/SP-2010-576 Version 1, NASA Risk-Informed Decision Making Handbook*).

CRM is then used to manage risks over the course of the development and implementation phases of the life cycle to assure that requirements related to safety, technical, cost, and schedule are met. In the past, RM was considered equivalent to the CRM process; now, RM is defined as comprising both the RIDM and CRM processes, which work together to assure proactive risk management as NASA programs and projects are conceived, developed, and executed. FIGURE 6.4-4 illustrates the concept.

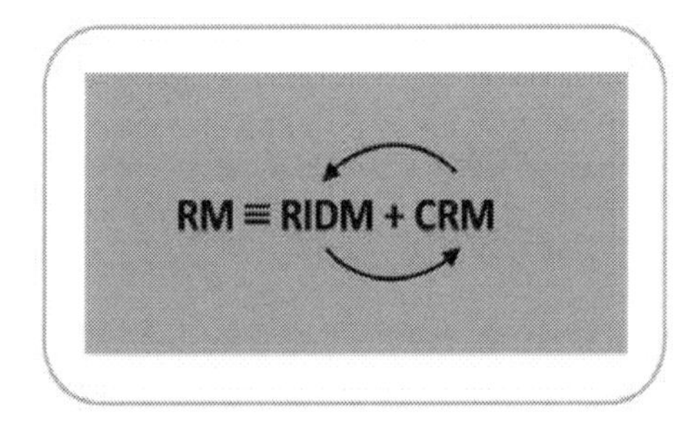

FIGURE 6.4-4 Risk Management as the Interaction of Risk-Informed Decision Making and Continuous Risk Management (Source: NASA/SP-2011-3422)

6.4.1.2.4 *Prepare for Technical Risk Mitigation*

This includes selecting the risks that will be mitigated and more closely monitored, identifying the risk level or threshold that will trigger a risk mitigation action plan, and identifying for each risk which stakeholders will need to be informed that a mitigation/contingency action is determined as well as which organizations will need to become involved to perform the mitigation/contingency action.

6.4.1.2.5 *Monitor the Status of Each Technical Risk Periodically*

Risk status will need to be monitored periodically at a frequency identified in the risk plan. Risks that are approaching the trigger thresholds will be monitored on a more frequent basis. Reports of the status are made to the appropriate program/project management or board for communication and for decisions whether to trigger a mitigation action early. Risk status will also be reported at most life cycle reviews.

6.4.1.2.6 *Implement Technical Risk Mitigation and Contingency Action Plans as Triggered*

When the applicable thresholds are triggered, the technical risk mitigation and contingency action plans are implemented. This includes monitoring the results of the action plan implementation and modifying them as necessary, continuing the mitigation until the residual risk and/or consequence impacts are acceptable, and communicating the actions and results to the identified stakeholders. Action plan reports are prepared and results reported at appropriate boards and at life cycle reviews.

6.4.1.2.7 *Capture Work Products*

Work products include the strategy and procedures for conducting technical risk management; the rationale for decisions made; assumptions made in prioritizing, handling, and reporting technical risks and action plan effectiveness; actions taken to correct action plan implementation anomalies; and lessons learned.

6.4.1.3 *Outputs*

Following are key risk outputs from activities:

- **Technical Risk Mitigation and/or Contingency Actions:** Actions taken to mitigate identified risks or contingency actions taken in case risks are realized.
- **Technical Risk Reports:** Reports of the technical risk policies, status, remaining residual risks, actions taken, etc. Output at the agreed-to frequency and recipients.
- **Work Products:** Includes the procedures for conducting technical risk management; rationale for decisions made; selected decision alternatives; assumptions made in prioritizing, handling, and reporting technical risks; and lessons learned.

6.4.2 Risk Management Process Guidance

For additional guidance on risk management, refer to *NASA/SP-2010-576, NASA RIDM Handbook* and *NASA/SP-2011-3422, NASA Risk Management Handbook.*

6.5 Configuration Management

Configuration management is a management discipline applied over the product's life cycle to provide visibility into and to control changes to performance and functional and physical characteristics. Additionally, according to SAE Electronic Industries Alliance (EIA) 649B, improper configuration management may result in incorrect, ineffective, and/or unsafe products being released. Therefore, in order to protect and ensure the integrity of NASA products, NASA has endorsed the implementation of the five configuration management functions and the associated 37 underlying principles defined within *SAE/EIA-649-2 Configuration Management Requirements for NASA Enterprises.*

Together, these standards address what configuration management activities are to be done, when they are to happen in the product life cycle, and what planning and resources are required. Configuration management is a key systems engineering practice that, when properly implemented, provides visibility of a true representation of a product and attains the product's integrity by controlling the changes made to the baseline configuration and tracking such changes. Configuration management ensures that the configuration of a product is known and reflected in product information, that any product change is beneficial and is effected without adverse consequences, and that changes are managed.

CM reduces technical risks by ensuring correct product configurations, distinguishes among product versions, ensures consistency between the product and information about the product, and avoids the embarrassment cost of stakeholder dissatisfaction and complaint. In general, NASA adopts the CM principles as defined by *SAE/EIA 649B, Configuration Management Standard*, in addition to implementation as defined by NASA CM professionals and as approved by NASA management.

When applied to the design, fabrication/assembly, system/subsystem testing, integration, and operational and sustaining activities of complex technology items, CM represents the "backbone" of the enterprise structure. It instills discipline and keeps the product attributes and documentation consistent. CM enables all stakeholders in the technical effort, at any given time in the life of a product, to use identical data for development activities and decision-making. CM principles are applied to keep the documentation consistent with the approved product, and to ensure that the product conforms to the functional and physical requirements of the approved design.

6.5.1 Process Description

FIGURE 6.5-1 provides a typical flow diagram for the Configuration Management Process and identifies typical inputs, outputs, and activities to consider in addressing CM.

6.5.1.1 *Inputs*

The inputs for this process are:

- **CM plan:** This plan would have been developed under the Technical Planning Process and serves as the overall guidance for this process for the program/project
- **Engineering change proposals:** These are the requests for changes to the established baselines in whatever form they may appear throughout the life cycle.
- **Expectation, requirements and interface documents:** These baselined documents or models are key to the design and development of the product.
- **Approved requirements baseline changes:** The approved requests for changes will authorize the update of the associated baselined document or model.
- **Designated configuration items to be controlled:** As part of technical planning, a list or philosophy would have been developed that identifies the types of items that will need to be placed under configuration control.

6.5.1.2 *Process Activities*

There are five elements of CM (see FIGURE 6.5-2):

- configuration planning and management
- configuration identification,

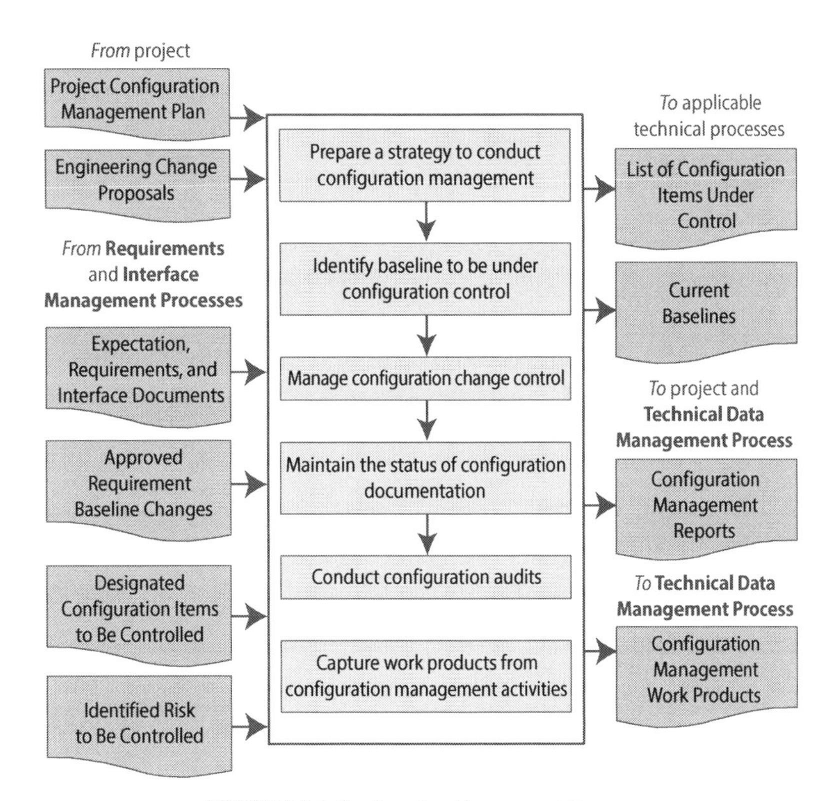

FIGURE 6.5-1 Configuration Management Process

- configuration change management,
- Configuration Status Accounting (CSA), and
- configuration verification.

6.5.1.2.1 *Prepare a Strategy to Conduct CM*

CM planning starts at a program's or project's inception. The CM office should carefully weigh the value of prioritizing resources into CM tools or into CM surveillance of the contractors. Reviews by the Center Configuration Management Organization (CMO) are warranted and will cost resources and time, but the correction of systemic CM problems before they erupt into losing configuration control are always preferable to explaining why incorrect or misidentified parts are causing major problems in the program/project.

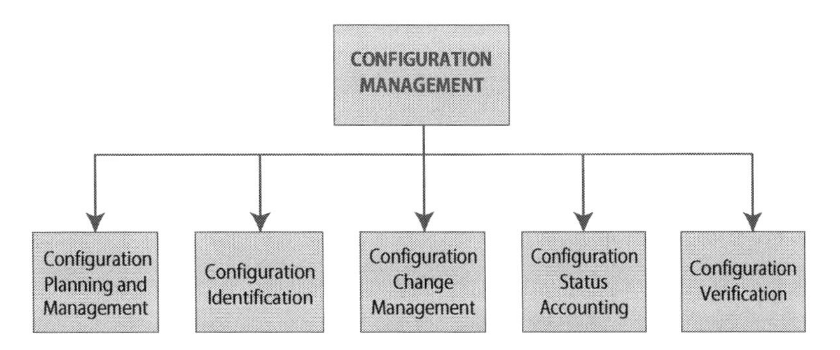

FIGURE 6.5-2 Five Elements of Configuration Management

One of the key inputs to preparing for CM implementation is a strategic plan for the project's complete CM process. This is typically contained in a CM plan. See *Appendix M* for an outline of a typical CM plan.

This plan has both internal and external uses:

- **Internal:** It is used within the program/project office to guide, monitor, and measure the overall CM process. It describes all the CM activities and the schedule for implementing those activities within the program/project.
- **External:** The CM plan is used to communicate the CM process to the contractors involved in the program/project. It establishes consistent CM processes and working relationships.

The CM plan may be a standalone document or it may be combined with other program/project planning documents. It should describe the criteria for each technical baseline creation, technical approvals, and audits.

6.5.1.2.2 *Identify Baseline to be Under Configuration Control*

Configuration identification is the systematic process of selecting, organizing, and stating the product attributes. Identification requires unique identifiers for a product and its configuration documentation. The CM activity associated with identification includes selecting the Configuration Items (CIs), determining the CIs' associated configuration documentation, determining the appropriate change control authority, issuing unique identifiers for both CIs and CI documentation, releasing configuration documentation, and establishing configuration baselines.

NASA has four baselines, each of which defines a distinct phase in the evolution of a product design. The baseline identifies an agreed-to description of attributes of a CI at a point in time and provides a known configuration to which changes are addressed. Baselines are established by agreeing to (and documenting) the stated definition of a CI's attributes. The approved "current" baseline defines the basis of the subsequent change. The system specification is typically finalized following the SRR. The functional baseline is established at the SDR and will usually transfer to NASA's control at that time for contracting efforts. For in-house efforts, the baseline is set/controlled by the NASA program/project.

The four baselines (see FIGURE 6.5-3) normally controlled by the program, project, or Center are the following:

- **Functional Baseline:** The functional baseline is the approved configuration documentation that describes a system's or top-level CI's performance requirements (functional, interoperability, and interface characteristics) and the verification required to demonstrate the achievement of those specified characteristics. The functional baseline is established at the SDR by the NASA program/project. The program/project will direct through contractual agreements, how the functional baselines are managed at the different functional levels. (Levels 1–4)
- **Allocated Baseline:** The allocated baseline is the approved performance-oriented configuration documentation for a CI to be developed that describes the functional, performance, and interface characteristics that are allocated from a higher level requirements document or a CI and the verification required to demonstrate achievement of those specified characteristics. The allocated baseline extends the top-level performance requirements of the functional baseline to sufficient detail for defining the functional and

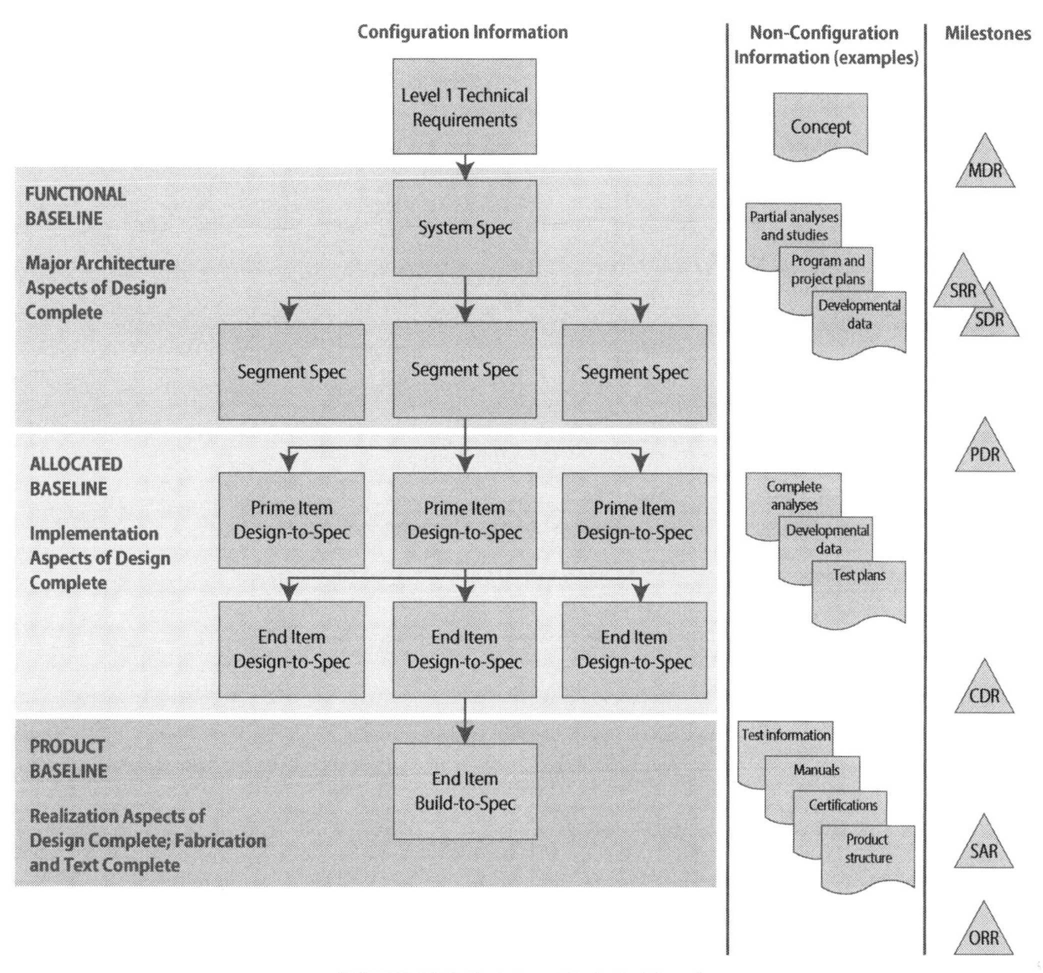

FIGURE 6.5-3 Evolution of Technical Baseline

performance characteristics and for initiating detailed design for a CI. The allocated baseline is usually controlled by the design organization until all design requirements have been verified. The allocated baseline is typically established at the successful completion of the PDR. Prior to CDR, NASA normally reviews design output for conformance to design requirements through incremental deliveries of engineering data. NASA control of the allocated baseline occurs through review of the engineering deliveries as data items.

- **Product Baseline:** The product baseline is the approved technical documentation that describes the configuration of a CI during the production, fielding/deployment, and operational support phases of its life cycle. The established product baseline is controlled as described in the configuration management plan that was developed during Phase A. The product baseline is typically established at the completion of the CDR. The product baseline describes:

  » Detailed physical or form, fit, and function characteristics of a CI;
  » The selected functional characteristics designated for production acceptance testing; and
  » The production acceptance test requirements.

- **As-Deployed Baseline:** The as-deployed baseline occurs at the ORR. At this point, the design is considered to be functional and ready for flight. All changes will have been incorporated into the documentation.

6.5.1.2.3 *Manage Configuration Change Control*

Configuration change management is a process to manage approved designs and the implementation of approved changes. Configuration change management is achieved through the systematic proposal, justification, and evaluation of proposed changes followed by incorporation of approved changes and verification of implementation. Implementing configuration change management in a given program/project requires unique knowledge of the program/project objectives and requirements. The first step establishes a robust and well-disciplined internal NASA Configuration Control Board (CCB) system, which is chaired by someone with program/project change authority. CCB members represent the stakeholders with authority to commit the team they represent. The second step creates configuration change management surveillance of the contractor's activity. The CM office advises the NASA program or project manager to achieve a balanced configuration change management implementation that suits the unique program/project situation. See FIGURE 6.5-4 for an example of a typical configuration change management control process.

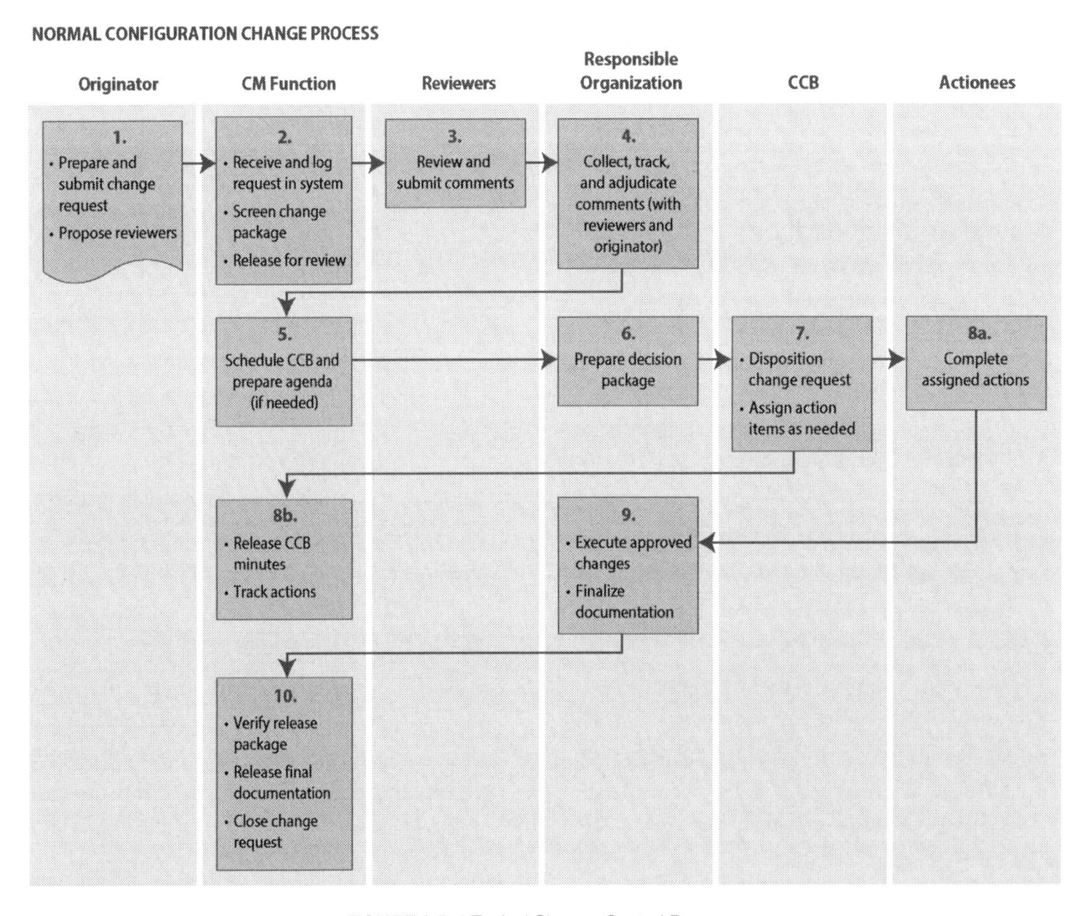

FIGURE 6.5-4 Typical Change Control Process

TYPES OF CONFIGURATION MANAGEMENT CHANGES

- **Engineering Change:** An engineering change is an iteration in the baseline. Changes can be major or minor. They may or may not include a specification change. Changes affecting an external interface must be coordinated and approved by all stakeholders affected.

    > A "major" change is a change to the baseline configuration documentation that has significant impact (i.e., requires retrofit of delivered products or affects the baseline specification, cost, safety, compatibility with interfacing products, or operator, or maintenance training).
    > A "minor" change corrects or modifies configuration documentation or processes without impact to the interchangeability of products or system elements in the system structure.

- **Waiver:** A waiver is a documented agreement intentionally releasing a program or project from meeting a requirement. (Some Centers use deviations prior to Implementation and waivers during Implementation.) Authorized waivers do not constitute a change to a baseline.

6.5.1.2.4 *Maintain the Status of Configuration Documentation*

Configuration Status Accounting (CSA) is the recording and reporting of configuration data necessary to manage CIs effectively. An effective CSA system provides timely and accurate configuration information such as:

- Complete current and historical configuration documentation and unique identifiers.
- Status of proposed changes, deviations, and waivers from initiation to implementation.
- Status and final disposition of identified discrepancies and actions identified during each configuration audit.

Some useful purposes of the CSA data include:

- An aid for proposed change evaluations, change decisions, investigations of design problems, warranties, and shelf-life calculations
- Historical traceability
- Software trouble reporting
- Performance measurement data

The following are critical functions or attributes to consider if designing or purchasing software to assist with the task of managing configuration.

- Ability to share data real time with internal and external stakeholders securely;
- Version control and comparison (track history of an object or product);
- Secure user checkout and check in;
- Tracking capabilities for gathering metrics (i.e., time, date, who, time in phases, etc.);
- Web based;
- Notification capability via e-mail;
- Integration with other databases or legacy systems;
- Compatible with required support contractors and/or suppliers (i.e., can accept data from a third party as required);
- Integration with drafting and modeling programs as required;

- Provide neutral format viewer for users;
- License agreement allows for multiple users within an agreed-to number;
- Workflow and life cycle management;
- Limited customization;
- Migration support for software upgrades;
- User friendly;
- Consideration for users with limited access;
- Ability to attach standard format files from desktop
- Workflow capability (i.e., route a CI as required based on a specific set of criteria); and
- Capable of acting as the one and only source for released information.

6.4.1.2.5 *Conduct Configuration Audits*

Configuration verification is accomplished by inspecting documents, products, and records; reviewing procedures, processes, and systems of operations to verify that the product has achieved its required performance requirements and functional attributes; and verifying that the product's design is documented. This is sometimes divided into functional and physical configuration audits. (See Section 6.7.2.3 for more on technical reviews.)

6.4.1.2.6 *Capture work Products*

These include the strategy and procedures for configuration management, the list of identified configuration items, descriptions of the configuration items, change requests, disposition of the requests, rational for dispositions, reports, and audit results.

6.5.1.3 *Outputs*

NPR 7120.5 defines a project's life cycle in progressive phases. Beginning with Pre-Phase A, these steps in turn are grouped under the headings of Formulation and Implementation. Approval is required to transition between these phases. Key Decision Points (KDPs) define transitions between the phases. CM plays an important role in determining whether a KDP has been met. Major outputs of CM are:

- **List of configuration items under control (Configuration Status Accounting (CSA) reports):** This output is the list of all the items, documents, hardware, software, models, etc., that were identified as needing to be placed under configuration control. CSA reports are updated and maintained throughout the program and project life cycle.

- **Current baselines:** Baselines of the current configurations of all items that are on the CM list are made available to all technical teams and stakeholders.

- **CM reports:** Periodic reports on the status of the CM items should be available to all stakeholders on an agreed-to frequency and at key life cycle reviews.

- **Other CM work products:** Other work products include the strategy and procedures used for CM; descriptions, drawings and/or models of the CM items; change requests and their disposition and accompanying rationale; reports; audit results as well as any corrective actions needed.

6.5.2 CM Guidance

Refer to Section 6.5.2 in the NASA Expanded Guidance for Systems Engineering at *https://nen.nasa.gov/web/se/doc-repository* for additional guidance on:

- the impact of not doing CM,
- warning signs when you know you are in trouble, and
- when it is acceptable to use redline drawings.

6.6 Technical Data Management

The Technical Data Management Process is used to plan for, acquire, access, manage, protect, and use data of a technical nature to support the total life cycle of a system. Data Management (DM) includes the development, deployment, operations and support, eventual retirement, and retention of appropriate technical, to include mission and science, data beyond system retirement as required by NPR 1441.1, NASA Records Retention Schedules.

DM is illustrated in FIGURE 6.6-1. Key aspects of DM for systems engineering include:

- application of policies and procedures for data identification and control,
- timely and economical acquisition of technical data,
- assurance of the adequacy of data and its protection,
- facilitating access to and distribution of the data to the point of use,
- analysis of data use,
- evaluation of data for future value to other programs/projects, and
- process access to information written in legacy software.

The Technical Data Management and Configuration Management Processes work side-by-side to ensure all information about the project is safe, known, and accessible. Changes to information under configuration control require a Change Request (CR) and are typically approved by a Configuration Control Board. Changes to information under Technical Data Management do not need a CR but still need to be managed by identifying who can make changes to each type of technical data.

6.6.1 Process Description

FIGURE 6.6-1 provides a typical flow diagram for the Technical Data Management Process and identifies typical inputs, outputs, and activities to consider in addressing technical data management.

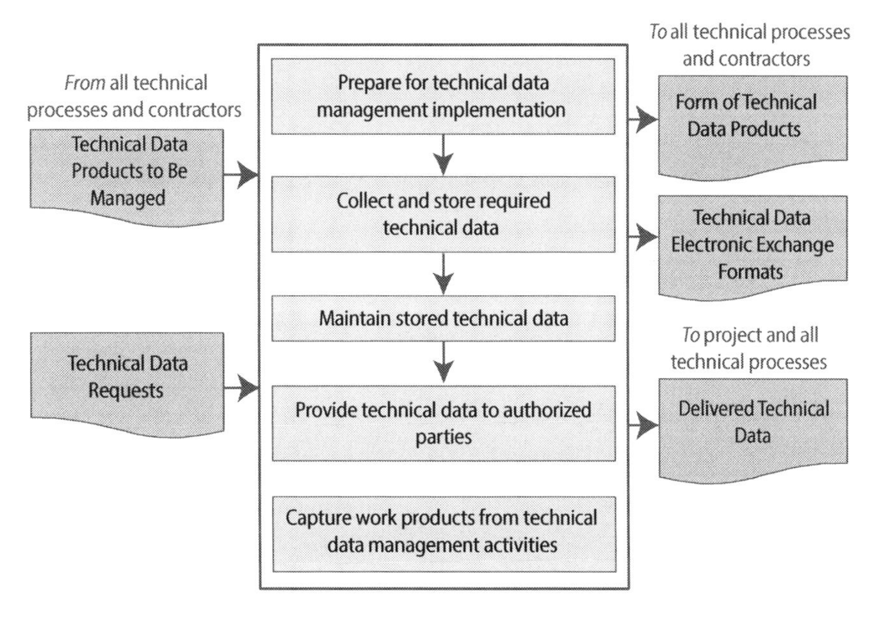

FIGURE 6.6-1 Technical Data Management Process

6.6.1.1 *Inputs*

The inputs for this process are:

- **Technical data products to be managed:** Technical data, regardless of the form or method of recording and whether the data are generated by the contractor or Government during the life cycle of the system being developed. (Electronic technical data should be stored with sufficient metadata to enable easy retrieval and sorting.)
- **Technical data requests:** External or internal requests for any of the technical data generated by the program/project.

6.6.1.2 *Process Activities*

Each Center is responsible for policies and procedures for technical DM. NPR 7120.5 and NPR 7123.1 define the need to manage data, but leave specifics to the individual Centers. However, NPR 7120.5 does require that DM planning be provided as either a section in the program/project plan, CM plan, or as a separate document. The program or project manager is responsible for ensuring that the data required are captured and stored, data integrity is maintained, and data are disseminated as required.

Other NASA policies address the acquisition and storage of data and not just the technical data used in the life cycle of a system.

6.6.1.2.1 *Prepare for Technical Data Management Implementation*

The recommended procedure is that the DM plan be a separate plan apart from the program/project plan. DM issues are usually of sufficient magnitude to justify a separate plan. The plan should cover the following major DM topics:

- Identification/definition/management of data sets.
- Control procedures—receipt, modification, review, and approval.
- Guidance on how to access/search for data for users.
- Data exchange formats that promote data reuse and help to ensure that data can be used consistently throughout the system, family of systems, or system of systems.
- Data rights and distribution limitations such as export-control Sensitive But Unclassified (SBU).
- Storage and maintenance of data, including master lists where documents and records are maintained and managed.

Prepare a technical data management strategy. This strategy can document how the program/project data management plan will be implemented by the technical effort or, in the absence of such a program-level plan, be used as the basis for preparing a detailed technical data management plan, including:

- Items of data that will be managed according to program/project or organizational policy, agreements, or legislation;
- The data content and format;
- A framework for data flow within the program/project and to/from contractors including the language(s) to be employed in technical effort information exchanges;
- Technical data management responsibilities and authorities regarding the origin, generation, capture, archiving, security, privacy, and disposal of data products;

- Establishing the rights, obligations, and commitments regarding the retention of, transmission of, and access to data items; and
- Relevant data storage, transformation, transmission, and presentation standards and conventions to be used according to program/project or organizational policy, agreements, or legislative constraints.
- Obtain strategy/plan commitment from relevant stakeholders.
- Prepare procedures for implementing the technical data management strategy for the technical effort and/or for implementing the activities of the technical data management plan.
- Establish a technical database(s) to use for technical data maintenance and storage or work with the program/project staff to arrange use of the program/project database(s) for managing technical data.
- Establish data collection tools, as appropriate to the technical data management scope and available resources.
- Establish electronic data exchange interfaces in accordance with international standards/agreements and applicable NASA standards.

Train appropriate stakeholders and other technical personnel in the established technical data management strategy/plan, procedures, and data collection tools, as applicable

Data Identification/Definition

Each program/project determines data needs during the life cycle. Data types may be defined in standard documents. Center and Agency directives sometimes specify content of documents and are appropriately used for in-house data preparation. The standard description is modified to suit program/project-specific needs, and appropriate language is included in SOWs to implement actions resulting from the data evaluation. "Data suppliers" may be contractors, academia, or the Government. Procurement of data from an outside supplier is a formal procurement action that requires a procurement document; in-house requirements may be handled using a less formal method. Below are the different types of data that might be utilized within a program/project:

- **Data**
  - "Data" is defined in general as "recorded information regardless of the form or method of recording." However, the terms "data" and "information" are frequently used interchangeably. To be more precise, data generally should be processed in some manner to generate useful, actionable information.
  - "Data," as used in SE DM, includes technical data; computer software documentation; and representation of facts, numbers, or data of any nature that can be communicated, stored, and processed to form information required by a contract or agreement to be delivered to, or accessed by, the Government.
  - Data include that associated with system development, modeling and simulation used in development or test, test and evaluation, installation, parts, spares, repairs, usage data required for product sustainability, and source and/or supplier data.
  - Data specifically not included in Technical Data Management would be data relating to general NASA workforce operations information, communications information (except where related to a specific requirement), financial transactions, personnel data,

transactional data, and other data of a purely business nature.

- **Data Call:** Solicitation from Government stakeholders (specifically Integrated Product Team (IPT) leads and functional managers) identifies and justifies their data requirements from a proposed contracted procurement. Since data provided by contractors have a cost to the Government, a data call (or an equivalent activity) is a common control mechanism used to ensure that the requested data are truly needed. If approved by the data call, a description of each data item needed is then developed and placed on contract.

- **Information:** Information is generally considered as processed data. The form of the processed data is dependent on the documentation, report, review formats, or templates that are applicable.

- **Technical Data Package:** A technical data package is a technical description of an item adequate for supporting an acquisition strategy, production, engineering, and logistics support. The package defines the required design configuration and procedures to ensure adequacy of item performance. It consists of all applicable items such as drawings, associated lists, specifications, standards, performance requirements, quality assurance provisions, and packaging details.

- **Technical Data Management System:** The strategies, plans, procedures, tools, people, data formats, data exchange rules, databases, and other entities and descriptions required to manage the technical data of a program/project.

6.6.1.2.2 *Collect and Store Data*

Subsequent activities collect, store, and maintain technical data and provide it to authorized parties as required. Some considerations that impact these activities for implementing Technical Data Management include:

- Requirements relating to the flow/delivery of data to or from a contractor should be specified in the technical data management plan and included in the Request for Proposal (RFP) and contractor agreement.

- NASA should not impose changes on existing contractor data management systems unless the program/project technical data management requirements, including data exchange requirements, cannot otherwise be met.

- Responsibility for data inputs into the technical data management system lies solely with the originator or generator of the data.

- The availability/access of technical data lies with the author, originator, or generator of the data in conjunction with the manager of the technical data management system.

- The established availability/access description and list should be baselined and placed under configuration control.

- For new programs/projects, a digital generation and delivery medium is desired. Existing programs/projects should weigh the cost/benefit trades of digitizing hard copy data.

TABLE 6.6-1 defines the tasks required to capture technical data.

6.6.1.2.3 *Provide Data to Authorized Parties*

All data deliverables should include distribution statements and procedures to protect all data that contain critical technology information, as well as to ensure that limited distribution data, intellectual property data, or proprietary data are properly handled during

DATA COLLECTION CHECKLIST

- Have the frequency of collection and the points in the technical and technical management processes when data inputs will be available been determined?
- Has the timeline that is required to move data from the point of origin to storage repositories or stakeholders been established?
- Who is responsible for the input of the data?
- Who is responsible for data storage, retrieval, and security?
- Have necessary supporting tools been developed or acquired?

systems engineering activities. This injunction applies whether the data are hard copy or digital.

As part of overall asset protection planning, NASA has established special procedures for the protection of Critical Program Information (CPI). CPI may include components; engineering, design, or manufacturing processes; technologies; system capabilities, and vulnerabilities; and any other information that gives a system its distinctive operational capability.

CPI protection should be a key consideration for the technical data management effort and is part of the asset protection planning process.

6.6.1.3 *Outputs*

Outputs include timely, secure availability of needed data in various representations to those authorized to receive it. Major outputs from the Technical Data Management Process include the following (see FIGURE 6.6-1):

- **Form of Technical Data Products:** How each type of data is held and stored such as textual, graphic, video, etc.

- **Technical Data Electronic Exchange Formats:** Description and perhaps templates, models or other ways to capture the formats used for the various data exchanges.

- **Delivered Technical Data:** The data that were delivered to the requester.

Other work products generated as part of this process include the strategy and procedures used for technical data management, request dispositions, decisions, and assumptions.

6.6.2 Technical Data Management Guidance

Refer to Section 6.6.2 in the NASA Expanded Guidance for Systems Engineering at *https://nen.nasa.gov/web/se/doc-repository* for additional guidance on:

- data security and
- ITAR.

6.7 Technical Assessment

Technical assessment is the crosscutting process used to help monitor technical progress of a program/project through periodic technical reviews and through monitoring of technical indicators such as MOEs, MOPs, Key Performance Parameters (KPPs), and TPMs. The reviews and metrics also provide status

TABLE 6.6-1 Technical Data Tasks

Description	Tasks	Expected Outcomes
Technical data capture	Collect and store inputs and technical effort outcomes from the technical and technical management processes, including: • results from technical assessments; • descriptions of methods, tools, and metrics used; • recommendations, decisions, assumptions, and impacts of technical efforts and decisions; • lessons learned; • deviations from plan; • anomalies and out-of-tolerances relative to requirements; and • other data for tracking requirements. Perform data integrity checks on collected data to ensure compliance with content and format as well as technical data checks to ensure there are no errors in specifying or recording the data. Report integrity check anomalies or variances to the authors or generators of the data for correction. Prioritize, review, and update data collection and storage procedures as part of regularly scheduled maintenance.	Sharable data needed to perform and control the technical and technical management processes is collected and stored. Stored data inventory.
Technical data maintenance	Implement technical management roles and responsibilities with technical data products received. Manage database(s) to ensure that collected data have proper quality and integrity; and are properly retained, secure, and available to those with access authority. Periodically review technical data management activities to ensure consistency and identify anomalies and variances. Review stored data to ensure completeness, integrity, validity, availability, accuracy, currency, and traceability. Perform technical data maintenance, as required. Identify and document significant issues, their impacts, and changes made to technical data to correct issues and mitigate impacts. Maintain, control, and prevent the stored data from being used inappropriately. Store data in a manner that enables easy and speedy retrieval. Maintain stored data in a manner that protects the technical data against foreseeable hazards, e.g., fire, flood, earthquake, etc.	Records of technical data maintenance. Technical effort data, including captured work products, contractor-delivered documents, and acquirer-provided documents are controlled and maintained. Status of data stored is maintained to include: version description, timeline, and security classification.
Technical data/ information distribution	Maintain an information library or reference index to provide technical data availability and access instructions. Receive and evaluate requests to determine data requirements and delivery instructions. Process special requests for technical effort data or information according to established procedures for handling such requests. Ensure that required and requested data are appropriately distributed to satisfy the needs of the acquirer and requesters in accordance with the agreement, program/project directives, and technical data management plans and procedures. Ensure that electronic access rules are followed before database access is allowed or any requested data are electronically released/transferred to the requester. Provide proof of correctness, reliability, and security of technical data provided to internal and external recipients.	Access information (e.g., available data, access means, security procedures, time period for availability, and personnel cleared for access) is readily available. Technical data are provided to authorize requesters in the appropriate format, with the appropriate content, and by a secure mode of delivery, as applicable.

(continued)

Description	Tasks	Expected Outcomes
Data management system maintenance	Implement safeguards to ensure protection of the technical database and of *en route* technical data from unauthorized access or intrusion. Establish proof of coherence of the overall technical dataset to facilitate effective and efficient use. Maintain, as applicable, backups of each technical database. Evaluate the technical data management system to identify collection and storage performance issues and problems; satisfaction of data users; risks associated with delayed or corrupted data, unauthorized access, or survivability of information from hazards such as fire, flood, earthquake, etc. Review systematically the technical data management system, including the database capacity, to determine its appropriateness for successive phases of the Defense Acquisition Framework. Recommend improvements for discovered risks and problems: Handle risks identified as part of technical risk management. Control recommended changes through established program/project change management activities.	Current technical data management system. Technical data are appropriately and regularly backed up to prevent data loss.

information to support assessing system design, product realization, and technical management decisions.

NASA has multiple review cycle processes for both space flight programs and projects (see NPR 7120.5), and research and technology programs and projects. (See NPR 7120.8, NASA Research and Technology Program and Project Management Requirements.) These different review cycles all support the same basic goals but with differing formats and formalities based on the particular program or project needs.

6.7.1 Process Description

FIGURE 6.7-1 provides a typical flow diagram for the Technical Assessment Process and identifies typical inputs, outputs, and activities to consider in addressing technical assessment. Technical assessment is focused on providing a periodic assessment of the program/project's technical and programmatic status and health at key points in the life cycle. There are 6 criteria considered in this assessment process: alignment with and contribution to Agency strategic goals; adequacy of management approach; adequacy of technical approach; adequacy of the integrated cost and schedule estimates and funding strategy; adequacy and availability of non-budgetary resources, and adequacy of the risk management approach.

6.7.1.1 *Inputs*

Typical inputs needed for the Technical Assessment Process would include the following:

- **Technical Plans:** These are the planning documents that will outline the technical reviews/assessment process as well as identify the technical product/process measures that will be tracked and assessed to determine technical progress. Examples of these plans are the program (or project) plan, SEMP (if applicable), review plans (which may be part of the program or project plan), ILS plan, and EVM plan (if applicable). These plans contain the information and descriptions of the program/project's alignment with and contribution to Agency strategic goals, its management approach, its technical approach, its integrated cost and schedule, its budget, resource allocations, and its risk management approach.

- **Technical Process and Product Measures:** These are the identified technical measures that will be

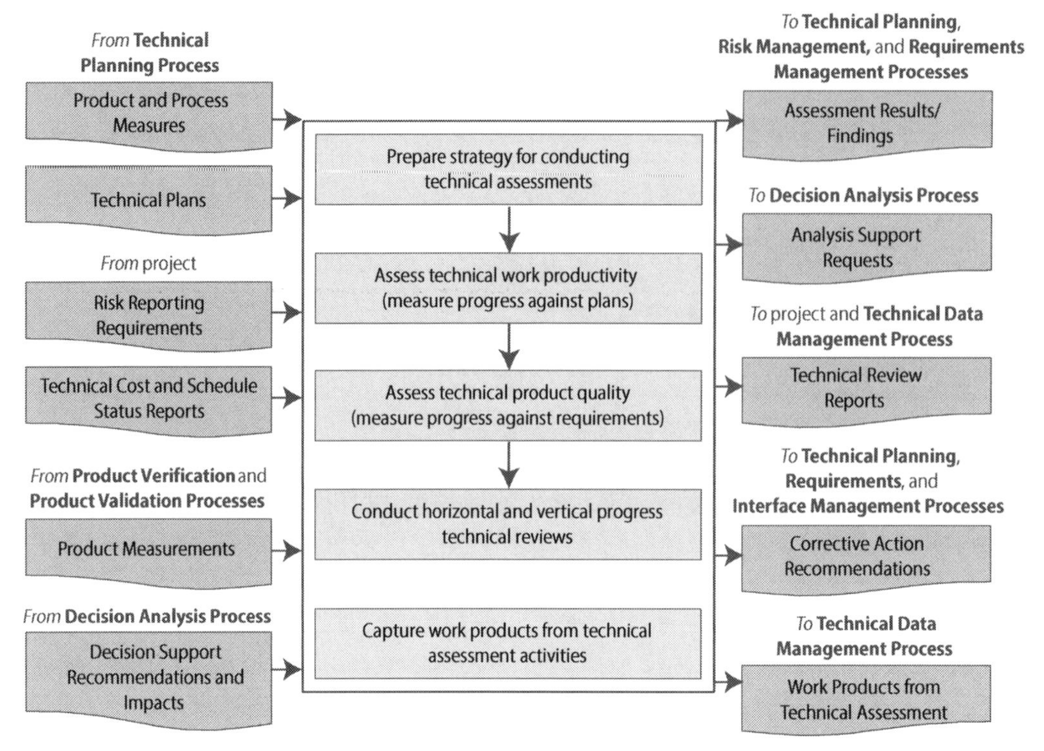

FIGURE 6.7-1 Technical Assessment Process

assessed or tracked to determine technical progress. These measures are also referred to as MOEs, MOPs, KPPs, and TPMs. (See Section 6.7.2.6.2 in the NASA Expanded Guidance for Systems Engineering at *https://nen.nasa.gov/web/se/doc-repository*.) They provide indications of the program/project's performance in key management, technical, cost (budget), schedule, and risk areas.

- **Reporting Requirements:** These are the requirements on the methodology in which the status of the technical measures will be reported with regard to management, technical cost (budget), schedule, and risk. The requirements apply internally to the program/project and are used externally by the Centers and Mission Directorates to assess the performance of the program or project. The methodology and tools used for reporting the status will be established on a project-by-project basis.

6.7.1.2 *Process Activities*

6.7.1.2.1 *Prepare Strategy for Conducting Technical Assessments*

As outlined in FIGURE 6.7-1, the technical plans provide the initial inputs into the Technical Assessment Process. These documents outline the technical reviews/assessment approach as well as identify the technical measures that will be tracked and assessed to determine technical progress. An important part of the technical planning is determining what is needed in time, resources, and performance to complete a system that meets desired goals and objectives. Project managers need visibility into the progress of those plans in order to exercise proper management control. Typical activities in determining progress against the identified technical measures include status reporting and assessing the data. Status reporting will identify where the project stands with regard to a particular technical measure. Assessing will analytically convert the output of the status reporting into a more useful form from which trends can be determined and variances from expected results can be understood. Results of the assessment activity then feed into the Decision Analysis Process (see Section 6.8) where potential corrective action may be necessary.

These activities together form the feedback loop depicted in FIGURE 6.7-2.

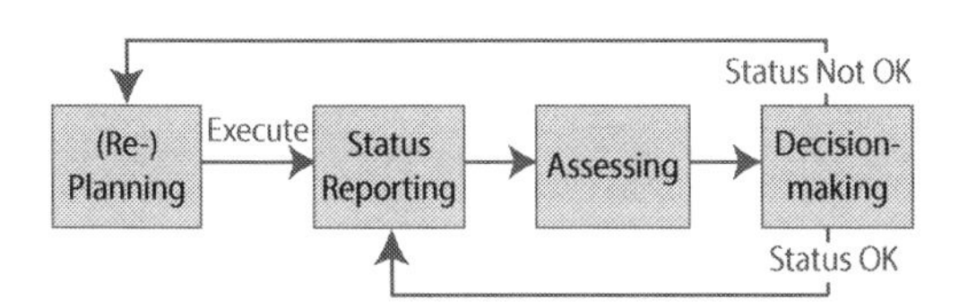

FIGURE 6.7-2 Planning and Status Reporting Feedback Loop

This loop takes place on a continual basis throughout the project life cycle. This loop is applicable at each level of the project hierarchy. Planning data, status reporting data, and assessments flow up the hierarchy with appropriate aggregation at each level; decisions cause actions to be taken down the hierarchy. Managers at each level determine (consistent with policies established at the next higher level of the project hierarchy) how often and in what form status data should be reported and assessments should be made. In establishing these status reporting and assessment requirements, some principles of good practice are as follows:

- Use an agreed-upon set of well-defined technical measures. (See Section 6.7.2.6.2 in the NASA Expanded Guidance for Systems Engineering at *https://nen.nasa.gov/web/se/doc-repository*.)
- Report these technical measures in a consistent format at all project levels.
- Maintain historical data for both trend identification and cross-project analyses.
- Encourage a logical process of rolling up technical measures (e.g., use the WBS or PBS for project progress status).
- Support assessments with quantitative risk measures.
- Summarize the condition of the project by using color-coded (red, yellow, and green) alert zones for all technical measures.

6.7.1.2.2 *Assess Technical Work Productivity and Product Quality and Conduct Progress Reviews*

Regular, periodic (e.g., monthly) tracking of the technical measures is recommended, although some measures should be tracked more often when there is rapid change or cause for concern. Key reviews, such as PDRs and CDRs, or status reviews are points at which technical measures and their trends should be carefully scrutinized for early warning signs of potential problems. Should there be indications that existing trends, if allowed to continue, will yield an unfavorable outcome, corrective action should begin as soon as practical. Section 6.7.2.6.1 in the NASA Expanded Guidance for Systems Engineering at *https://nen.nasa.gov/web/se/doc-repository* provides additional information on status reporting and assessment techniques for costs and schedules (including EVM), technical performance, and systems engineering process metrics.

The measures are predominantly assessed during the program and project technical reviews. Typical activities performed for technical reviews include (1) identifying, planning, and conducting phase-to-phase technical reviews; (2) establishing each review's purpose, objective, and entry and success criteria; (3)

establishing the makeup of the review team; and (4) identifying and resolving action items resulting from the review. Section 6.7.2.3 summarizes the types of technical reviews typically conducted on a program/project and the role of these reviews in supporting management decision processes. This section address the types of technical reviews typically conducted for both space flight and research and technology programs/projects and the role of these reviews in supporting management decision processes. It also identifies some general principles for holding reviews, but leaves explicit direction for executing a review to the program/project team to define.

The process of executing technical assessment has close relationships to other areas, such as risk management, decision analysis, and technical planning. These areas may provide input into the Technical Assessment Process or be the benefactor of outputs from the process.

TABLE 6.7-1 provides a summary of the types of reviews for a space flight project, their purpose, and timing.

6.7.1.2.3 *Capture Work Products*

The work products generated during these activities should be captured along with key decisions made, supporting decision rationale and assumptions, and lessons learned in performing the Technical Assessment Process.

6.7.1.3 *Outputs*

Typical outputs of the Technical Assessment Process would include the following:

- **Assessment Results, Findings, and Recommendations:** This is the collective data on the established measures from which trends can be determined and variances from expected results can be understood. Results then feed into the Decision Analysis Process where corrective action may be necessary.
- **Technical Review Reports/Minutes:** This is the collective information coming out of each review that captures the results, recommendations, and actions with regard to meeting the review's success criteria.
- **Other Work Products:** These would include strategies and procedures for technical assessment, key decisions and associated rationale, assumptions, and lessons learned.

6.7.2 Technical Assessment Guidance

Refer to Section 6.7.2 in the NASA Expanded Guidance for Systems Engineering at *https://nen.nasa.gov/web/se/doc-repository* for additional guidance on:

- the basis of technical reviews,
- audits,
- Key Decision Points,
- required technical reviews for space flight projects,
- other reviews,
- status reporting and assessment (including MOEs, MOPs, KPPs, TPMs, EVM and other metrics,

Additional information is also available in NASA/SP-2014-3705, *NASA Space Flight Program and Project Management Handbook.*

6.8 Decision Analysis

The purpose of this section is to provide an overview of the Decision Analysis Process, highlighting selected tools and methodologies. Decision Analysis is a framework within which analyses of diverse types are applied to the formulation and characterization of decision alternatives that best implement the decision-maker's priorities given the decision-maker's state of knowledge.

TABLE 6.7-1 Purpose and Results for Life-Cycle Reviews for Spaceflight Projects

Name of Review	Purpose	Timing	Entrance/Success Criteria	Results of Review
Mission Concept Review (MCR)	The MCR will affirm the mission need and evaluates the proposed objectives and the concept for meeting those objectives.	The MCR should be completed prior to entering the concept development phase (Phase A)	The MCR entrance and success criteria are defined in Table G-3 of NPR 7123.1.	A successful MCR supports the determination that the proposed mission meets the customer need and has sufficient quality and merit to support a field Center management decision to propose further study to the cognizant NASA program Associate Administrator as a candidate Phase A effort.
System Requirements Review (SRR)	The SRR evaluates the functional and performance requirements defined for the system and the preliminary program or project plan and ensures that the requirements and selected concept will satisfy the mission.	The SRR is conducted during the concept development phase (Phase A) and before conducting the SDR or MDR.	The SRR entrance and success criteria for a program are defined in Table G-1 of NPR 7123.1. The SRR entrance and success criteria for projects and single-project programs are defined in Table G-4 of NPR 7123.1.	Successful completion of the SRR freezes program/project requirements and leads to a formal decision by the cognizant program Associate Administrator to proceed with proposal request preparations for project implementation
Mission Definition Review (MDR)/ System Definition Review (SDR)	Sometimes called the MDR by robotic projects and SDR for human flight projects, this review evaluates whether the proposed architecture is responsive to the functional and performance requirements and that the requirements have been allocated to all functional elements of the mission/system.	The MDR/SDR is conducted during the concept development phase (Phase A) prior to KDP B and the start of preliminary design.	The MDR/SDR entrance and success criteria for a program are defined in Table G-2 of NPR 7123.1. The MDR/SDR entrance and success criteria for projects and single-project programs are defined in Table G-5 of NPR 7123.1.	A successful MDR/SDR supports the decision to further develop the system architecture/design and any technology needed to accomplish the mission. The results reinforce the mission/system's merit and provide a basis for the system acquisition strategy. As a result of successful completion, the mission/system and its operation are well enough understood to warrant design and acquisition of the end items.
Preliminary Design Review (PDR)	The PDR demonstrates that the preliminary design meets all system requirements with acceptable risk and within the cost and schedule constraints and establishes the basis for proceeding with detailed design. It shows that the correct design options have been selected, interfaces have been identified, and verification methods have been described. The PDR should address and resolve critical, system-wide issues and show that work can begin on detailed design.	PDR occurs near the completion of the preliminary design phase (Phase B) as the last review in the Formulation Phase.	The entrance and success criteria for the PDR are defined in Table G-6 of NPR 7123.1.	As a result of successful completion of the PDR, the design-to baseline is approved. A successful review result also authorizes the project to proceed into the Implementation Phase and toward final design.

(continued)

Name of Review	Purpose	Timing	Entrance/Success Criteria	Results of Review
Critical Design Review (CDR)	The CDR demonstrates that the maturity of the design is appropriate to support proceeding with full scale fabrication, assembly, integration, and test. CDR determines if the technical effort is on track to complete the system development, meeting mission performance requirements within the identified cost and schedule constraints.	CDR occurs during the final design phase (Phase C).	The entrance and success criteria for the CDR are defined in Table G-7 of NPR 7123.1.	As a result of successful completion of the CDR, the build-to baseline, production, and verification plans are approved. A successful review result also authorizes coding of deliverable software (according to the build-to baseline and coding standards presented in the review) and system qualification testing and integration. All open issues should be resolved with closure actions and schedules.
Production Readiness Review (PRR)	A PRR is held for projects developing or acquiring multiple or similar systems greater than three or as determined by the project. The PRR determines the readiness of the system developers to efficiently produce the required number of systems. It ensures that the production plans; fabrication, assembly, and integration-enabling products; and personnel are in place and ready to begin production.	PRR occurs during the final design phase (Phase C).	The entrance and success criteria for the PRR are defined in Table G-8 of NPR 7123.1.	As a result of successful completion of the PRR, the final production build-to baseline, production, and verification plans are approved. Approved drawings are released and authorized for production. A successful review result also authorizes coding of deliverable software (according to the build-to baseline and coding standards presented in the review) and system qualification testing and integration. All open issues should be resolved with closure actions and schedules.
System Integration Review (SIR)	An SIR ensures segments, components, and subsystems are on schedule to be integrated into the system. Integration facilities, support personnel, and integration plans and procedures are on schedule to support integration.	SIR occurs at the end of the final design phase (Phase C) and before the systems assembly, integration, and test phase (Phase D) begins.	The entrance and success criteria for the SIR are defined in Table G-9 of NPR 7123.1.	As a result of successful completion of the SIR, the final as-built baseline and verification plans are approved. Approved drawings are released and authorized to support integration. All open issues should be resolved with closure actions and schedules. The subsystems/systems integration procedures, ground support equipment, facilities, logistical needs, and support personnel are planned for and are ready to support integration.
System Acceptance Review (SAR)	The SAR verifies the completeness of the specific end products in relation to their expected maturity level and assesses compliance to stakeholder expectations. It also ensures that the system has sufficient technical maturity to authorize its shipment to the designated operational facility or launch site.		The entrance and success criteria for the SAR are defined in Table G-11 of NPR 7123.1.	As a result of successful completion of the SAR, the system is accepted by the buyer, and authorization is given to ship the hardware to the launch site or operational facility and to install software and hardware for operational use.

(continued)

Name of Review	Purpose	Timing	Entrance/Success Criteria	Results of Review
Operational Readiness Review (ORR)	The ORR examines the actual system characteristics and procedures used in the system or end product's operation. It ensures that all system and support (flight and ground) hardware, software, personnel, procedures, and user documentation accurately reflect the deployed state of the system.		The entrance and success criteria for the ORR are defined in Table G-12 of NPR 7123.1.	As a result of successful ORR completion, the system is ready to assume normal operations.
Flight Readiness Review (FRR)	The FRR examines tests, demonstrations, analyses, and audits that determine the system's readiness for a safe and successful flight or launch and for subsequent flight operations. It also ensures that all flight and ground hardware, software, personnel, and procedures are operationally ready.		The entrance and success criteria for the FRR are defined in Table G-13 of NPR 7123.1.	As a result of successful FRR completion, technical and procedural maturity exists for system launch and flight authorization and, in some cases, initiation of system operations.
Post-Launch Assessment Review (PLAR)	A PLAR is a post-deployment evaluation of the readiness of the spacecraft systems to proceed with full, routine operations. The review evaluates the status, performance, and capabilities of the project evident from the flight operations experience since launch. This can also mean assessing readiness to transfer responsibility from the development organization to the operations organization. The review also evaluates the status of the project plans and the capability to conduct the mission with emphasis on near-term operations and mission-critical events.	This review is typically held after the early flight operations and initial checkout.	The entrance and success criteria for the PLAR are defined in Table G-14 of NPR 7123.1.	As a result of successful PLAR completion, the system is ready to assume in-space operations.
Critical Event Readiness Review (CERR)	A CERR confirms the project's readiness to execute the mission's critical activities during flight operation. These include orbital insertion, rendezvous and docking, re-entry, scientific observations/encounters, etc.		The CERR entrance and success criteria for a program are defined in Table G-15 of NPR 7123.1.	As a result of successful CER completion, the system is ready to assume (or resume) in-space operations.
Post-Flight Assessment Review (PFAR)	The PFAR evaluates the activities from the flight after recovery. The review identifies all anomalies that occurred during the flight and mission and determines the actions necessary to mitigate or resolve the anomalies for future flights.		The entrance and success criteria for the PFAR are defined in Table G-16 of NPR 7123.1.	As a result of successful PFAR completion, the report documenting flight performance and recommendations for future missions is complete and all anomalies have been documented and dispositioned.

(continued)

Name of Review	Purpose	Timing	Entrance/Success Criteria	Results of Review
Decommissioning Review (DR)	The DR confirms the decision to terminate or decommission the system and assesses the readiness of the system for the safe decommissioning and disposal of system assets.	The DR is normally held near the end of routine mission operations upon accomplishment of planned mission objectives. It may be advanced if some unplanned event gives rise to a need to prematurely terminate the mission, or delayed if operational life is extended to permit additional investigations.	The entrance and success criteria for the DR are defined in Table G-17 of NPR 7123.1.	A successful DR completion ensures that the decommissioning and disposal of system items and processes are appropriate and effective.
Disposal Readiness Review (DRR)	A DRR confirms the readiness for the final disposal of the system assets.	The DRR is held as major assets are ready for final disposal.	The DRR entrance and success criteria for a program are defined in Table G-18 of NPR 7123.1.	A successful DRR completion ensures that the disposal of system items and processes are appropriate and effective.

The Decision Analysis Process is used in support of decision making bodies to help evaluate technical, cost, and schedule issues, alternatives, and their uncertainties. Decision models have the capacity for accepting and quantifying human subjective inputs: judgments of experts and preferences of decision makers.

The outputs from this process support the decision authority's difficult task of deciding among competing alternatives without complete knowledge; therefore, it is critical to understand and document the assumptions and limitation of any tool or methodology and integrate them with other factors when deciding among viable options.

6.8.1 Process Description

A typical process flow diagram is provided in FIGURE 6.8-1, including inputs, activities, and outputs. The first step in the process is understanding the decision to be made in the context of the system/mission. Understanding the decision needed requires knowledge of the intended outcome in terms of technical performance, cost, and schedule. For an issue that follows the decision analysis process, the definition of the decision criteria or the measures that are important to characterize the options for making a decision should be the next step in the process. With this defined, a set of alternative solutions can be defined for evaluation. These solutions should cover the full decision space as defined by the understanding of the decision and definition of the decision criteria. The need for specific decision analysis tools (defined in Section 6.8.3 in the NASA Expanded Guidance for Systems Engineering at *https://nen.nasa.gov/web/se/doc-repository*) can then be determined and employed to support the formulation of a solution. Following completion of the analysis, a description of how each alternative compares with the decision criteria can be captured for submission to the decision-making body or authority. A recommendation is typically provided from the decision analysis, but is not always required depending on the discretion of the decision-making body. A decision analysis report should be generated including: decision to be made,

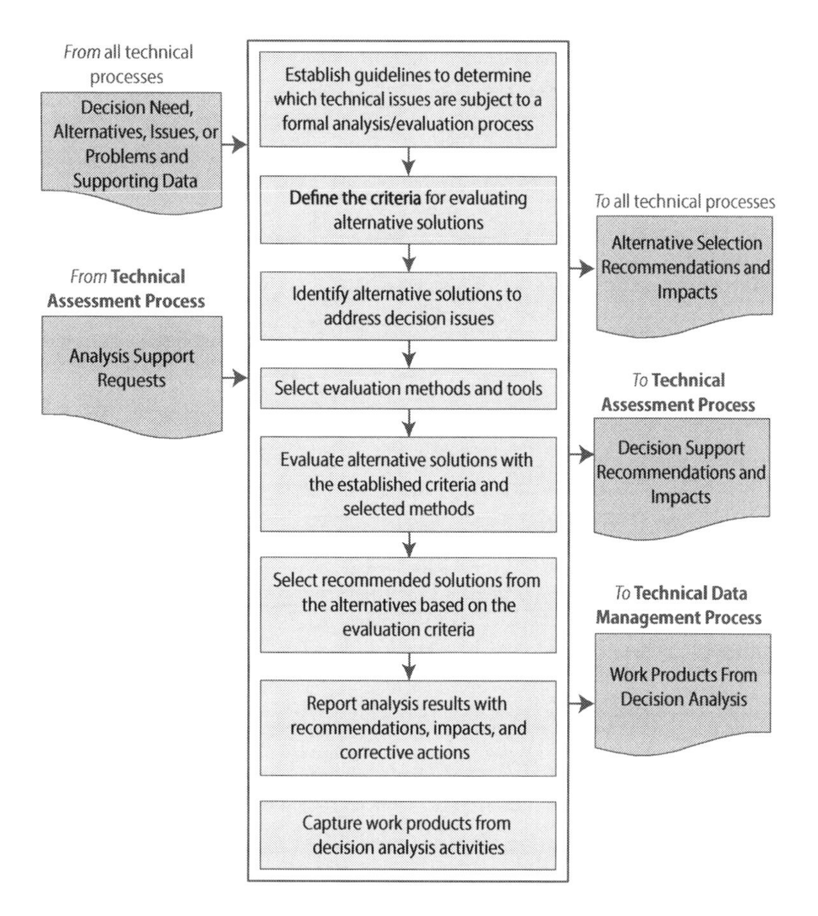

FIGURE 6.8-1 Decision Analysis Process

decision criteria, alternatives, evaluation methods, evaluation process and results, recommendation, and final decision.

Decision analysis covers a wide range of timeframes. Complex, strategic decisions may require weeks or months to fully assess all alternatives and potential outcomes. Decisions can also be made in hours or in a few days, especially for smaller projects or activities. Decisions are also made in emergency situations. Under such conditions, process steps, procedures, and meetings may be combined. In these cases, the focus of the systems engineer is on obtaining accurate decisions quickly. Once the decision is made, the report can be generated. The report is usually generated in an ongoing fashion during the decision analysis process. However, for quick or emergency decisions, the report information may be captured after the decision has been made.

Not all decisions require the same amount of analysis effort. The level and rigor required in a specific situation depend essentially on how clear-cut the decision is. If there is enough uncertainty in the alternatives' performance that the decision might change if that uncertainty were to be reduced, then consideration needs to be given to reducing that uncertainty. A robust decision is one that is based on sufficient technical evidence and characterization of uncertainties to determine that the selected alternative best reflects

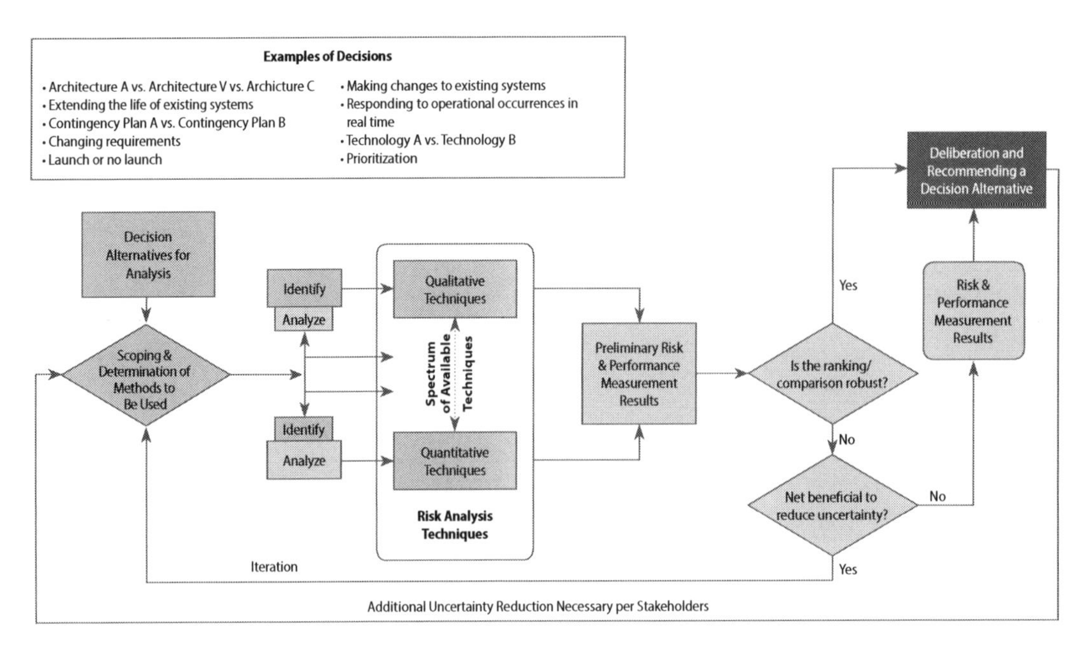

FIGURE 6.8-2 Risk Analysis of Decision Alternatives

decision-maker preferences and values given the state of knowledge at the time of the decision. This is suggested in FIGURE 6.8-2.

Note that in FIGURE 6.8-2, the phrase "net beneficial" in the decision node "Net beneficial to reduce uncertainty?" is meant to imply consideration of all factors, including whether the project can afford any schedule slip that might be caused by additional information collection and additional analysis.

6.8.1.1 *Inputs*

The technical, cost, and schedule inputs need to be comprehensively understood as part of the general decision definition. Based on this understanding, decision making can be addressed from a simple meeting to a formal analytical analysis. As illustrated in FIGURE 6.8-2, many decisions do not require extensive analysis and can be readily made with clear input from the responsible engineering and programmatic disciplines. Complex decisions may require more formal decision analysis when contributing factors have complicated or not well defined relationships. Due to this complexity, formal decision analysis has the potential to consume significant resources and time. Typically, its application to a specific decision is warranted only when some of the following conditions are met:

- **Complexity:** The actual ramifications of alternatives are difficult to understand without detailed analysis;
- **Uncertainty:** Uncertainty in key inputs creates substantial uncertainty in the ranking of alternatives and points to risks that may need to be managed;
- **Multiple Attributes:** Greater numbers of attributes cause a greater need for formal analysis; and
- **Diversity of Stakeholders:** Extra attention is warranted to clarify objectives and formulate TPMs

when the set of stakeholders reflects a diversity of values, preferences, and perspectives.

Satisfaction of all of these conditions is not a requirement for initiating decision analysis. The point is, rather, that the need for decision analysis increases as a function of the above conditions. In addition, often these decisions have the potential to result in high stakes impacts to cost, safety, or mission success criteria, which should be identified and addressed in the process. When the Decision Analysis Process is triggered, the following are inputs:

- **Decision need, identified alternatives, issues, or problems and supporting data:** This information would come from all technical, cost, and schedule management processes. It may also include high-level objectives and constraints (from the program/project).

- **Analysis support requests:** Requests will arise from the technical, cost, and schedule assessment processes.

6.8.1.2 *Process Activities*

For the Decision Analysis Process, the following activities are typically performed.

It is important to understand the decision needed in the context of the mission and system, which requires knowledge of the intended outcome in terms of technical performance, cost, and schedule. A part of this understanding is the definition of the decision criteria, or the measures that are important to characterize the options for making a decision. The specific decision-making body, whether the program/project manager, chief engineer, line management, or control board should also be well defined. Based on this understanding, then the specific approach to decision-making can be defined.

Decisions are based on facts, qualitative and quantitative data, engineering judgment, and open communications to facilitate the flow of information throughout the hierarchy of forums where technical analyses and evaluations are presented and assessed and where decisions are made. The extent of technical analysis and evaluation required should be commensurate with the consequences of the issue requiring a decision. The work required to conduct a formal evaluation is significant and applicability should be based on the nature of the problem to be resolved. Guidelines for use can be determined by the magnitude of the possible consequences of the decision to be made.

6.8.1.2.1 *Define the Criteria for Evaluating Alternative Solutions*

This step includes identifying the following:

- The types of criteria to consider, such as customer expectations and requirements, technology limitations, environmental impact, safety, risks, total ownership and life cycle costs, and schedule impact;

- The acceptable range and scale of the criteria; and

- The rank of each criterion by its importance.

Decision criteria are requirements for individually assessing the options and alternatives being considered. Typical decision criteria include cost, schedule, risk, safety, mission success, and supportability. However, considerations should also include technical criteria specific to the decision being made. Criteria should be objective and measurable. Criteria should also permit differentiating among options or alternatives. Some criteria may not be meaningful to a decision; however, they should be documented as having been considered. Criteria may be mandatory (i.e., "shall have") or enhancing. An option that does not meet mandatory criteria should be disregarded. For complex decisions, criteria can be grouped into categories or objectives.

6.8.1.2.2 *Identify Alternative Solutions to Address Decision Issues*

With the decision need well understood, alternatives can be identified that fit the mission and system context. There may be several alternatives that could potentially satisfy the decision criteria. Alternatives can be found from design options, operational options, cost options, and/or schedule options.

Almost every decision will have options to choose from. These options are often fairly clear within the mission and system context once the decision need is understood. In cases where the approach has uncertainty, there are several methods to help generate various options. Brainstorming decision options with those knowledgeable of the context and decision can provide a good list of candidate alternatives. A literature search of related systems and approaches to identify options may also provide some possible options. All possible options should be considered. This can get unwieldy if a large number of variations is possible. A "trade tree" (discussed later) is an excellent way to prune the set of variations before extensive analysis is undertaken, and to convey to other stakeholders the basis for that pruning.

A good understanding of decision need and criteria will include the definition of primary and secondary factors. Options should be focused on primary factors in the decision as defined by the decision criteria. Non-primary factors (i.e., secondary, tertiary) can be included in evaluations but should not, in general, define separate alternatives. This will require some engineering judgment that is based on the mission and system context as well as the identified decision criteria. Some options may quickly drop out of consideration as the analysis is conducted. It is important to document the fact that these options were considered. A few decisions might only have one option. It is a best practice to document a decision matrix for a major decision even if only one alternative is determined to be viable. (Sometimes doing nothing or not making a decision is an option.)

6.8.1.2.3 *Select Evaluation Methods and Tools*

Based on the decision to be made, various approaches can be taken to evaluate identified alternatives. These can range from simple discussion meetings with contributing and affected stakeholders to more formal evaluation methods. In selecting the approach, the mission and system context should be kept in mind and the complexity of the decision analysis should fit the complexity of the mission, system, and corresponding decision.

Evaluation methods and tools/techniques to be used should be selected based on the purpose for analyzing a decision and on the availability of the information used to support the method and/or tool. Typical evaluation methods include: simulations; weighted tradeoff matrices; engineering, manufacturing, cost, and technical opportunity trade studies; surveys; human-in-the-loop testing; extrapolations based on field experience and prototypes; user review and comment; and testing. *Section 6.8.2* provides several options.

6.8.1.2.4 *Evaluate Alternative Solutions with the Established Criteria and Selected Methods*

The performance of each alternative with respect to each chosen performance measure is evaluated. In all but the simplest cases, some consideration of uncertainty is warranted. Uncertainty matters in a particular analysis only if there is a non-zero probability that uncertainty reduction could alter the ranking of alternatives. If this condition is obtained, then it is necessary to consider the value of reducing that uncertainty, and act accordingly.

Regardless of the methods or tools used, results should include the following:

- Evaluation of assumptions related to evaluation criteria and of the evidence that supports the assumptions; and
- Evaluation of whether uncertainty in the values for alternative solutions affects the evaluation.

When decision criteria have different measurement bases (e.g., numbers, money, weight, dates), normalization can be used to establish a common base for mathematical operations. The process of "normalization" is making a scale so that all different kinds of criteria can be compared or added together. This can be done informally (e.g., low, medium, high), on a scale (e.g., 1-3-9), or more formally with a tool. No matter how normalization is done, the most important thing to remember is to have operational definitions of the scale. An operational definition is a repeatable, measurable number. For example, "high" could mean "a probability of 67 percent and above." "Low" could mean "a probability of 33 percent and below." For complex decisions, decision tools usually provide an automated way to normalize. It is important to question and understand the operational definitions for the weights and scales of the tool.

NOTE: Completing the decision matrix can be thought of as a default evaluation method. Completing the decision matrix is iterative. Each cell for each criterion and each option needs to be completed by the team. Use evaluation methods as needed to complete the entire decision matrix.

6.8.1.2.5 *Select Recommended Solutions from the Alternatives Based on the Evaluation Criteria and Report to the Decision-Maker*

Once the decision alternative evaluation is completed, recommendations should be brought back to the decision maker including an assessment of the robustness of the ranking (i.e., whether the uncertainties are such that reducing them could credibly change the ranking of the alternatives). Generally, a single alternative should be recommended. However, if the alternatives do not significantly differ, or if uncertainty reduction could credibly alter the ranking, the recommendation should include all closely ranked alternatives for a final selection by the decision-maker. In any case, the decision-maker is always free to select any alternative or ask for additional alternatives to be assessed (often with updated guidance on selection criteria). This step includes documenting the information, including assumptions and limitations of the evaluation methods used, and analysis of the uncertainty in the analysis of the alternatives' performance that justifies the recommendations made and gives the impacts of taking the recommended course of action, including whether further uncertainty reduction would be justifiable.

The highest score (e.g., percentage, total score) is typically the option that is recommended to management. If a different option is recommended, an explanation should be provided as to why the lower score is preferred. Usually, if an alternative having a lower score is recommended, the "risks" or "disadvantages" were too great for the highest ranking alternative indicating the scoring methods did not properly rank the alternatives. Sometimes the benefits and advantages of a lower or close score outweigh the highest score. If this occurs, the decision criteria should be reevaluated, not only the weights, but the basic definitions of what is being measured for each alternative. The criteria should be updated, with concurrence from the decision-maker, to more correctly reflect the suitability of each alternative.

6.8.1.2.6 *Report Analysis Results*

These results are reported to the appropriate stakeholders with recommendations, impacts, and corrective actions

6.8.1.2.7 *Capture Work Products*

These work products may include the decision analysis guidelines, strategy, and procedures that were used; analysis/evaluation approach; criteria, methods, and tools used; analysis/evaluation assumptions made in arriving at recommendations; uncertainties; sensitivities of the recommended actions or corrective actions; and lessons learned.

6.8.1.3 *Outputs*

6.8.1.3.1 *Alternative Selection and Decision Support Recommendations and Impacts*

Once the technical team recommends an alternative to a NASA decision-maker (e.g., a NASA board, forum, or panel), all decision analysis information should be documented. The team should produce a report to document all major recommendations to serve as a backup to any presentation materials used. A report in conjunction with a decision matrix provides clearly documented rationale for the presentation materials (especially for complex decisions). Decisions are typically captured in meeting minutes and should be captured in the report as well. Based on the mission and system context and the decision made, the report may be a simple white paper or a more formally formatted document. The important characteristic of the report is the content, which fully documents the decision needed, assessments done, recommendations, and decision finally made.

This report includes the following:

- mission and system context for the decision
- decision needed and intended outcomes
- decision criteria
- identified alternative solutions
- decision evaluation methods and tools employed
- assumptions, uncertainties, and sensitivities in the evaluations and recommendations
- results of all alternative evaluations
- alternative recommendations
- final decision made with rationale
- lessons learned

Typical information captured in a decision report is shown in TABLE 6.8-1.

6.8.2 Decision Analysis Guidance

Refer to Section 6.8.2 in the NASA Expanded Guidance for Systems Engineering at *https://nen.nasa.gov/web/se/doc-repository* for additional guidance on decision analysis methods supporting all SE processes and phases including:

- trade studies,
- cost-benefit analysis,
- influence diagrams,
- decision trees,
- analytic hierarchy process,
- Borda counting, and
- utility analysis,

Additional information on tools for decision making can be found in NASA Reference Publication 1358, *System Engineering "Toolbox" for Design-Oriented Engineers* located at *https://nen.nasa.gov/web/se/doc-repository*.

TABLE 6.8-1 Typical Information to Capture in a Decision Report

#	Section	Section Description
1	Executive Summary	Provide a short half-page executive summary of the report: • Recommendation (short summary—1 sentence) • Problem/issue requiring a decision (short summary—1 sentence)
2	Problem/Issue Description	Describe the problem/issue that requires a decision. Provide background, history, the decision maker(s) (e.g., board, panel, forum, council), and decision recommendation team, etc.
3	Decision Matrix Setup Rationale	Provide the rationale for setting up the decision matrix: • Criteria selected • Options selected • Weights selected • Evaluation methods selected Provide a copy of the setup decision matrix.
4	Decision Matrix Scoring Rationale	Provide the rationale for the scoring of the decision matrix. Provide the results of populating the scores of the matrix using the evaluation methods selected.
5	Final Decision Matrix	Cut and paste the final spreadsheet into the document. Also include any important snapshots of the decision matrix.
6	Risk/Benefits	For the final options being considered, document the risks and benefits of each option.
7	Recommendation and/or Final Decision	Describe the recommendation that is being made to the decision maker(s) and the rationale for why the option was selected. Can also document the final decision in this section.
8	Dissent	If applicable, document any dissent with the recommendation. Document how dissent was addressed (e.g., decision matrix, risk).
9	References	Provide any references.
A	Appendices	Provide the results of the literature search, including lessons learned, previous related decisions, and previous related dissent. Also document any detailed data analysis and risk analysis used for the decision. Can also document any decision metrics.

Appendix A: Acronyms

AADL	Architecture Analysis and Design Language
AD^2	Advancement Degree of Difficulty Assessment
AIAA	American Institute of Aeronautics and Astronautics
AO	Announcement of Opportunity
AS9100	Aerospace Quality Management Standard
ASME	American Society of Mechanical Engineers
ASQ	American Society for Quality
CAIB	Columbia Accident Investigation Board
CCB	Configuration Control Board
CDR	Critical Design Review
CE	Concurrent Engineering or Chief Engineer
CEQ	Council on Environmental Quality
CERR	Critical Event Readiness Review
CHSIP	Commercial Human Systems Integration Processes
CI	Configuration Item
CM	Configuration Management
CMO	Configuration Management Organization
ConOps	Concept of Operations
COSPAR	Committee on Space Research
COTS	Commercial Off-The-Shelf
CPI	Critical Program Information
CR	Change Request
CRM	Continuous Risk Management
CSA	Configuration Status Accounting
D&C	Design and Construction
DDT&E	Design, Development, Test, and Evaluation
DM	Data Management
DOD	(U.S.) Department of Defense
DODAF	DOD Architecture Framework
DR	Decommissioning Review
DRM	Design Reference Mission
DRR	Disposal Readiness Review
EDL	Entry, Descent, and Landing
EEE	Electrical, Electronic, and Electromechanical
EFFBD	Enhanced Functional Flow Block Diagram
EIA	Electronic Industries Alliance
EMC	Electromagnetic Compatibility
EMI	Electromagnetic Interference
EO	(U.S.) Executive Order
EOM	End of Mission
EVM	Earned Value Management
FA	Formulation Agreement
FAD	Formulation Authorization Document
FAR	Federal Acquisition Regulation
FCA	Functional Configuration Audit
FFBD	Functional Flow Block Diagram
FIPS	Federal Information Processing Standard
FM	Fault Management
FMEA	Failure Modes and Effects Analysis
FMR	Financial Management Requirements
FRR	Flight Readiness Review
FTE	Full Time Equivalent
GEO	Geostationary
GOTS	Government Off-The-Shelf
GSE	Government-Supplied Equipment or Ground Support Equipment
GSFC	Goddard Space Flight Center
HCD	Human-Centered Design
HF	Human Factors
HITL	Human-In-The-Loop
HQ	Headquarters
HSI	Human Systems Integration
HSIP	Human System Integration Plan
HWIL	HardWare-In-the-Loop
I&T	Integration and Test
ICD	Interface Control Document/Drawing
ICP	Interface Control Plan
IDD	Interface Definition Document

IDEF0	Integration Definition (for functional modeling)
IEEE	Institute of Electrical and Electronics Engineers
ILS	Integrated Logistics Support
INCOSE	International Council on Systems Engineering
IPT	Integrated Product Team
IRD	Interface Requirements Document
ISO	International Organization for Standardization
IT	Information Technology
ITA	Internal Task Agreement
ITAR	International Traffic in Arms Regulation
IV&V	Independent Verification and Validation
IVHM	Integrated Vehicle Health Management
IWG	Interface Working Group
JCL	Joint (cost and schedule) Confidence Level
JPL	Jet Propulsion Laboratory
KBSI	Knowledge Based Systems, Inc.
KDP	Key Decision Point
KDR	Key Driving Requirement
KPP	Key Performance Parameter
KSC	Kennedy Space Center
LCC	Life Cycle Cost
LEO	Low Earth Orbit or Low Earth Orbiting
M&S	Modeling and Simulation or Models and Simulations
MBSE	Model-Based Systems Engineering
MCR	Mission Concept Review
MDAA	Mission Directorate Associate Administrator
MDR	Mission Definition Review
MEL	Master Equipment List
MODAF	(U.K.) Ministry of Defense Architecture Framework
MOE	Measure of Effectiveness
MOP	Measure of Performance
MOTS	Modified Off-The-Shelf
MOU	Memorandum of Understanding
MRB	Material Review Board
MRR	Mission Readiness Review
MSFC	Marshall Space Flight Center
NASA	(U.S.) National Aeronautics and Space Administration
NEN	NASA Engineering Network
NEPA	National Environmental Policy Act
NFS	NASA FAR Supplement
NGO	Needs, Goals, and Objectives
NIAT	NASA Integrated Action Team
NID	NASA Interim Directive
NOA	New Obligation Authority
NOAA	(U.S.) National Oceanic and Atmospheric Administration
NODIS	NASA Online Directives Information System
NPD	NASA Policy Directive
NPR	NASA Procedural Requirements
NRC	(U.S.) Nuclear Regulatory Commission
NSTS	National Space Transportation System
OCE	(NASA) Office of the Chief Engineer
OCIO	(NASA) Office of the Chief Information Officer
OCL	Object Constraint Language
OMB	(U.S.) Office of Management and Budget
ORR	Operational Readiness Review
OTS	Off-the-Shelf
OWL	Web Ontology Language
PBS	Product Breakdown Structure
PCA	Physical Configuration Audit or Program Commitment Agreement
PD/NSC	(U.S.) Presidential Directive/National Security Council
PDR	Preliminary Design Review
PFAR	Post-Flight Assessment Review
PI	Performance Index or Principal Investigator
PIR	Program Implementation Review
PKI	Public Key Infrastructure
PLAR	Post-Launch Assessment Review
PM	Program Manager or Project Manager
PMC	Program Management Council
PPD	(U.S.) Presidential Policy Directive
PRA	Probabilistic Risk Assessment
PRD	Project Requirements Document
PRR	Production Readiness Review
QA	Quality Assurance
QVT	Query View Transformations
R&M	Reliability and Maintainability
R&T	Research and Technology
RACI	Responsible, Accountable, Consulted, Informed
REC	Record of Environmental Consideration
RF	Radio Frequency

RFA	Requests for Action
RFP	Request for Proposal
RID	Review Item Discrepancy or Review Item Disposition
RIDM	Risk-Informed Decision-Making
RM	Risk Management
RMA	Rapid Mission Architecture
RUL	Remaining Useful Life
SAR	System Acceptance Review or Safety Analysis Report (DOE)
SBU	Sensitive But Unclassified
SDR	Program/System Definition Review
SE	Systems Engineering
SECoP	Systems Engineering Community of Practice
SEMP	Systems Engineering Management Plan
SI	International System of Units (French: Système international d'unités)
SIR	System Integration Review
SMA	Safety and Mission Assurance
SME	Subject Matter Expert
SOW	Statement Of Work
SP	Special Publication
SRD	System Requirements Document
SRR	Program/System Requirements Review
SRS	Software Requirements Specification
STI	Scientific and Technical Information
STS	Space Transportation System
SysML	System Modeling Language
T&E	Test and Evaluation
TA	Technical Authority
TBD	To Be Determined
TBR	To Be Resolved
ToR	Terms of Reference
TPM	Technical Performance Measure
TRL	Technology Readiness Level
TRR	Test Readiness Review
TVC	Thrust Vector Controller
UFE	Unallocated Future Expenses
UML	Unified Modeling Language
V&V	Verification and Validation
WBS	Work Breakdown Structure
WYE	Work Year Equivalent
XMI	XML Metadata Interchange
XML	Extensible Markup Language

Appendix B: Glossary

Acceptable Risk: The risk that is understood and agreed to by the program/project, governing authority, mission directorate, and other customer(s) such that no further specific mitigating action is required.

Acquisition: The process for obtaining the systems, research, services, construction, and supplies that NASA needs to fulfill its missions. Acquisition, which may include procurement (contracting for products and services), begins with an idea or proposal that aligns with the NASA Strategic Plan and fulfills an identified need and ends with the completion of the program or project or the final disposition of the product or service.

Activity: A set of tasks that describe the technical effort to accomplish a process and help generate expected outcomes.

Advancement Degree of Difficulty Assessment (AD²): The process to develop an understanding of what is required to advance the level of system maturity.

Allocated Baseline (Phase C): The allocated baseline is the approved performance-oriented configuration documentation for a CI to be developed that describes the functional and interface characteristics that are allocated from a higher level requirements document or a CI and the verification required to demonstrate achievement of those specified characteristics. The allocated baseline extends the top-level performance requirements of the functional baseline to sufficient detail for initiating manufacturing or coding of a CI. The allocated baseline is controlled by NASA. The allocated baseline(s) is typically established at the Preliminary Design Review.

Analysis: Use of mathematical modeling and analytical techniques to predict the compliance of a design to its requirements based on calculated data or data derived from lower system structure end product validations.

Analysis of Alternatives: A formal analysis method that compares alternative approaches by estimating their ability to satisfy mission requirements through an effectiveness analysis and by estimating their life cycle costs through a cost analysis. The results of these two analyses are used together to produce a cost-effectiveness comparison that allows decision makers to assess the relative value or potential programmatic returns of the alternatives. An analysis of alternatives broadly examines multiple elements of program or project alternatives (including technical performance, risk, LCC, and programmatic aspects).

Analytic Hierarchy Process: A multi-attribute methodology that provides a proven, effective means to deal with complex decision- making and can assist with identifying and weighting selection criteria, analyzing the data collected for the criteria, and expediting the decision-making process.

Anomaly: The unexpected performance of intended function.

Approval: Authorization by a required management official to proceed with a proposed course of action. Approvals are documented.

Approval (for Implementation): The acknowledgment by the decision authority that the program/project has met stakeholder expectations and formulation requirements, and is ready to proceed to implementation. By approving a program/project, the decision authority commits the budget resources necessary to continue into implementation. Approval (for Implementation) is documented.

Architecture (System): Architecture is the high-level unifying structure that defines a system. It provides a set of rules, guidelines, and constraints that defines a cohesive and coherent structure consisting of constituent parts, relationships and connections that establish how those parts fit and work together. It addresses the concepts, properties and characteristics of the system and is represented by entities such as functions, functional flows, interfaces, relationships, resource flow items, physical elements, containers, modes, links, communication resources, etc. The entities are not independent but interrelated in the architecture through the relationships between them (NASA HQ).

Architecture (ISO Definition): Fundamental concepts or properties of a system in its environment embodied in its elements, relationships, and in the principles of its design and evolution (ISO 42010).

As-Deployed Baseline: The as-deployed baseline occurs at the Operational Readiness Review. At this point, the design is considered to be functional and ready for flight. All changes will have been incorporated into the documentation.

Automated: Automation refers to the allocation of system functions to machines (hardware or software) versus humans.

Autonomous: Autonomy refers to the relative locations and scope of decision-making and control functions between two locations within a system or across the system boundary.

Baseline : An agreed-to set of requirements, designs, or documents that will have changes controlled through a formal approval and monitoring process.

Bidirectional Traceability : The ability to trace any given requirement/expectation to its parent requirement/expectation and to its allocated children requirements/expectations.

Brassboard: A medium fidelity functional unit that typically tries to make use of as much operational hardware/software as possible and begins to address scaling issues associated with the operational system. It does not have the engineering pedigree in all aspects, but is structured to be able to operate in simulated operational environments in order to assess performance of critical functions.

Breadboard: A low fidelity unit that demonstrates function only, without respect to form or fit in the case of hardware, or platform in the case of software. It often uses commercial and/or ad hoc components and is not intended to provide definitive information regarding operational performance.

Component Facilities: Complexes that are geographically separated from the NASA Center or institution to which they are assigned, but are still part of the Agency.

Concept of Operations (ConOps) (Concept Documentation): Developed early in Pre-Phase A, the ConOps describes the overall high-level concept of how the system will be used to meet stakeholder expectations, usually in a time-sequenced manner. It describes the system from an operational perspective and helps facilitate an understanding of the system goals. It stimulates the development of the requirements and architecture related to the user elements of the system. It serves as the basis for subsequent definition documents and provides the foundation for the long-range operational planning activities.

Concurrence: A documented agreement by a management official that a proposed course of action is acceptable.

Concurrent Engineering: Design in parallel rather than serial engineering fashion. It is an approach to product development that brings manufacturing, testing, assurance, operations and other disciplines into the design cycle to ensure all aspects are incorporated into the design and thus reduce overall product development time.

Configuration Items (CI): Any hardware, software, or combination of both that satisfies an end use function and is designated for separate configuration management. For example, configuration items can be referred to by an alphanumeric identifier which also serves as the unchanging base for the assignment of serial numbers to uniquely identify individual units of the CI.

Configuration Management Process: A management discipline that is applied over a product's life cycle to provide visibility into and to control changes to performance and functional and physical characteristics. It ensures that the configuration of a product is known and reflected in product information, that any product change is beneficial and is effected without adverse consequences, and that changes are managed.

Context Diagram: A diagram that shows external systems that impact the system being designed.

Continuous Risk Management: A systematic and iterative process that efficiently identifies, analyzes, plans, tracks, controls, communicates, and documents risks associated with implementation of designs, plans, and processes.

Contract: A mutually binding legal relationship obligating the seller to furnish the supplies or services (including construction) and the buyer to pay for them. It includes all types of commitments that obligate the Government to an expenditure of appropriated funds and that, except as otherwise authorized, are in writing. In addition to bilateral instruments, contracts include (but are not limited to) awards and notices of awards; job orders or task letters issued under basic ordering agreements; letter contracts; orders, such as purchase orders under which the contract becomes effective by written acceptance or performance; and bilateral contract modifications. Contracts do not include grants and cooperative agreements.

Contractor: An individual, partnership, company, corporation, association, or other service having a contract with the Agency for the design, development, manufacture, maintenance, modification, operation, or supply of items or services under the terms of a contract to a program or project. Research grantees, research contractors, and research subcontractors are excluded from this definition.

Control Account Manager: A manager responsible for a control account and for the planning, development, and execution of the budget content for those accounts.

Control Gate (or milestone): A defined point in the program/project life cycle where the decision authority can evaluate progress and determine next actions. These may include a key decision point, life cycle review, or other milestones identified by the program/project.

Cost-Benefit Analysis: A methodology to determine the advantage of one alternative over another in terms of equivalent cost or benefits. It relies on totaling positive factors and subtracting negative factors to determine a net result.

Cost-Effectiveness Analysis: A systematic quantitative method for comparing the costs of alternative means of achieving the same equivalent benefit for a specific objective.

Critical Design Review: A review that demonstrates that the maturity of the design is appropriate to support proceeding with full-scale fabrication, assembly, integration, and test, and that the technical effort is on track to complete the system development meeting performance requirements within the identified cost and schedule constraints.

Critical Event (or key event): An event in the operations phase of the mission that is time-sensitive and is required to be accomplished successfully in order to achieve mission success. These events should be considered early in the life cycle as drivers for system design.

Critical Event Readiness Review: A review that evaluates the readiness of a project's flight system to execute the critical event during flight operation.

Customer: The organization or individual that has requested a product and will receive the product to be delivered. The customer may be an end user of the product, the acquiring agent for the end user, or the requestor of the work products from a technical effort. Each product within the system hierarchy has a customer.

Data Management: DM is used to plan for, acquire, access, manage, protect, and use data of a technical nature to support the total life cycle of a system.

Decision Analysis Process: A methodology for making decisions that offers techniques for modeling decision problems mathematically and finding optimal decisions numerically. The methodology entails identifying alternatives, one of which should be decided upon; possible events, one of which occurs thereafter; and outcomes, each of which results from a combination of decision and event.

Decision Authority: The individual authorized by the Agency to make important decisions for programs and projects under his or her authority.

Decision Matrix: A methodology for evaluating alternatives in which valuation criteria are typically displayed in rows on the left side of the matrix and alternatives are the column headings of the matrix. A "weight" is typically assigned to each criterion.

Decision Support Package: Documentation submitted in conjunction with formal reviews and change requests.

Decision Tree: A decision model that displays the expected consequences of all decision alternatives by making discreet all "chance" nodes, and, based on this, calculating and appropriately weighting the possible consequences of all alternatives.

Decommissioning Review: A review that confirms the decision to terminate or decommission a system and assess the readiness for the safe decommissioning and disposal of system assets. The DR is normally held near the end of routine mission operations upon accomplishment of planned mission objectives. It may be advanced if some unplanned event gives rise to a need to prematurely terminate the mission, or delayed if operational life is extended to permit additional investigations.

Deliverable Data Item: Consists of technical data, such as requirements specifications, design documents, management data plans, and metrics reports, that have been identified as items to be delivered with an end product.

Demonstration: Showing that the use of an end product achieves the individual specified requirement (verification) or stakeholder expectation (validation). It is generally a basic confirmation of performance capability, differentiated from testing by the lack of detailed data gathering. Demonstrations can involve the use of physical models or mock-ups; for example, a requirement that all controls shall be reachable by the pilot could be verified by having a pilot perform flight-related tasks in a cockpit mock-up or simulator. A demonstration could also be the actual operation of the end product by highly qualified personnel, such as test pilots, who perform a one-time event that demonstrates a capability to operate at extreme limits of system performance.

Derived Requirements: Requirements arising from constraints, consideration of issues implied but not explicitly stated in the high-level direction provided by NASA Headquarters and Center institutional requirements, factors introduced by the selected architecture, and the design. These requirements are finalized through requirements analysis as part of the overall systems engineering process and become part of the program or project requirements baseline. Requirements arising from constraints, consideration of issues implied but not explicitly stated in the high-level direction provided by NASA Headquarters and Center institutional requirements, factors introduced by the selected architecture, and the design. These requirements are finalized through requirements analysis as part of the overall systems engineering process and become part of the program or project requirements baseline.

Descope: As a verb, take out of (or remove from) the scope of a project. As a noun, as in "performance descope," it indicates the process or the result of the process of narrowing the scope; i.e., removing part of the original scope.

Design Solution Definition Process: The process used to translate the outputs of the logical decomposition process into a design solution definition. It includes transforming the defined logical decomposition models and their associated sets of derived technical requirements into alternative solutions and analyzing each alternative to be able to select a preferred alternative and fully define that alternative into a final design solution that will satisfy the technical requirements.

Designated Governing Authority: For the technical effort, this is the Center Director or the person that has been designated by the Center Director to ensure the appropriate level of technical management oversight. For large programs, this will typically be the Engineering Technical Authority. For smaller projects, this function can be delegated to line managers.

Detection: Determination that system state or behavior is different from expected performance.

Diagnosis: Determining the possible locations and/or causes of an anomaly or a failure.

Discrepancy: Any observed variance from, lack of agreement with, or contradiction to the required or expected outcome, configuration, or result.

Earned Value: The sum of the budgeted cost for tasks and products that have actually been produced (completed or in progress) at a given time in the schedule.

Earned Value Management: A tool for measuring and assessing project performance through the integration of technical scope with schedule and cost objectives during the execution of the project. EVM provides quantification of technical progress, enabling management to gain insight into project status and project completion costs and schedules. Two essential characteristics of successful EVM are EVM system data integrity and carefully targeted monthly EVM data analyses (i.e., risky WBS elements).

Emergent Behavior: An unanticipated behavior shown by a system due to interactions between large numbers of simple components of that system.

End Product: The hardware/software or other product that performs the operational functions. This product is to be delivered to the next product layer or to the final customer.

Enabling Products: The life cycle support products and services (e.g., production, test, deployment, training, maintenance, and disposal) that facilitate the progression and use of the operational end product through its life cycle. Since the end product and its enabling products are interdependent, they are viewed as a system. Project responsibility thus extends to acquiring services from the relevant enabling products in each life cycle phase. When a suitable enabling product does not already exist, the project that is responsible for the end product may also be responsible for creating and using the enabling product.

Engineering Unit: A high fidelity unit that demonstrates critical aspects of the engineering processes involved in the development of the operational unit. Engineering test units are intended to closely resemble the final product (hardware/software) to the maximum extent possible and are built and tested so as to establish confidence that the design will function in the expected environments. In some cases, the engineering unit will become the final product, assuming that proper traceability has been exercised over the components and hardware handling.

Enhanced Functional Flow Block Diagram: A block diagram that represents control flows and data flows as well as system functions and flow.

Entrance Criteria: Guidance for minimum accomplishments each project needs to fulfill prior to a life cycle review.

Environmental Impact: The direct, indirect, or cumulative beneficial or adverse effect of an action on the environment.

Environmental Management: The activity of ensuring that program and project actions and decisions that potentially impact or damage the environment are assessed and evaluated during the formulation and planning phase and reevaluated throughout implementation. This activity is performed according to all NASA policy and Federal, state, and local environmental laws and regulations.

Establish (with respect to processes): The act of developing policy, work instructions, or procedures to implement process activities.

Evaluation: The continual self- and independent assessment of the performance of a program or project and incorporation of the evaluation findings to ensure adequacy of planning and execution according to plan.

Extensibility: The ability of a decision to be extended to other applications.

Failure: The inability of a system, subsystem, component, or part to perform its required function within specified limits (Source: NPR 8715.3 and Avizienis 2004).

Failure Tolerance: The ability to sustain a certain number of failures and still retain capability (Source:

NPR 8705.2). A function should be preserved despite the presence of any of a specified number of coincident, independent failure causes of specified types.

Fault: A physical or logical cause, which explains a failure (Source: Avizienis 2004).

Fault Identification: Determining the possible locations of a failure or anomaly cause(s), to a defined level of granularity.

Fault Isolation: The act of containing the effects of a fault to limit the extent of failure.

Fault Management: A specialty engineering discipline that encompasses practices that enable an operational system to contain, prevent, detect, diagnose, identify, respond to, and recover from conditions that may interfere with nominal mission operations.

Fault Tolerance: See "Failure Tolerance."

Feasible: Initial evaluations show that the concept credibly falls within the technical cost and schedule constraints for the project.

Flexibility: The ability of a decision to support more than one current application.

Flight Readiness Review: A review that examines tests, demonstrations, analyses, and audits that determine the system's readiness for a safe and successful flight/launch and for subsequent flight operations. It also ensures that all flight and ground hardware, software, personnel, and procedures are operationally ready.

Float: The amount of time that a task in a project network schedule can be delayed without causing a delay to subsequent tasks or the project completion date.

Formulation Phase: The first part of the NASA management life cycle defined in NPR 7120.5 where system requirements are baselined, feasible concepts are determined, a system definition is baselined for the selected concept(s), and preparation is made for progressing to the Implementation Phase.

Functional Analysis: The process of identifying, describing, and relating the functions a system should perform to fulfill its goals and objectives.

Functional Baseline (Phase B): The functional baseline is the approved configuration documentation that describes a system's or top-level CIs' performance requirements (functional, interoperability, and interface characteristics) and the verification required to demonstrate the achievement of those specified characteristics.

Functional Configuration Audit (FCA): Examines the functional characteristics of the configured product and verifies that the product has met, via test results, the requirements specified in its functional baseline documentation approved at the PDR and CDR plus any approved changes thereafter. FCAs will be conducted on both hardware- and software-configured products and will precede the PCA of the configured product.

Functional Decomposition: A subfunction under logical decomposition and design solution definition, it is the examination of a function to identify subfunctions necessary for the accomplishment of that function and functional relationships and interfaces.

Functional Flow Block Diagram: A block diagram that defines system functions and the time sequence of functional events.

Gantt Chart: A bar chart depicting start and finish dates of activities and products in the WBS.

Goal: Goals elaborate on the need and constitute a specific set of expectations for the system. They further define what we hope to accomplish by addressing the critical issues identified during the problem assessment. Goals need not be in a quantitative or measurable form, but they must allow us to assess whether the system has achieved them.

Government Mandatory Inspection Points: Inspection points required by Federal regulations to ensure 100 percent compliance with safety/mission-critical attributes when noncompliance can result in loss of life or loss of mission.

Health Assessment: The activity under Fault Management that carries out detection, diagnosis, and identification of faults and prediction of fault propagation states into the future.

Health Monitoring: The activity under Fault Management that implements system state data collection, storage, and reporting though sensing and communication.

Heritage (or legacy): Refers to the original manufacturer's level of quality and reliability that is built into the parts, which have been proven by (1) time in service, (2) number of units in service, (3) mean time between failure performance, and (4) number of use cycles.

Human-Centered Design: An approach to the development of interactive systems that focuses on making systems usable by ensuring that the needs, abilities, and limitations of the human user are met throughout the system's life cycle.

Human Factors Engineering: The discipline that studies human-system interfaces and provides requirements, standards, and guidelines to ensure the human component of an integrated system is able to function as intended.

Human Systems Integration: An interdisciplinary and comprehensive management and technical process that focuses on the integration of human considerations into the system acquisition and development processes to enhance human system design, reduce life cycle ownership cost, and optimize total system performance.

Implementation Phase: The part of the NASA management life cycle defined in NPR 7120.5 where the detailed design of system products is completed and the products to be deployed are fabricated, assembled, integrated, and tested and the products are deployed to their customers or users for their assigned use or mission.

Incommensurable Costs: Costs that cannot be easily measured, such as controlling pollution on launch or mitigating debris.

Influence Diagram: A compact graphical and mathematical representation of a decision state. Its elements are decision nodes, chance nodes, value nodes, and arrows to indicate the relationships among these elements.

Inspection: The visual examination of a realized end product. Inspection is generally used to verify physical design features or specific manufacturer identification. For example, if there is a requirement that the safety arming pin has a red flag with the words "Remove Before Flight" stenciled on the flag in black letters, a visual inspection of the arming pin flag can be used to determine if this requirement was met.

Integrated Logistics Support: The management, engineering activities, analysis, and information management associated with design requirements definition, material procurement and distribution, maintenance, supply replacement, transportation, and disposal that are identified by space flight and ground systems supportability objectives.

Interface Management Process: The process to assist in controlling product development when efforts are divided among parties (e.g., Government, contractors, geographically diverse technical teams) and/or to define and maintain compliance among the products that should interoperate.

Iterative: Application of a process to the same product or set of products to correct a discovered discrepancy or other variation from requirements. (See "recursive" and "repeatable.")

Key Decision Point: The event at which the decision authority determines the readiness of a program/project to progress to the next phase of the life cycle (or to the next KDP).

Key Event (or Critical Event): See "Critical Event."

Key Performance Parameter: Those capabilities or characteristics (typically engineering-based or related to health and safety or operational performance) considered most essential for successful mission accomplishment. They characterize the major drivers of operational performance, supportability, and interoperability.

Knowledge Management: A collection of policies, processes, and practices relating to the use of intellectual- and knowledge-based assets in an organization.

Least-Cost Analysis: A methodology that identifies the least-cost project option for meeting the technical requirements.

Liens: Requirements or tasks not satisfied that have to be resolved within a certain assigned time to allow passage through a control gate to proceed.

Life Cycle Cost (LCC): The total of the direct, indirect, recurring, nonrecurring, and other related expenses both incurred and estimated to be incurred in the design, development, verification, production, deployment, prime mission operation, maintenance, support, and disposal of a project, including closeout, but not extended operations. The LCC of a project or system can also be defined as the total cost of ownership over the project or system's planned life cycle from Formulation (excluding Pre–Phase A) through Implementation (excluding extended operations). The LCC includes the cost of the launch vehicle.

Logical Decomposition Models: Mathematical or visual representations of the relationships between requirements as identified in the Logical Decomposition Process.

Logical Decomposition Process: A process used to improve understanding of the defined technical requirements and the relationships among the requirements (e.g., functional, behavioral, performance, and temporal) and to transform the defined set of technical requirements into a set of logical decomposition models and their associated set of derived technical requirements for lower levels of the system and for input to the Design Solution Definition Process.

Logistics (or Integrated Logistics Support): See "Integrated Logistics Support."

Loosely Coupled Program: Programs that address specific objectives through multiple space flight projects of varied scope. While each individual project has an assigned set of mission objectives, architectural and technological synergies and strategies that benefit the program as a whole are explored during the formulation process. For instance, Mars orbiters designed for more than one Mars year in orbit are required to carry a communication system to support present and future landers.

Maintain (with respect to establishment of processes): The act of planning the process, providing resources, assigning responsibilities, training people, managing configurations, identifying and involving stakeholders, and monitoring process effectiveness.

Maintainability: The measure of the ability of an item to be retained in or restored to specified conditions when maintenance is performed by personnel having specified skill levels, using prescribed procedures and resources, at each prescribed level of maintenance.

Margin: The allowances carried in budget, projected schedules, and technical performance parameters (e.g., weight, power, or memory) to account for uncertainties and risks. Margins are allocated in the formulation process based on assessments of risks and are typically consumed as the program/project proceeds through the life cycle.

Master Equipment List (MEL): The MEL is a listing of all the parts of a system and includes pertinent information such as serial numbers, model numbers, manufacturer, equipment type, system/element it is located within, etc.

Measure of Effectiveness (MOE): A measure by which a stakeholder's expectations are judged in assessing satisfaction with products or systems produced and delivered in accordance with the associated technical effort. The MOE is deemed to be critical to not only the acceptability of the product by the stakeholder but also critical to operational/mission usage. A MOE is typically qualitative in nature or not able to be used directly as a design-to requirement.

Measure of Performance (MOP): A quantitative measure that, when met by the design solution, helps ensure that a MOE for a product or system will be satisfied. These MOPs are given special attention during design to ensure that the MOEs to which they are associated are met. There are generally two or more measures of performance for each MOE.

Metric: The result of a measurement taken over a period of time that communicates vital information about the status or performance of a system, process, or activity. A metric should drive appropriate action.

Mission: A major activity required to accomplish an Agency goal or to effectively pursue a scientific, technological, or engineering opportunity directly related to an Agency goal. Mission needs are independent of any particular system or technological solution.

Mission Concept Review: A review that affirms the mission/project need and examines the proposed mission's objectives and the ability of the concept to fulfill those objectives.

Mission Definition Review: A life cycle review that evaluates whether the proposed mission/system architecture is responsive to the program mission/system functional and performance requirements and requirements have been allocated to all functional elements of the mission/system.

Mitigation: An action taken to mitigate the effects of a fault towards achieving existing or redefined system goals.

Model: A model is a physical, mathematical, or logical representation of reality.

Need: A single statement that drives everything else. It should relate to the problem that the system is supposed to solve, but not be the solution.

Nonconforming product: Software, hardware, or combination, either produced, acquired, or in some combination that is identified as not meeting documented requirements.

Objective: Specific target levels of outputs the system must achieve. Each objective should relate to a particular goal. Generally, objectives should meet four criteria:

1. **Specific:** Objectives should aim at results and reflect what the system needs to do, but they don't outline how to implement the solution. They need to be specific enough to provide clear direction, so developers, customers, and testers can understand them.

2. **Measurable:** Objectives need to be quantifiable and verifiable. The project needs to monitor the system's success in achieving each objective.

3. **Aggressive, but attainable:** Objectives need to be challenging but reachable, and targets need to be realistic. At first, objectives "To Be Determined" (TBD) may be included until trade studies occur, operations concepts solidify, or technology matures. But objectives need to be feasible before starting to write requirements and design systems.

4. **Results-oriented:** Objectives need to focus on desired outputs and outcomes, not on the methods used to achieve the target (what, not how).

Objective Function (sometimes Cost Function): A mathematical expression of the values of combinations of possible outcomes as a single measure of cost-effectiveness.

Operational Environment: The environment in which the final product will be operated. In the case of space flight hardware/software, it is space. In the case of ground-based or airborne systems that are not directed toward space flight, it is the environments defined by the scope of operations. For software, the environment is defined by the operational platform.

Operational Readiness Review: A review that examines the actual system characteristics and the procedures used in the system or product's operation and ensures that all system and support (flight and ground) hardware, software, personnel, procedures, and user documentation accurately reflects the deployed state of the system and are operationally ready.

Operations Concept: A description of how the flight system and the ground system are used together to ensure that the concept of operation is reasonable. This might include how mission data of interest, such as engineering or scientific data, are captured, returned to Earth, processed, made available to users, and archived for future reference. (Source: NPR 7120.5)

Optimal Solution: A feasible solution that best meets criteria when balanced at a system level.

Other Interested Parties (Stakeholders): A subset of "stakeholders," other interested parties are groups or individuals who are not customers of a planned technical effort but may be affected by the resulting product, the manner in which the product is realized or used, or have a responsibility for providing life cycle support services.

Peer Review: Independent evaluation by internal or external subject matter experts who do not have a vested interest in the work product under review. Peer reviews can be planned, focused reviews conducted on selected work products by the producer's peers to identify defects and issues prior to that work product moving into a milestone review or approval cycle.

Performance Standards: Defines what constitutes acceptable performance by the provider. Common metrics for use in performance standards include cost and schedule.

Physical Configuration Audits (PCA) or configuration inspection: The PCA examines the physical configuration of the configured product and verifies that the product corresponds to the build-to (or code-to) product baseline documentation previously approved at the CDR plus the approved changes thereafter. PCAs are conducted on both hardware-and software-configured products.

Post-Flight Assessment Review: Evaluates how well mission objectives were met during a mission and identifies all flight and ground system anomalies that occurred during the flight and determines the actions necessary to mitigate or resolve the anomalies for future flights of the same spacecraft design.

Post-Launch Assessment Review: A review that evaluates the readiness of the spacecraft systems to proceed with full, routine operations after post-launch deployment. The review also evaluates the status of the project plans and the capability to conduct the mission with emphasis on near-term operations and mission-critical events.

Precedence Diagram: Workflow diagram that places activities in boxes connected by dependency arrows; typical of a Gantt chart.

Preliminary Design Review: A review that demonstrates that the preliminary design meets all system requirements with acceptable risk and within the cost and schedule constraints and establishes the basis for proceeding with detailed design. It will show that the correct design option has been selected, interfaces have been identified, and verification methods have been described.

Process: A set of activities used to convert inputs into desired outputs to generate expected outcomes and satisfy a purpose.

Producibility: A system characteristic associated with the ease and economy with which a completed design can be transformed (i.e., fabricated, manufactured, or coded) into a hardware and/or software realization.

Product: A part of a system consisting of end products that perform operational functions and enabling products that perform life cycle services related to the end product or a result of the technical efforts in the form of a work product (e.g., plan, baseline, or test result).

Product Baseline (Phase D/E): The product baseline is the approved technical documentation that describes the configuration of a CI during the production, fielding/deployment, and operational support phases of its life cycle. The product baseline describes detailed physical or form, fit, and function characteristics of a CI; the selected functional characteristics designated for production acceptance testing; and the production acceptance test requirements.

Product Breakdown Structure: A hierarchical breakdown of the hardware and software products of a program/project.

Product Implementation Process: A process used to generate a specified product of a product layer through buying, making, or reusing in a form consistent with the product life cycle phase exit (success) criteria and that satisfies the design solution definition-specified requirements (e.g., drawings, specifications).

Product Integration Process: A process used to transform the design solution definition into the desired end product of the product layer through assembly and integration of lower-level validated end products in a form that is consistent with the product life cycle phase exit (success) criteria and that satisfies the design solution definition requirements (e.g., drawings, specifications).

Product Realization: The act of making, buying, or reusing a product, or the assembly and integration of lower-level realized products into a new product, as well as the verification and validation that the product satisfies its appropriate set of requirements and the transition of the product to its customer.

Product Transition Process: A process used to transition a verified and validated end product that has been generated by product implementation or product integration to the customer at the next level in the system structure for integration into an end product or, for the top-level end product, transitioned to the intended end user.

Product Validation Process: A process used to confirm that a verified end product generated by product implementation or product integration fulfills (satisfies) its intended use when placed in its intended environment and to assure that any anomalies discovered during validation are appropriately resolved prior to delivery of the product (if validation is done by the supplier of the product) or prior to integration with other products into a higher-level assembled product (if validation is done by the receiver of the product). The validation is done against the set of baselined stakeholder expectations.

Product Verification Process: A process used to demonstrate that an end product generated from product implementation or product integration conforms to its design solution definition requirements as a function of the product life cycle phase and the location of the product layer end product in the system structure.

Production Readiness Review (PRR): A review for projects developing or acquiring multiple or similar systems greater than three or as determined by the project. The PRR determines the readiness of the system developers to efficiently produce the required number of systems. It ensures that the production plans, fabrication, assembly, integration-enabling products, operational support, and personnel are in place and ready to begin production.

Prognosis: The prediction of a system's future health states, degradation, and Remaining Useful Life (RUL).

Program: A strategic investment by a mission directorate or mission support office that has a defined architecture and/or technical approach, requirements, funding level, and a management structure that initiates and directs one or more projects. A program defines a strategic direction that the Agency has identified as critical.

Program/System Definition Review: A review that examines the proposed program architecture and the flowdown to the functional elements of the system. The proposed program's objectives and the concept for meeting those objectives are evaluated. Key technologies and other risks are identified and assessed. The baseline program plan, budgets, and schedules are presented.

Program Requirements: The set of requirements imposed on the program office, which are typically found in the program plan plus derived requirements that the program imposes on itself.

Program System Requirements Review: A review that evaluates the credibility and responsiveness of a proposed program requirements/architecture to the mission directorate requirements, the allocation of program requirements to the projects, and the maturity of the program's mission/system definition.

Programmatic Requirements: Requirements set by the mission directorate, program, project, and PI, if applicable. These include strategic scientific and exploration requirements, system performance requirements, and schedule, cost, and similar nontechnical constraints.

Project: A specific investment having defined goals, objectives, requirements, life cycle cost, a beginning, and an end. A project yields new or revised products or services that directly address NASA's strategic needs. The products may be produced or the services performed wholly in-house; by partnerships with Government, industry, or academia; or through contracts with private industry.

Project Plan: The document that establishes the project's baseline for implementation, signed by the responsible program manager, Center Director, project manager, and the MDAA, if required.

Project Requirements: The set of requirements imposed on the project and developer, which are typically found in the project plan plus derived requirements that the project imposes on itself. It includes identification of activities and deliverables (end products and work products) and outputs of the development and operations.

Phase Product: An end product that is to be provided as a result of the activities of a given life cycle phase. The form depends on the phase—a product of early phases might be a simulation or model; a product of later phases may be the (final) end product itself.

Product Form: A representation of a product that depends on the development phase, current use, and maturity. Examples include mock-up, model, engineering unit, prototype unit, and flight unit.

Product Realization: The desired output from the application of the four product realization processes. The form of this product is dependent on the phase of the product life cycle and the phase exit (success) criteria.

Prototype: The prototype unit demonstrates form, fit, and function at a scale deemed to be representative of the final product operating in its operational environment. A subscale test article provides fidelity sufficient to permit validation of analytical models capable of predicting the behavior of full-scale systems in an operational environment. The prototype is used to "wring out" the design solution so that experience gained from the prototype can be fed back into design changes that will improve the manufacture, integration, and maintainability of a single flight item or the production run of several flight items.

Quality Assurance: An independent assessment performed throughout a product's life cycle in order to acquire confidence that the system actually produced and delivered is in accordance with its functional, performance, and design requirements.

Realized Product: The end product that has been implemented/integrated, verified, validated, and transitioned to the next product layer.

Recovery: An action taken to restore the functions necessary to achieve existing or redefined system goals after a fault/failure occurs.

Recursive: Value is added to the system by the repeated application of processes to design next lower-layer system products or to realize next upper-layer end products within the system structure. This also applies to repeating the application of the same processes to the system structure in the next life cycle phase to mature the system definition and satisfy phase exit (success) criteria.

Relevant Stakeholder: A subset of the term "stakeholder" that applies to people or roles that are designated in a plan for stakeholder involvement. Since "stakeholder" may describe a very large number of people, a lot of time and effort would be consumed by attempting to deal with all of them. For this reason, "relevant stakeholder" is used in most practice statements to describe the people identified to contribute to a specific task.

Relevant Environment: Not all systems, subsystems, and/or components need to be operated in the operational environment in order to satisfactorily address performance margin requirements or stakeholder expectations. Consequently, the relevant environment is the specific subset of the operational environment that is required to demonstrate critical "at risk" aspects of the final product performance in an operational environment.

Reliability: The measure of the degree to which a system ensures mission success by functioning properly over its intended life. It has a low and acceptable probability of failure, achieved through simplicity, proper design, and proper application of reliable parts and materials. In addition to long life, a reliable system is robust and fault tolerant.

Repeatable: A characteristic of a process that can be applied to products at any level of the system structure or within any life cycle phase.

Requirement: The agreed-upon need, desire, want, capability, capacity, or demand for personnel, equipment, facilities, or other resources or services by specified quantities for specific periods of time or at a specified time expressed as a "shall" statement. Acceptable form for a requirement statement is individually clear, correct, feasible to obtain, unambiguous in meaning, and can be validated at the level of the system structure at which it is stated. In pairs of requirement statements or as a set, collectively, they are not redundant, are adequately related with respect to terms used, and are not in conflict with one another.

Requirements Allocation Sheet: Documents the connection between allocated functions, allocated performance, and the physical system.

Requirements Management Process: A process used to manage the product requirements identified, baselined, and used in the definition of the products of each product layer during system design. It provides bidirectional traceability back to the top product layer requirements and manages the changes to established requirement baselines over the life cycle of the system products.

Risk: In the context of mission execution, risk is the potential for performance shortfalls that may be realized in the future with respect to achieving explicitly established and stated performance requirements. The performance shortfalls may be related to any one or more of the following mission execution domains: (1) safety, (2) technical, (3) cost, and (4) schedule. (Source: NPR 8000.4, Agency Risk Management Procedural Requirements)

Risk Assessment: An evaluation of a risk item that determines (1) what can go wrong, (2) how likely it is to occur, (3) what the consequences are, and (4) what the uncertainties associated with the likelihood and consequences are, and 5) what the mitigation plans are.

Risk-Informed Decision Analysis Process: A five-step process focusing first on objectives and next on developing decision alternatives with those objectives clearly in mind and/or using decision alternatives that have been developed under other systems engineering processes. The later steps of the process interrelate heavily with the Technical Risk Management Process.

Risk Management: Risk management includes Risk-Informed Decision-Making (RIDM) and Continuous Risk Management (CRM) in an integrated framework. RIDM informs systems engineering decisions through better use of risk and uncertainty information in selecting alternatives and establishing baseline requirements. CRM manages risks over the course of the development and the Implementation Phase of the life cycle to ensure that safety, technical, cost, and schedule requirements are met. This is done to foster proactive risk management, to better inform decision-making through better use of risk information, and then to more effectively manage Implementation risks by focusing the CRM process on the baseline performance requirements emerging from the RIDM process. (Source: NPR 8000.4, Agency Risk Management Procedural Requirements) These processes are applied at a level of rigor commensurate with the complexity, cost, and criticality of the program.

Safety: Freedom from those conditions that can cause death, injury, occupational illness, damage to or loss of equipment or property, or damage to the environment.

Search Space (or Alternative Space): The envelope of concept possibilities defined by design constraints and parameters within which alternative concepts can be developed and traded off.

Single-Project Programs: Programs that tend to have long development and/or operational lifetimes, represent a large investment of Agency resources, and have contributions from multiple organizations/agencies. These programs frequently combine program and project management approaches, which they document through tailoring.

Software: Computer programs, procedures, rules, and associated documentation and data pertaining to the development and operation of a computer system. Software also includes Commercial Off-The-Shelf (COTS), Government Off-The-Shelf (GOTS), Modified Off-The-Shelf (MOTS), embedded software, reuse, heritage, legacy, autogenerated code, firmware, and open source software components.

Note 1: For purposes of the NASA Software Release program only, the term "software," as redefined in NPR 2210.1, Release of NASA Software, does not include computer databases or software documentation.

Note 2: Definitions for the terms COTS, GOTS, heritage software, MOTS, legacy software, software reuse, and classes of software are provided in NPR 7150.2, NASA Software Engineering Requirements. (Source: NPD 7120.4, NASA Engineering and Program/Project Management Policy)

Solicitation: The vehicle by which information is solicited from contractors for the purpose of awarding a contract for products or services. Any request to submit offers or quotations to the Government. Solicitations under sealed bid procedures are called "invitations for bids." Solicitations under negotiated procedures are called "requests for proposals." Solicitations under simplified acquisition procedures may require submission of either a quotation or an offer.

Specification: A document that prescribes completely, precisely, and verifiably the requirements, design, behavior, or characteristics of a system or system component. In NPR 7123.1, "specification" is treated as a "requirement."

Stakeholder: A group or individual who is affected by or has an interest or stake in a program or project. There are two main classes of stakeholders. See "customers" and "other interested parties."

Stakeholder Expectations: A statement of needs, desires, capabilities, and wants that are not expressed as a requirement (not expressed as a "shall" statement) is referred to as an "expectation." Once the set of expectations from applicable stakeholders is collected, analyzed, and converted into a "shall" statement, the expectation becomes a requirement. Expectations can be stated in either qualitative (non-measurable) or quantitative (measurable) terms. Requirements are always stated in quantitative terms. Expectations can be stated in terms of functions, behaviors, or constraints with respect to the product being engineered or the process used to engineer the product.

Stakeholder Expectations Definition Process: A process used to elicit and define use cases, scenarios, concept of operations, and stakeholder expectations for the applicable product life cycle phases and product layer. The baselined stakeholder expectations are used for validation of the product layer end product.

Standing Review Board: The board responsible for conducting independent reviews (life-cycle and special) of a program or project and providing objective, expert judgments to the convening authorities. The reviews are conducted in accordance with approved Terms of Reference (ToR) and life cycle requirements per NPR 7123.1.

State Diagram: A diagram that shows the flow in the system in response to varying inputs in order to characterize the behavior of the system.

Success Criteria: Specific accomplishments that need to be satisfactorily demonstrated to meet the objectives of a technical review so that a technical effort can progress further in the life cycle. Success criteria are documented in the corresponding technical review plan. Formerly referred to as "exit" criteria, a term still used in some NPDs/NPRs.

Surveillance : The monitoring of a contractor's activities (e.g., status meetings, reviews, audits, site visits) for progress and production and to demonstrate fiscal responsibility, ensure crew safety and mission success, and determine award fees for extraordinary (or penalty fees for substandard) contract execution.

System: (1) The combination of elements that function together to produce the capability to meet a need. The elements include all hardware, software, equipment, facilities, personnel, processes, and procedures needed for this purpose. (2) The end product (which performs operational functions) and enabling products (which provide life cycle support services to the operational end products) that make up a system.

System Acceptance Review: The SAR verifies the completeness of the specific end products in relation to their expected maturity level, assesses compliance to stakeholder expectations, and ensures that the system has sufficient technical maturity to authorize its shipment to the designated operational facility or launch site.

System Definition Review: The Mission/System Definition Review (MDR/SDR) evaluates whether the proposed mission/system architecture is responsive to the program mission/system functional and performance requirements and requirements have been allocated to all functional elements of the mission/system. This review is used for projects and for single-project programs.

System Integration Review: A SIR ensures that segments, components, and subsystems are on schedule to be integrated into the system and that integration facilities, support personnel, and integration plans and procedures are on schedule to support integration.

System Requirements Review: For a program, the SRR is used to ensure that its functional and performance requirements are properly formulated and correlated with the Agency and mission directorate strategic objectives.

For a system/project, the SRR evaluates whether the functional and performance requirements defined for the system are responsive to the program's requirements and ensures that the preliminary project plan and requirements will satisfy the mission.

System Safety Engineering: The application of engineering and management principles, criteria, and techniques to achieve acceptable mishap risk within the constraints of operational effectiveness and suitability, time, and cost throughout all phases of the system life cycle.

System Structure: A system structure is made up of a layered structure of product-based WBS models. (See "Work Breakdown Structure" and Product Breakdown Structure.")

Systems Approach: The application of a systematic, disciplined engineering approach that is quantifiable, recursive, iterative, and repeatable for the development, operation, and maintenance of systems integrated into a whole throughout the life cycle of a project or program.

Systems Engineering (SE) Engine: The SE model shown in Figure 2.1-1 that provides the 17 technical processes and their relationships with each other. The model is called an "SE engine" in that the appropriate set of processes is applied to the products being engineered to drive the technical effort.

Systems Engineering Management Plan (SEMP): The SEMP identifies the roles and responsibility interfaces of the technical effort and specifies how those interfaces will be managed. The SEMP is the vehicle that documents and communicates the technical approach, including the application of the common technical processes; resources to be used; and the key technical tasks, activities, and events along with their metrics and success criteria.

Tailoring: A process used to adjust or seek relief from a prescribed requirement to accommodate the needs of a specific task or activity (e.g., program or project). The tailoring process results in the generation of deviations and waivers depending on the timing of the request.

OR

The process used to seek relief from NPR 7123.1 requirements consistent with program or project objectives, allowable risk, and constraints.

Technical Assessment Process: A process used to help monitor progress of the technical effort and provide status information for support of the system design, product realization, and technical management processes. A key aspect of the process is conducting life cycle and technical reviews throughout the system life cycle.

Technical Cost Estimate: The cost estimate of the technical work on a project created by the technical team based on its understanding of the system requirements and operational concepts and its vision of the system architecture.

Technical Data Management Process: A process used to plan for, acquire, access, manage, protect, and use data of a technical nature to support the total life cycle of a system. This process is used to capture trade studies, cost estimates, technical analyses, reports, and other important information.

Technical Data Package: An output of the Design Solution Definition Process, it evolves from phase to phase, starting with conceptual sketches or models and ending with complete drawings, parts list, and other details needed for product implementation or product integration.

Technical Measures: An established set of measures based on the expectations and requirements that will be tracked and assessed to determine overall system or product effectiveness and customer satisfaction. Common terms for these measures are Measures Of Effectiveness (MOEs), Measures Of Performance (MOPs), and Technical Performance Measures (TPMs).

Technical Performance Measures: A set of performance measures that are monitored by comparing the current actual achievement of the parameters with that anticipated at the current time and on future dates. TPMs are used to confirm progress and identify deficiencies that might jeopardize meeting a system requirement. Assessed parameter values that fall outside an expected range around the anticipated values indicate a need for evaluation and corrective action. Technical performance measures are typically selected from the defined set of Measures Of Performance (MOPs).

Technical Planning Process: A process used to plan for the application and management of each common technical process. It is also used to identify, define, and plan the technical effort applicable to the product life cycle phase for product layer location within the system structure and to meet project objectives and product life cycle phase exit (success) criteria. A key document generated by this process is the SEMP.

Technical Requirements: A set of requirements imposed on the end products of the system, including the system itself. Also referred to as "product requirements."

Technical Requirements Definition Process: A process used to transform the stakeholder expectations into a complete set of validated technical requirements expressed as "shall" statements that can be used for defining a design solution for the Product Breakdown Structure (PBS) model and related enabling products.

Technical Risk: Risk associated with the achievement of a technical goal, criterion, or objective. It applies to undesired consequences related to technical performance, human safety, mission assets, or environment.

Technical Risk Management Process: A process used to make risk-informed decisions and examine, on a continuing basis, the potential for deviations from the project plan and the consequences that could result should they occur.

Technical Team: A group of multidisciplinary individuals with appropriate domain knowledge, experience, competencies, and skills who are assigned to a specific technical task.

Technology Readiness Assessment Report: A document required for transition from Phase B to Phase C/D demonstrating that all systems, subsystems, and components have achieved a level of technological maturity with demonstrated evidence of qualification in a relevant environment.

Technology Assessment: A systematic process that ascertains the need to develop or infuse technological advances into a system. The technology assessment process makes use of basic systems engineering principles and processes within the framework of the Product Breakdown Structure (PBS). It is a two-step process comprised of (1) the determination of the current technological maturity in terms of Technology Readiness Levels (TRLs) and (2) the determination of the difficulty associated with moving a technology from one TRL to the next through the use of the Advancement Degree of Difficulty Assessment (AD^2).

Technology Development Plan: A document required for transition from Phase A to Phase B identifying technologies to be developed, heritage systems to be modified, alternative paths to be pursued, fallback positions and corresponding performance descopes, milestones, metrics, and key decision points. It is incorporated in the preliminary project plan.

Technology Maturity Assessment: A process to determine a system's technological maturity based on Technology Readiness Levels (TRLs).

Technology Readiness Level: Provides a scale against which to measure the maturity of a technology. TRLs range from 1, basic technology research, to 9, systems test, launch, and operations. Typically, a TRL of 6 (i.e., technology demonstrated in a relevant environment) is required for a technology to be integrated into an SE process.

Test: The use of a realized end product to obtain detailed data to verify or validate performance or to provide sufficient information to verify or validate performance through further analysis.

Test Readiness Review: A review that ensures that the test article (hardware/software), test facility, support personnel, and test procedures are ready for testing and data acquisition, reduction, and control.

Threshold Requirements: A minimum acceptable set of technical and project requirements; the set could represent the descope position of the project.

Tightly Coupled Programs: Programs with multiple projects that execute portions of a mission(s). No single project is capable of implementing a complete mission. Typically, multiple NASA Centers contribute to the program. Individual projects may be managed at different Centers. The program may also include contributions from other agencies or international partners.

Traceability: A discernible association among two or more logical entities such as requirements, system elements, verifications, or tasks.

Trade Study: A means of evaluating system designs by devising alternative means to meet functional requirements, evaluating these alternatives in terms of the measures of effectiveness and system cost, ranking the alternatives according to appropriate selection criteria, dropping less promising alternatives, and proceeding to the next level of resolution, if needed.

Trade Study Report: A report written to document a trade study. It should include: the system under analysis; system goals, objectives (or requirements, as appropriate to the level of resolution), and constraints; measures and measurement methods (models) used; all data sources used; the alternatives chosen for analysis; computational results, including uncertainty ranges and sensitivity analyses performed; the selection rule used; and the recommended alternative.

Trade Tree: A representation of trade study alternatives in which each layer represents some system aspect that needs to be treated in a trade study to determine the best alternative.

Transition: The act of delivery or moving of a product from one location to another. This act can include packaging, handling, storing, moving, transporting, installing, and sustainment activities.

Uncoupled Programs: Programs implemented under a broad theme and/or a common program implementation concept, such as providing frequent flight opportunities for cost-capped projects selected through AO or NASA Research Announcements. Each such project is independent of the other projects within the program.

Utility: A measure of the relative value gained from an alternative. The theoretical unit of measurement for utility is the "util."

Validated Requirements: A set of requirements that are well formed (clear and unambiguous), complete (agree with customer and stakeholder needs and expectations), consistent (conflict free), and individually verifiable and traceable to a higher level requirement or goal.

Validation (of a product): The process of showing proof that the product accomplishes the intended purpose based on stakeholder expectations and the Concept of Operations. May be determined by a combination of test, analysis, demonstration, and inspection. (Answers the question, "Am I building the right product?")

Variance: In program control terminology, a difference between actual performance and planned costs or schedule status.

Verification (of a product): Proof of compliance with specifications. Verification may be determined by test, analysis, demonstration, or inspection or a combination thereof. (Answers the question, "Did I build the product right?")

Waiver: A documented authorization releasing a program or project from meeting a requirement after the requirement is put under configuration control at the level the requirement will be implemented.

Work Breakdown Structure (WBS): A product-oriented hierarchical division of the hardware, software, services, and data required to produce the program/project's end product(s) structured according to the way the work will be performed, reflecting the way in which program/project costs, schedule, technical, and risk data are to be accumulated, summarized, and reported.

WBS Model: A WBS model describes a system that consists of end products and their subsystems (which perform the operational functions of the system), the supporting or enabling products, and any other work products (plans, baselines) required for the development of the system.

Workflow Diagram: A scheduling chart that shows activities, dependencies among activities, and milestones.

Appendix C: How to Write a Good Requirement—Checklist

C.1 Use of Correct Terms

- ☐ Shall = requirement
- ☐ Will = facts or declaration of purpose
- ☐ Should = goal

C.2 Editorial Checklist

Personnel Requirement

- ☐ The requirement is in the form "responsible party shall perform such and such." In other words, use the active, rather than the passive voice. A requirement should state who shall (do, perform, provide, weigh, or other verb) followed by a description of what should be performed.

Product Requirement

- ☐ The requirement is in the form "product ABC shall XYZ." A requirement should state "The product shall" (do, perform, provide, weigh, or other verb) followed by a description of what should be done.
- ☐ The requirement uses consistent terminology to refer to the product and its lower-level entities.
- ☐ Complete with tolerances for qualitative/performance values (e.g., less than, greater than or equal to, plus or minus, 3 sigma root sum squares).
- ☐ Is the requirement free of implementation? (Requirements should state WHAT is needed, NOT HOW to provide it; i.e., state the problem not the solution. Ask, "Why do you need the requirement?" The answer may point to the real requirement.)
- ☐ Free of descriptions of operations? (Is this a need the product should satisfy or an activity involving the product? Sentences like "The operator shall..." are almost always operational statements not requirements.)

Example Product Requirements

- ☐ The system shall operate at a power level of...
- ☐ The software shall acquire data from the...
- ☐ The structure shall withstand loads of...
- ☐ The hardware shall have a mass of...

C.3 General Goodness Checklist

- ☐ The requirement is grammatically correct.
- ☐ The requirement is free of typos, misspellings, and punctuation errors.
- ☐ The requirement complies with the project's template and style rules.
- ☐ The requirement is stated positively (as opposed to negatively, i.e., "shall not").
- ☐ The use of "To Be Determined" (TBD) values should be minimized. It is better to use a best

estimate for a value and mark it "To Be Resolved" (TBR) with the rationale along with what should be done to eliminate the TBR, who is responsible for its elimination, and by when it should be eliminated.

☐ The requirement is accompanied by an intelligible rationale, including any assumptions. Can you validate (concur with) the assumptions? Assumptions should be confirmed before baselining.

☐ The requirement is located in the proper section of the document (e.g., not in an appendix).

C.4 Requirements Validation Checklist

Clarity

☐ Are the requirements clear and unambiguous? (Are all aspects of the requirement understandable and not subject to misinterpretation? Is the requirement free from indefinite pronouns (this, these) and ambiguous terms (e.g., "as appropriate," "etc.," "and/or," "but not limited to")?)

☐ Are the requirements concise and simple?

☐ Do the requirements express only one thought per requirement statement, a stand-alone statement as opposed to multiple requirements in a single statement, or a paragraph that contains both requirements and rationale?

☐ Does the requirement statement have one subject and one predicate?

Completeness

☐ Are requirements stated as completely as possible? Have all incomplete requirements been captured as TBDs or TBRs and a complete listing of them maintained with the requirements?

☐ Are any requirements missing? For example, have any of the following requirements areas been overlooked: functional, performance, interface, environment (development, manufacturing, test, transport, storage, and operations), facility (manufacturing, test, storage, and operations), transportation (among areas for manufacturing, assembling, delivery points, within storage facilities, loading), training, personnel, operability, safety, security, appearance and physical characteristics, and design.

☐ Have all assumptions been explicitly stated?

Compliance

☐ Are all requirements at the correct level (e.g., system, segment, element, subsystem)?

☐ Are requirements free of implementation specifics? (Requirements should state what is needed, not how to provide it.)

☐ Are requirements free of descriptions of operations? (Don't mix operation with requirements: update the ConOps instead.)

☐ Are requirements free of personnel or task assignments? (Don't mix personnel/task with product requirements: update the SOW or Task Order instead.)

Consistency

☐ Are the requirements stated consistently without contradicting themselves or the requirements of related systems?

☐ Is the terminology consistent with the user and sponsor's terminology? With the project glossary?

- ☐ Is the terminology consistently used throughout the document? Are the key terms included in the project's glossary?

Traceability

- ☐ Are all requirements needed? Is each requirement necessary to meet the parent requirement? Is each requirement a needed function or characteristic? Distinguish between needs and wants. If it is not necessary, it is not a requirement. Ask, "What is the worst that could happen if the requirement was not included?"
- ☐ Are all requirements (functions, structures, and constraints) bidirectionally traceable to higher-level requirements or mission or system-of-interest scope (i.e., need(s), goals, objectives, constraints, or concept of operations)?
- ☐ Is each requirement stated in such a manner that it can be uniquely referenced (e.g., each requirement is uniquely numbered) in subordinate documents?

Correctness

- ☐ Is each requirement correct?
- ☐ Is each stated assumption correct? Assumptions should be confirmed before the document can be baselined.
- ☐ Are the requirements technically feasible?

Functionality

- ☐ Are all described functions necessary and together sufficient to meet mission and system goals and objectives?

Performance

- ☐ Are all required performance specifications and margins listed (e.g., consider timing, throughput, storage size, latency, accuracy and precision)?
- ☐ Is each performance requirement realistic?
- ☐ Are the tolerances overly tight? Are the tolerances defendable and cost-effective? Ask, "What is the worst thing that could happen if the tolerance was doubled or tripled?"

Interfaces

- ☐ Are all external interfaces clearly defined?
- ☐ Are all internal interfaces clearly defined?
- ☐ Are all interfaces necessary, sufficient, and consistent with each other?

Maintainability

- ☐ Have the requirements for maintainability of the system been specified in a measurable, verifiable manner?
- ☐ Are requirements written so that ripple effects from changes are minimized (i.e., requirements are as weakly coupled as possible)?

Reliability

- ☐ Are clearly defined, measurable, and verifiable reliability requirements specified?
- ☐ Are there error detection, reporting, handling, and recovery requirements?
- ☐ Are undesired events (e.g., single-event upset, data loss or scrambling, operator error) considered and their required responses specified?
- ☐ Have assumptions about the intended sequence of functions been stated? Are these sequences required?
- ☐ Do these requirements adequately address the survivability after a software or hardware fault of

the system from the point of view of hardware, software, operations, personnel and procedures?

Verifiability/Testability

- ☐ Can the system be tested, demonstrated, inspected, or analyzed to show that it satisfies requirements? Can this be done at the level of the system at which the requirement is stated? Does a means exist to measure the accomplishment of the requirement and verify compliance? Can the criteria for verification be stated?

- ☐ Are the requirements stated precisely to facilitate specification of system test success criteria and requirements?

- ☐ Are the requirements free of unverifiable terms (e.g., flexible, easy, sufficient, safe, ad hoc, adequate, accommodate, user-friendly, usable, when required, if required, appropriate, fast, portable, light-weight, small, large, maximize, minimize, sufficient, robust, quickly, easily, clearly, other "ly" words, other "ize" words)?

Data Usage

- ☐ Where applicable, are "don't care" conditions truly "don't care"? ("Don't care" values identify cases when the value of a condition or flag is irrelevant, even though the value may be important for other cases.) Are "don't care" conditions values explicitly stated? (Correct identification of "don't care" values may improve a design's portability.)

Appendix D: Requirements Verification Matrix

When developing requirements, it is important to identify an approach for verifying the requirements. This appendix provides an example matrix that defines how all the requirements are verified. Only "shall" requirements should be included in these matrices. The matrix should identify each "shall" by unique identifier and be definitive as to the source, i.e., document from which the requirement is taken. This matrix could be divided into multiple matrices (e.g., one for each requirements document) to delineate sources of requirements depending on the project. The example is shown to provide suggested guidelines for the minimum information that should be included in the verification matrix.

NOTE: See *Appendix I* for an outline of the Verification and Validation Plan. The matrix shown here (TABLE D-1) is *Appendix C* in that outline.

TABLE D-1 Requirements Verification Matrix

Requirement No.	Document	Paragraph	Shall Statement	Verification Success Criteria	Verification Method	Facility or Lab	Phase[a]	Acceptance Requirement?	Preflight Acceptance?	Performing Organization	Results
Unique identifier or each requirement	*Document number the requirement is contained within*	*Paragraph number of the requirement*	*Text (within reason) of the requirement, i.e., the "shall"*	*Success criteria for the requirement*	*Verification method for the requirement (analysis, inspection, demonstration, test)*	*Facility or laboratory used to perform the verification and validation.*	*Phase in which the verification and validation will be performed.*	*Indicate whether this requirement is also verified during initial acceptance testing of each unit.*	*Indicate whether this requirement is also verified during any pre-flight or recurring acceptance testing of each unit*	*Organization responsible for performing the verification*	*Indicate documents that contain the objective evidence that requirement was satisfied*
P-1	xxx	3.2.1.1 Capability: Support Uplinked Data (LDR)	System X shall provide a max. ground-to-station uplink of…	1. System X locks to forward link at the min and max data rate tolerances 2. System X locks to the forward link at the min and max operating frequency tolerances	Test	xxx	5	Yes	No	xxx	TPS xxxx
P-i	xxx	Other paragraphs	Other "shalls" in PTRS	Other criteria	xxx	xxx	xxx	Yes/No	Yes/No	xxx	Memo xxx
S-i or other unique designator	xxxxx (other specs, ICDs, etc.)	Other paragraphs	Other "shalls" in specs, ICDs, etc.	Other criteria	xxx	xxx	xxx	Yes/No	Yes/No	xxx	Report xxx

[a] Phases defined as: (1) Pre-Declared Development, (2) Formal Box-Level Functional, (3) Formal Box-Level Environmental, (4) Formal System-Level Environmental, (5) Formal System-Level Functional, (6) Formal End-to-End Functional, (7) Integrated Vehicle Functional, (8) On-Orbit Functional.

Appendix E: Creating the Validation Plan with a Validation Requirements Matrix

NOTE: See *Appendix I* for an outline of the Verification and Validation Plan. The matrix shown here (TABLE E-1) is *Appendix D* in that outline.

When developing requirements, it is important to identify a validation approach for how additional validation evaluation, testing, analysis, or other demonstrations will be performed to ensure customer/sponsor satisfaction.

There are a number of sources to draw from for creating the validation plan:

- ConOps
- Stakeholder/customer needs, goals, and objectives documentation
- Rationale statements for requirements and in verification requirements
- Lessons learned database
- System architecture modeling
- Test-as-you-fly design goals and constraints
- SEMP, HSIP, V&V plans

Validation products can take the form of a wide range of deliverables, including:

- Stakeholder evaluation and feedback
- Peer reviews
- Physical models of all fidelities
- Simulations
- Virtual modeling
- Tests
- Fit-checks
- Procedure dry-runs
- Integration activities (to inform on-orbit maintenance procedures)
- Phase-level review solicitation and feedback

Particular attention should be paid to the planning for life cycle phase since early validation can have a profound impact on the design and cost in the later life cycle phases.

TABLE E-1 shows an example validation matrix.

TABLE E-1 Validation Requirements Matrix

Validation Product #	Activity	Objective	Validation Method	Facility or Lab	Phase	Performing Organization	Results
Unique identifier for validation product	*Describe evaluation by the customer/ sponsor that will be performed*	*What is to be accomplished by the customer/ sponsor evaluation*	*Validation method for the requirement (analysis, inspection, demonstration, or test)*	*Facility or laboratory used to perform the validation*	*Phase in which the verification/ validation will be performed*[a]	*Organization responsible for coordinating the validation activity*	*Indicate the objective evidence that validation activity occurred*
1	Customer/ sponsor will evaluate the candidate displays	1. Ensure legibility is acceptable 2. Ensure overall appearance is acceptable	Test	xxx	Phase A	xxx	TPS 123456

a. Example: (1) during product selection process, (2) prior to final product selection (if COTS) or prior to PDR, (3) prior to CDR, (4) during box-level functional, (5) during system-level functional, (6) during end-to-end functional, (7) during integrated vehicle functional, (8) during on-orbit functional.

Appendix F: Functional, Timing, and State Analysis

This appendix was removed. For additional guidance on functional flow block diagrams, requirements allocation sheets/models, N-squared diagrams, timing analysis, and state analysis refer to Appendix F in the NASA Expanded Guidance for Systems Engineering at *https://nen.nasa.gov/web/se/doc-repository*.

Appendix G: Technology Assessment/Insertion

G.1 Introduction, Purpose, and Scope

In 2014, the Headquarters Office of Chief Engineer and Office of Chief Technologist conducted an Agency-wide study on Technical Readiness Level (TRL) usage and Technology Readiness Assessment (TRA) implementation. Numerous findings, observations, and recommendations were identified, as was a wealth of new guidance, best practices, and clarifications on how to interpret TRL and perform TRAs. These are presently being collected into a NASA TRA Handbook (in work), which will replace this appendix. In the interim, contact HQ/Steven Hirshorn on any specific questions on interpretation and application of TRL/TRA. Although the information contained in this appendix may change, it does provide some information until the TRA Handbook can be completed.

Agency programs and projects frequently require the development and infusion of new technological advances to meet mission goals, objectives, and resulting requirements. Sometimes the new technological advancement being infused is actually a heritage system that is being incorporated into a different architecture and operated in a different environment from that for which it was originally designed. It is important to recognize that the adaptation of heritage systems frequently requires technological advancement. Failure to account for this requirement can result in key steps of the development process being given short shrift—often to the detriment of the program/project. In both contexts of technological advancement (new and adapted heritage), infusion is a complex process that is often dealt with in an ad hoc manner differing greatly from project to project with varying degrees of success.

Technology infusion frequently results in schedule slips, cost overruns, and occasionally even in cancellations or failures. In post mortem, the root cause of such events is often attributed to "inadequate definition of requirements." If such is indeed the root cause, then correcting the situation is simply a matter of defining better requirements, but this may not be the case—at least not totally.

In fact, there are many contributors to schedule slip, cost overrun, and project cancellation and failure—among them lack of adequate requirements definition. The case can be made that most of these contributors are related to the degree of uncertainty at the outset of the project and that a dominant factor in the degree of uncertainty is the lack of understanding of the maturity of the technology required to bring the project to fruition and a concomitant lack of understanding of the cost and schedule reserves required to advance the technology from its present state to a point where it can be qualified and successfully infused with a high degree of confidence. Although this uncertainty cannot be eliminated, it can be substantially reduced through the early application of good systems engineering practices focused on understanding the technological requirements; the maturity of the required technology; and the technological advancement required to meet program/project goals, objectives, and requirements.

TABLE G.1-1 Products Provided by the TA as a Function of Program/Project Phase

Gate	Product
KDP A: Transition from Pre-Phase A to Phase A	Requires an assessment of potential technology needs versus current and planned technology readiness levels, as well as potential opportunities to use commercial, academic, and other government agency sources of technology. Included as part of the draft integrated baseline. Technology Development Plan is baselined that identifies technologies to be developed, heritage systems to be modified, alternative paths to be pursued, fallback positions and corresponding performance descopes, milestones, metrics, and key decision points. Initial Technology Readiness Assessment (TRA) is available.
KDP B: Transition from Phase A to Phase B	Technology Development Plan and Technology Readiness Assessment (TRA) are updated. Incorporated in the preliminary project plan.
KDP C: Transition from Phase B to Phase C/D	Requires a TRAR demonstrating that all systems, subsystems, and components have achieved a level of technological maturity with demonstrated evidence of qualification in a relevant environment.

Source: NPR 7120.5.

A number of processes can be used to develop the appropriate level of understanding required for successful technology insertion. The intent of this appendix is to describe a systematic process that can be used as an example of how to apply standard systems engineering practices to perform a comprehensive Technology Assessment (TA). The TA comprises two parts, a Technology Maturity Assessment (TMA) and an Advancement Degree of Difficulty Assessment (AD^2). The process begins with the TMA which is used to determine technological maturity via NASA's Technology Readiness Level (TRL) scale. It then proceeds to develop an understanding of what is required to advance the level of maturity through the AD^2. It is necessary to conduct TAs at various stages throughout a program/project to provide the Key Decision Point (KDP) products required for transition between phases. (See TABLE G.1-1.)

The initial TMA provides the baseline maturity of the system's required technologies at program/project outset and allows monitoring progress throughout development. The final TMA is performed just prior to the Preliminary Design Review (PDR). It forms the basis for the Technology Readiness Assessment Report (TRAR), which documents the maturity of the technological advancement required by the systems, subsystems, and components demonstrated through test and analysis. The initial AD^2 provides the material necessary to develop preliminary cost and to schedule plans and preliminary risk assessments. In subsequent assessments, the information is used to build the Technology Development Plan and in the process, identify alternative paths, fallback positions, and performance descope options. The information is also vital to preparing milestones and metrics for subsequent Earned Value Management (EVM).

The TMA is performed against the hierarchical breakdown of the hardware and software products of the program/project PBS to achieve a systematic, overall understanding at the system, subsystem, and component levels. (See FIGURE G.1-1.)

G.2 Inputs/Entry Criteria

It is extremely important that a TA process be defined at the beginning of the program/project and that it be performed at the earliest possible stage (concept development) and throughout the program/project through PDR. Inputs to the process will vary in level of detail according to the phase of the program/project, and even though there is a lack of detail in Pre-Phase A, the TA will drive out the major critical

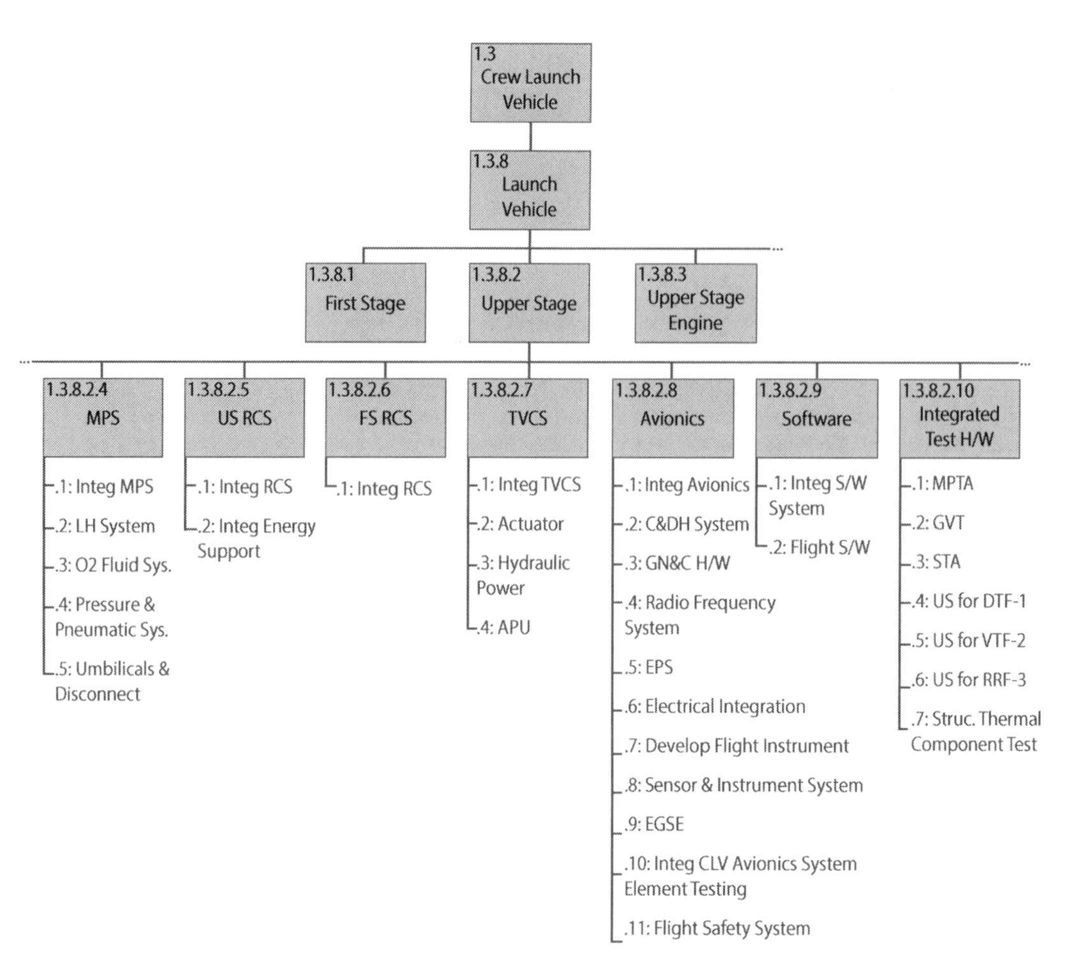

FIGURE G.1-1 PBS Example

technological advancements required. Therefore, at the beginning of Pre-Phase A, the following should be provided:

- Refinement of TRL definitions.
- Definition of AD^2.
- Definition of terms to be used in the assessment process.
- Establishment of meaningful evaluation criteria and metrics that will allow for clear identification of gaps and shortfalls in performance.
- Establishment of the TA team.
- Establishment of an independent TA review team.

G.3 How to Do Technology Assessment

The technology assessment process makes use of basic systems engineering principles and processes. As mentioned previously, it is structured to occur within the framework of the Product Breakdown Structure (PBS) to facilitate incorporation of the results. Using the PBS as a framework has a twofold benefit—it breaks the “problem” down into systems, subsystems, and components that can be more accurately assessed; and it provides the results of the assessment in a format that can be readily used in the generation of program costs and schedules. It can also be highly beneficial in providing milestones and metrics for progress tracking using EVM. As discussed above, it is a two-step process comprised of (1) the determination of the current technological maturity in terms of TRLs and (2) the determination of the difficulty associated with moving a technology from one TRL to the next through the use of the AD^2.

Conceptual Level Activities

The overall process is iterative, starting at the conceptual level during program Formulation, establishing the initial identification of critical technologies, and establishing the preliminary cost, schedule, and risk mitigation plans. Continuing on into Phase A, the process is used to establish the baseline maturity, the Technology Development Plan, and the associated costs and schedule. The final TA consists only of the TMA and is used to develop the TRAR, which validates that all elements are at the requisite maturity level. (See FIGURE G.3-1.)

Even at the conceptual level, it is important to use the formalism of a PBS to avoid allowing important technologies to slip through the cracks. Because of the preliminary nature of the concept, the systems, subsystems, and components will be defined at a level that will not permit detailed assessments to be made. The process of performing the assessment, however, is the same as that used for subsequent, more detailed steps that occur later in the program/project where systems are defined in greater detail.

Architectural Studies

Once the concept has been formulated and the initial identification of critical technologies made, it is

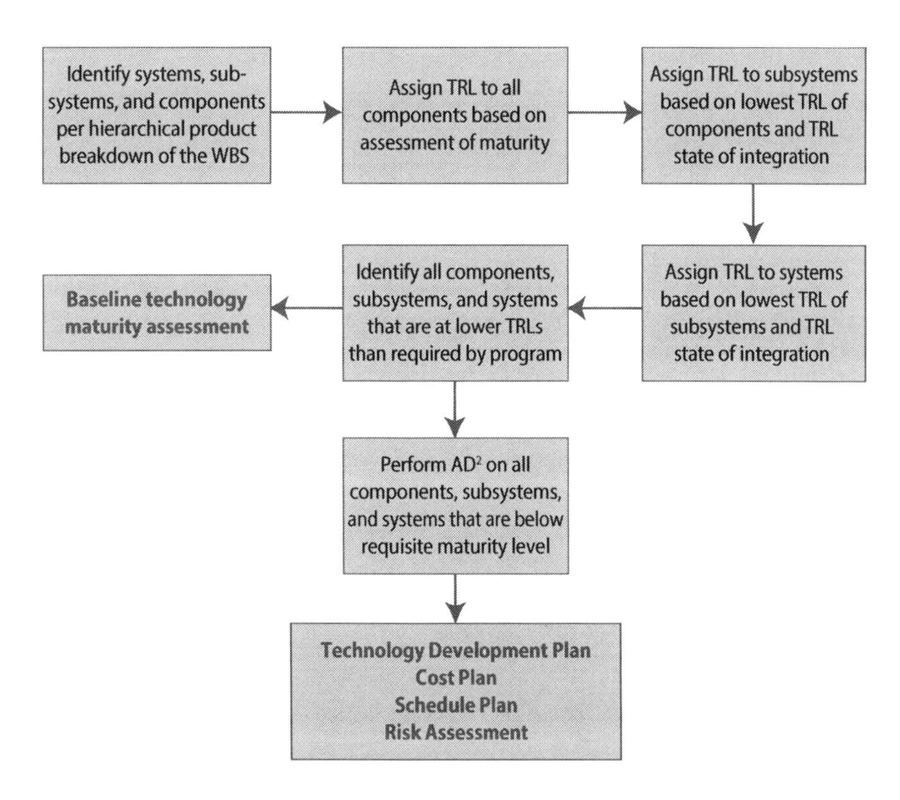

FIGURE G.3-1 Technology Assessment Process

necessary to perform detailed architecture studies with the Technology Assessment Process intimately interwoven. (See FIGURE G.3-2.)

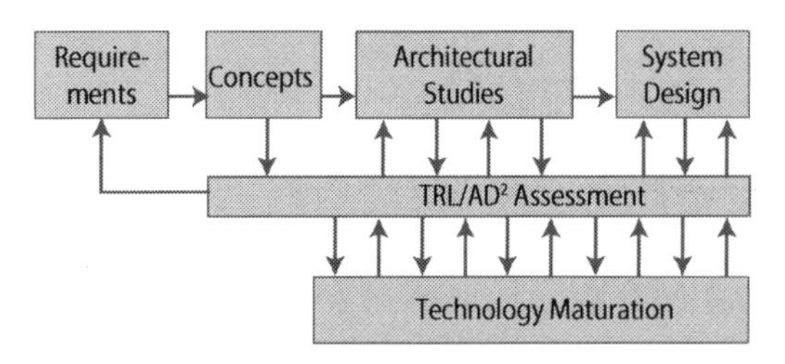

FIGURE G.3-2 Architectural Studies and Technology Development

The purpose of the architecture studies is to refine end-item system design to meet the overall scientific requirements of the mission. It is imperative that there be a continuous relationship between architectural studies and maturing technology advances. The architectural studies should incorporate the results of the technology maturation, planning for alternative paths and identifying new areas required for development as the architecture is refined. Similarly, it is incumbent upon the technology maturation process to identify requirements that are not feasible and development routes that are not fruitful and to transmit that information to the architecture studies in a timely manner. It is also incumbent upon the architecture studies to provide feedback to the technology development process relative to changes in requirements. Particular attention should be given to "heritage" systems in that they are often used in architectures and environments different from those in which they were designed to operate.

G.4 Establishing TRLs

A Technology Readiness Level (TRL) is, at its most basic, a description of the performance history of a given system, subsystem, or component relative to a set of levels first described at NASA HQ in the 1980s. The TRL essentially describes the state of a given technology and provides a baseline from which maturity is gauged and advancement defined. (See FIGURE G.4-1.)

Programs are often undertaken without fully understanding either the maturity of key technologies or what is needed to develop them to the required level. *It is impossible to understand the magnitude and scope of a development program without having a clear understanding of the baseline technological maturity of all elements of the system.* Establishing the TRL is a vital first step on the way to a successful program. A frequent misconception is that in practice, it is too difficult to determine TRLs and that when you do, it is not meaningful. On the contrary, identifying TRLs can be a straightforward systems engineering process of determining what was demonstrated and under what conditions it was demonstrated.

Terminology

At first glance, the TRL descriptions in FIGURE G.4-1 appear to be straightforward. It is in the process of trying to assign levels that problems arise. A primary cause of difficulty is in terminology; e.g., everyone knows what a breadboard is, but not everyone has the same definition. Also, what is a "relevant environment?" What is relevant to one application may or may not be relevant to another. Many of these terms originated in various branches of engineering and had, at the time, very specific meanings to that particular field. They have since become commonly used throughout the engineering field and often acquire differences in meaning from discipline to discipline, some differences subtle, some not so subtle. "Breadboard," for example, comes from electrical engineering where the original use referred to checking out the functional design of an electrical circuit by populating a "breadboard" with components to verify that the design operated as anticipated. Other terms come from mechanical engineering, referring

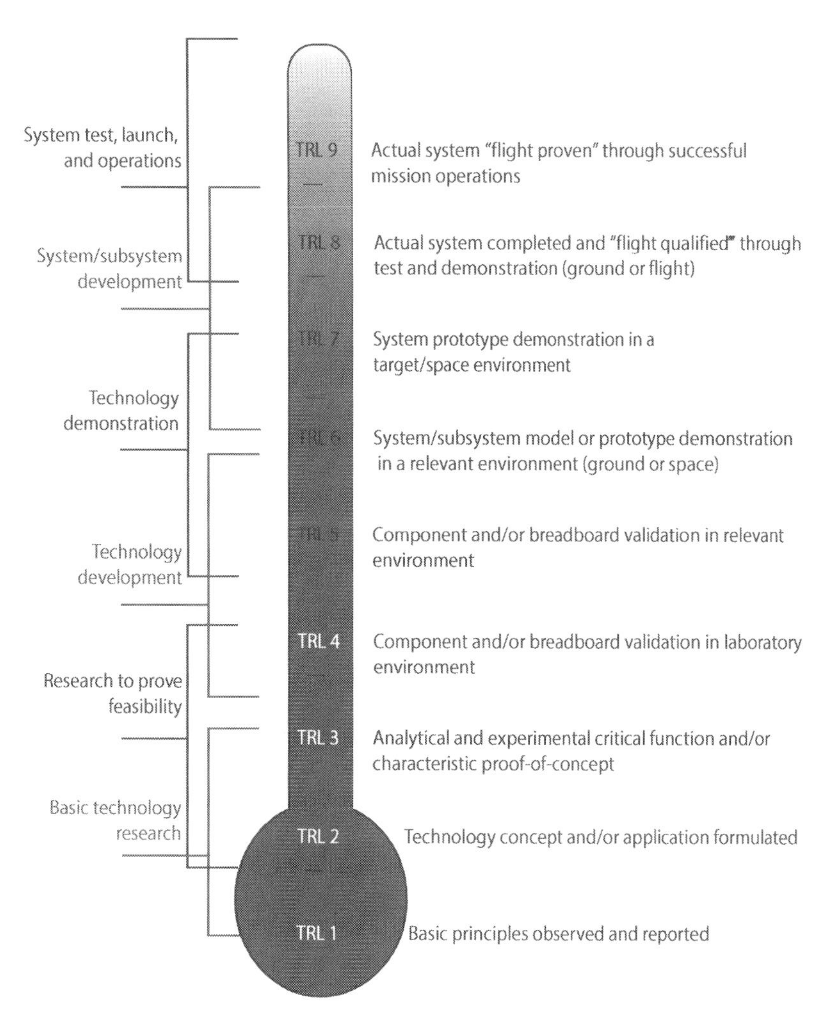

FIGURE G.4-1 Technology Readiness Levels

primarily to units that are subjected to different levels of stress under testing, e.g., qualification, protoflight, and flight units. The first step in developing a uniform TRL assessment (see FIGURE G.4-2) is to define the terms used. It is extremely important to develop and use a consistent set of definitions over the course of the program/project.

Judgment Calls

Having established a common set of terminology, it is necessary to proceed to the next step: quantifying "judgment calls" on the basis of past experience. Even with clear definitions, judgment calls will be required when it comes time to assess just how similar a given element is relative to what is needed (i.e., is it close enough to a prototype to be considered a prototype, or is it more like an engineering breadboard?). Describing what has been done in terms of form, fit, and function provides a means of quantifying an element based on its design intent and subsequent performance. The current definitions for software TRLs are contained in NPR 7123.1, NASA Systems Engineering Processes and Requirements.

Assessment Team

A third critical element of any assessment relates to the question of who is in the best position to make judgment calls relative to the status of the technology in question. For this step, it is extremely important to have a well-balanced, experienced assessment team. Team members do not necessarily have to be discipline experts. The primary expertise required for a TRL assessment is that the systems engineer/user understands the current state of the art in applications. User considerations are evaluated by HFE personnel who understand the challenges of technology insertions at various stages of the product life cycle. Having established a set of definitions, defined a process for quantifying judgment calls, and assembled an expert assessment team, the process primarily consists of asking the right questions. The flowchart depicted in FIGURE G.4-2 demonstrates the questions to ask to determine TRL at any level in the assessment.

Heritage Systems

Note the second box particularly refers to heritage systems. If the architecture and the environment have changed, then the TRL drops to TRL 5—at least initially. Additional testing may need to be done for heritage systems for the new use or new environment. If in subsequent analysis the new environment is sufficiently close to the old environment or the new architecture sufficiently close to the old architecture, then the resulting evaluation could be TRL 6 or 7, but the most important thing to realize is that it is no longer at TRL 9. Applying this process at the system level and then proceeding to lower levels of subsystem and component identifies those elements that require development and sets the stage for the subsequent phase, determining the AD².

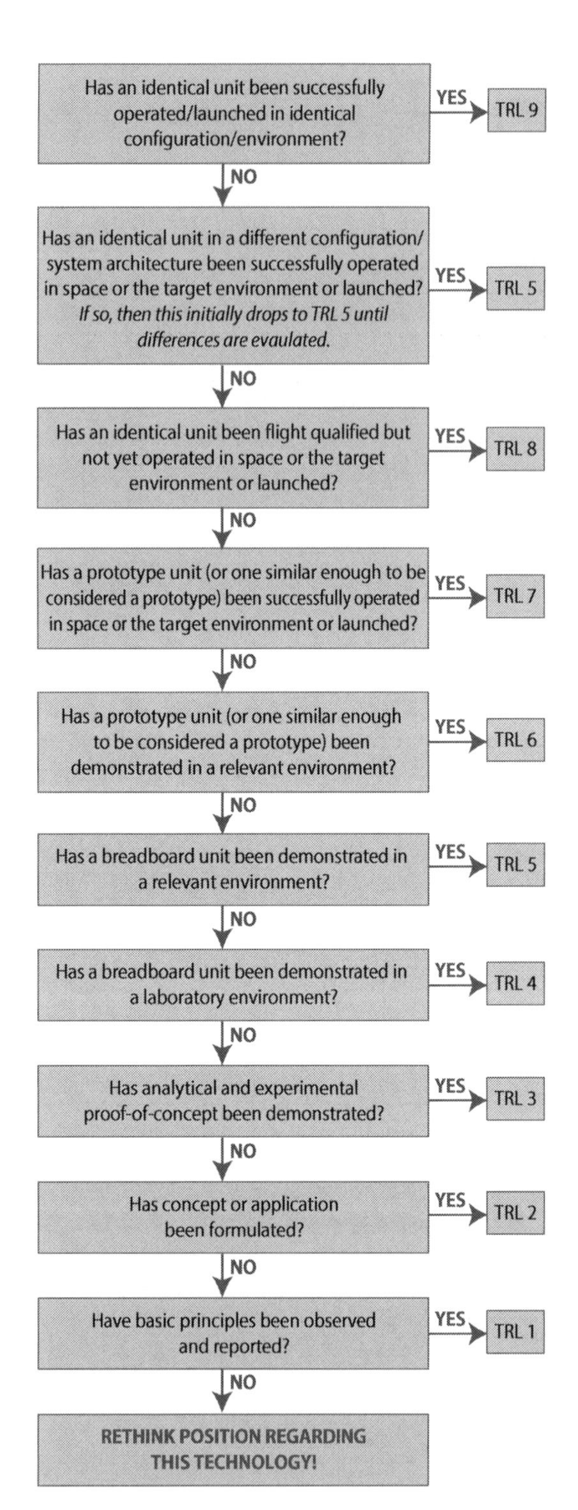

FIGURE G.4-2 TMA Thought Process

Formal Process for Determining TRLs

A method for formalizing this process is shown in FIGURE G.4-3. Here, the process has been set up as a table: the rows identify the systems, subsystems, and components that are under assessment. The columns identify the categories that will be used to determine the TRL; i.e., what units have been built, to what scale, and in what environment have they been tested. Answers to these questions determine the TRL of an item under consideration. The TRL of the system is determined by the lowest TRL present in the system; i.e., a system is at TRL 2 if any single element in the system is at TRL 2. The problem of multiple elements being at low TRLs is dealt with in the AD[2] process. Note that the issue of integration affects the TRL of every system, subsystem, and component. All of the elements can be at a higher TRL, but if they have never been integrated as a unit, the TRL will be lower for the unit. How much lower depends on the complexity of the integration. The assessed complexity depends upon the combined judgment of the engineers. It is important to have a good cross-section of senior people sitting in judgment.

Red = Below TRL 3
Yellow = TRL 3, 4 & 5
Green = TRL 6 and above
White = Unknown
X = Exists

	Demonstration Units						Environment				Unit Description				
	Concept	Breadboard	Brassboard	Developmental Model	Prototype	Flight Qualified	Laboratory Environment	Relevant Environment	Space Environment	Space Launch Operation	Form	Fit	Function	Appropriatae Scale	Overall TRL
1.0 System															
1.1 Subsystem X															
1.1.1 Mechanical Components															
1.1.2 Mechanical Systems															
1.1.3 Electrical Components					X				X		X	X	X		
1.1.4 Electrical Systems															
1.1.5 Control Systems															
1.1.6 Thermal Systems							X				X	X			
1.1.7 Fluid Systems		X													
1.1.8 Optical Systems															
1.1.9 Electro-Optical Systems															
1.1.10 Software Systems															
1.1.11 Mechanisms	X														
1.1.12 Integration															
1.2 Subsystem Y															
1.2.1 Mechanical Components															

FIGURE G.4-3 TRL Assessment Matrix

Appendix H: Integration Plan Outline

H.1 Purpose

The integration plan defines the integration and verification strategies for a project interface with the system design and decomposition into the lower-level elements.[1] The integration plan is structured to bring the elements together to assemble each subsystem and to bring all of the subsystems together to assemble the system/product. The primary purposes of the integration plan are: (1) to describe this coordinated integration effort that supports the implementation strategy, (2) to describe for the participants what needs to be done in each integration step, and (3) to identify the required resources and when and where they will be needed.

H.2 Questions/Checklist

- Does the integration plan include and cover integration of all of the components and subsystems of the project, either developed or purchased?
- Does the integration plan account for all external systems to be integrated with the system (for example, communications networks, field equipment, other complete systems owned by the government or owned by other government agencies)?
- Does the integration plan fully support the implementation strategy, for example, when and where the subsystems and system are to be used?
- Does the integration plan mesh with the verification plan?
- For each integration step, does the integration plan define what components and subsystems are to be integrated?
- For each integration step, does the integration plan identify all the needed participants and define what their roles and responsibilities are?
- Does the integration plan establish the sequence and schedule for every integration step?
- Does the integration plan spell out how integration problems are to be documented and resolved?

H.3 Integration Plan Contents

Title Page

The title page should follow the NASA procedures or style guide. At a minimum, it should contain the following information:

- INTEGRATION PLAN FOR THE *[insert name of project]* AND *[insert name of organization]*
- *Contract number*
- *Date that the document was formally approved*
- *The organization responsible for preparing the document*

1 The material in this appendix is adapted from Federal Highway Administration and CalTrans, *Systems Engineering Guidebook for ITS, Version 2.0.*

- *Internal document control number, if available*
- *Revision version and date issued*

1.0 Purpose of Document

This section gives a brief statement of the purpose of this document. It is the plan for integrating the components and subsystems of the project prior to verification.

2.0 Scope of Project

This section gives a brief description of the planned project and the purpose of the system to be built. Special emphasis is placed on the project's deployment complexities and challenges.

3.0 Integration Strategy

This section tells the reader what the high-level plan for integration is and, most importantly, why the integration plan is structured the way it is. The integration plan is subject to several, sometimes conflicting, constraints. Also, it is one part of the larger process of build, integrate, verify, and deploy, all of which should be synchronized to support the same project strategy. So, for even a moderately complex project, the integration strategy, which is based on a clear and concise statement of the project's goals and objectives, is described here at a high but all-inclusive level. It may also be necessary to describe the analysis of alternative strategies to make it clear why this particular strategy was selected.

The same strategy is the basis for the build plan, the verification plan, and the deployment plan. This section covers and describes each step in the integration process. It describes what components are integrated at each step and gives a general idea of what threads of the operational capabilities (requirements) are covered. It ties the plan to the previously identified goals and objectives so the stakeholders can understand the rationale for each integration step. This summary-level description also defines the schedule for all the integration efforts.

4.0 Phase 1 Integration

This and the following sections define and explain each step in the integration process. The intent here is to identify all the needed participants and to describe to them what they have to do. In general, the description of each integration step should identify the following:

- *The location of the activities.*
- *The project-developed equipment and software products to be integrated. Initially this is just a high-level list, but eventually the list should be exact and complete, showing part numbers and quantity.*
- *Any support equipment (special software, test hardware, software stubs, and drivers to simulate yet-to-be-integrated software components, external systems) needed for this integration step. The same support equipment is most likely needed for the subsequent verification step.*
- *All integration activities that need to be performed after installation, including integration with onsite systems and external systems at other sites.*
- *A description of the verification activities, as defined in the applicable verification plan, that occur after this integration step.*
- *The responsible parties for each activity in the integration step.*
- *The schedule for each activity.*

5.0 Multiple Phase Integration Steps (1 or N steps)

This and any needed additional sections follow the format for Section 3.0. Each covers each step in a multiple-step integration effort.

Appendix I: Verification and Validation Plan Outline

Sample Outline

The Verification and Validation (V&V) Plan needs to be baselined after the comments from PDR are incorporated. In this annotated outline, the use of the term "system" is indicative of the entire scope for which this plan is developed. This may be an entire spacecraft, just the avionics system, or a card within the avionics system. Likewise, the terms "end item," "subsystem," or "element" are meant to imply the lower-level products that, when integrated together, will produce the "system." The general term "end item" is used to encompass activities regardless of whether the end item is a hardware or software element.

The various sections are intended to move from the high-level generic descriptions to the more detailed. The sections also flow from the lower-level items in the product layer to larger and larger assemblies and to the completely integrated system. The sections also describe how that system may be integrated and further verified/validated with its externally interfacing elements. This progression will help build a complete understanding of the overall plans for verification and validation.

1.0 Introduction

1.1 Purpose and Scope

This section states the purpose of this Verification and Validation Plan and the scope (i.e., systems) to which it applies. The purpose of the V&V Plan is to identify the activities that will establish compliance with the requirements (verification) and to establish that the system will meet the customers' expectations (validation).

1.2 Responsibility and Change Authority

This section will identify who has responsibility for the maintenance of this plan and who or what board has the authority to approve any changes to it.

1.3 Definitions

This section will define any key terms used in the plan. The section may include the definitions of verification, validation, analysis, test, demonstration, and test. See appendix B *of this handbook for definitions of these and other terms that might be used.*

2.0 Applicable and Reference Documents

2.1 Applicable Documents

These are the documents that may impose additional requirements or from which some of the requirements have been taken.

2.2 Reference Documents

These are the documents that are referred to within the V&V Plan that do not impose requirements, but which may have additional useful information.

2.3 Order of Precedence

This section identifies which documents take precedence whenever there are conflicting requirements.

3.0 System Description

3.1 System Requirements Flowdown

This section describes where the requirements for this system come from and how they are flowed down to subsystems and lower-level elements. It should also indicate what method will be used to perform the flow-down and bidirectional traceability of the requirements: spreadsheet, model, or other means. It can point to the file, document, or spreadsheet that captures the actual requirements flowdown.

3.2 System Architecture

This section describes the system that is within the scope of this V&V Plan. The description should be enough so that the V&V activities will have the proper context and be understandable.

3.3 End Item Architectures

This section describes each of the major end items (subsystems, elements, units, modules, etc.) that when integrated together, will form the overall system that is the scope of this V&V Plan.

3.3.1 *System End Item A*

This section describes the first major end item/subsystem in more detail so that the V&V activities have context and are understandable.

3.3.n *System End Item n*

Each end item/subsystem is separately described in a similar manner as above.

3.4 Ground Support Equipment

This section describes any major ground-support equipment that will be used during the V&V activities. This may include carts for supplying power or fuel, special test fixtures, lifting aids, simulators, or other type of support.

3.5 Other Architecture Descriptions

This section describes any other items that are important for the V&V activities but which are not included in the sections above. This may be an existing control center, training facility, or other support.

4.0 Verification and Validation Process

This section describes the process that will be used to perform verification and validation.

4.1 Verification and Validation Management Responsibilities

This section describes the responsibilities of key players in the V&V activities. It may include identification and duty description for test directors/conductors, managers, facility owners, boards, and other key stakeholders.

4.2 Verification Methods

This section defines and describes the methods that will be used during the verification activities.

4.2.1 *Analysis*

Defines what this verification method means (See Appendix B of this handbook) and how it will be applied to this system.

4.2.2 *Inspection*

Defines what this verification method means (See Appendix B of this handbook) and how it will be applied to this system.

4.2.3 *Demonstration*

Defines what this verification method means (See Appendix B *of this handbook) and how it will be applied to this system.*

4.2.4 *Test*

Defines what this verification method means (See Appendix B *of this handbook) and how it will be applied to this system. This category may need to be broken down into further categories.*

4.2.4.1 *Qualification Testing*

This section describes the test philosophy for the environmental and other testing that is performed at higher than normal levels to ascertain margins and performance in worst-case scenarios. Includes descriptions of how the minimum and maximum extremes will be determined for various types of tests (thermal, vibration, etc.), whether it will be performed at a component, subsystem, or system level, and the pedigree (flight unit, qualification unit, engineering unit, etc.) of the units these tests will be performed on.

4.2.4.2 *Other Testing*

This section describes any other testing that will be used as part of the verification activities that are not part of the qualification testing. It includes any testing of requirements within the normal operating range of the end item. It may include some engineering tests that will form the foundation or provide dry runs for the official verification testing.

4.3 Validation Methods

This section defines and describes the methods to be used during the validation activities.

4.2.1 *Analysis*

Defines what this validation method means (See Appendix B *of this handbook) and how it will be applied to this system.*

4.2.2 *Inspection*

Defines what this validation method means (See Appendix B *of this handbook) and how it will be applied to this system.*

4.2.3 *Demonstration*

Defines what this validation method means (See Appendix B *of this handbook) and how it will be applied to this system.*

4.2.4 *Test*

Defines what this validation method means (See Appendix B *of this handbook) and how it will be applied to this system. This category may need to be broken down into further categories such as end-to-end testing, testing with humans, etc.)*

4.4 Certification Process

Describes the overall process by which the results of these verification and validation activities will be used to certify that the system meets its requirements and expectations and is ready to be put into the field or fly. In addition to the verification and validation results, the certification package may also include special forms, reports, safety documentation, drawings, waivers, or other supporting documentation.

4.5 Acceptance Testing

Describes the philosophy of how/which of the verification/validation activities will be performed on each of the operational units as they are manufactured/coded and are readied for flight/use. Includes how/if data packages will be developed and provided as part of the delivery.

5.0 Verification and Validation Implementation

5.1 System Design and Verification and Validation Flow

This section describes how the system units/modules will flow from manufacturing/coding through verification and validation. Includes whether each unit will be verified/validated separately, or assembled to some level and then evaluated or other statement of flow.

5.2 Test Articles

This section describes the pedigree of test articles that will be involved in the verification/validation activities. This may include descriptions of breadboards, prototypes, engineering units, qualification units, protoflight units, flight units, or other specially named units. A definition of what is meant by these terms needs to be included to ensure clear understanding of the expected pedigree of each type of test article. Descriptions of what kind of test/analysis activities will be performed on each type of test article is included.

5.3 Support Equipment

This section describes any special support equipment that will be needed to perform the verification/validation activities. This will be a more detailed description than is stated in Section 3.4 of this outline.

5.4 Facilities

This section identifies and describes major facilities that will be needed in order to accomplish the verification and validation activities. These may include environmental test facilities, computational facilities, simulation facilities, training facilities, test stands, and other facilities as needed.

6.0 End Item Verification and Validation

This section describes in detail the V&V activities that will be applied to the lower-level subsystems/elements/end items. It can point to other stand-alone descriptions of these tests if they will be generated as part of organizational responsibilities for the products at each product layer.

6.1 End Item A

This section focuses in on one of the lower-level end items and describes in detail what type of verification activities it will undergo.

6.1.1 Developmental/Engineering Unit Evaluations

This section describes what kind of testing, analysis, demonstrations, or inspections the prototype/engineering or other types of units/modules will undergo prior to performing official verification and validation.

6.1.2 Verification Activities

This section describes in detail the verification activities that will be performed on this end item.

6.1.2.1 Verification by Testing

This section describes all verification testing that will be performed on this end item.

6.1.2.1.1 Qualification Testing

This section describes the test environmental and other testing that is performed at higher than normal levels to ascertain margins and performance in worst-case scenarios. It includes what minimum and maximum extremes will be used on qualification tests (thermal, vibration, etc.) of this unit, whether it will be performed at a component, subsystem, or system level, and the pedigree (flight unit, qualification unit, engineering unit, etc.) of the units these tests will be performed on.

6.1.2.1.2 Other Testing

This section describes all other verification tests that are not performed as part of the qualification testing. These will include verification of requirements in the normal operating ranges.

6.1.2.2 Verification by Analysis

This section describes the verifications that will be performed by analysis (including verification by similarity). This may include thermal analysis, stress analysis, analysis of fracture control, materials analysis, Electrical, Electronic, and Electromechanical (EEE) parts analysis, and other analyses as needed for the verification of this end item.

6.1.2.3 *Verification by Inspection*

This section describes the verifications that will be performed for this end item by inspection.

6.1.2.4 *Verification Demonstration*

This section describes the verifications that will be performed for this end item by demonstration.

6.1.3 Validation Activities

6.1.3.1 *Validation by Testing*

This section describes what validation tests will be performed on this end item.

6.1.3.2 *Validation by Analysis*

This section describes the validation that will be performed for this end item through analysis.

6.1.3.3 *Validation by Inspection*

This section describes the validation that will be performed for this end item through inspection.

6.1.3.4 *Validation by Demonstration*

This section describes the validations that will be performed for this end item by demonstration.

6.1.4 Acceptance Testing

This section describes the set of tests, analysis, demonstrations, or inspections that will be performed on the flight/final version of the end item to show it has the same design as the one that is being verified, that the workmanship on this end item is good, and that it performs the identified functions properly.

6.n End Item n

In a similar manner as above, a description of how each end item that makes up the system will be verified and validated is made.

7.0 System Verification and Validation

7.1 End-Item Integration

This section describes how the various end items will be assembled/integrated together, verified and validated. For example, the avionics and power systems may be integrated and tested together to ensure their interfaces and performance is as required and expected prior to integration with a larger element. This section describes the verification and validation that will be performed on these major assemblies. Complete system integration will be described in later sections.

7.1.1 Developmental/Engineering Unit Evaluations

This section describes the unofficial (not the formal verification/validation) testing/analysis that will be performed on the various assemblies that will be tested together and the pedigree of the units that will be used. This may include system-level testing of configurations using engineering units, breadboard, simulators, or other forms or combination of forms.

7.1.2 Verification Activities

This section describes the verification activities that will be performed on the various assemblies.

7.1.2.1 *Verification by Testing*

This section describes all verification testing that will be performed on the various assemblies. The section may be broken up to describe qualification testing performed on the various assemblies and other types of testing.

7.1.2.2 *Verification by Analysis*

This section describes all verification analysis that will be performed on the various assemblies.

7.1.2.3 *Verification by Inspection*

This section describes all verification inspections that will be performed on the various assemblies.

7.1.2.4 *Verification by Demonstration*

This section describes all verification demonstrations that will be performed on the various assemblies.

7.1.3 Validation Activities

7.1.3.1 *Validation by Testing*

This section describes all validation testing that will be performed on the various assemblies.

7.1.3.2 *Validation by Analysis*

This section describes all validation analysis that will be performed on the various assemblies.

7.1.3.3 *Validation by Inspection*

This section describes all validation inspections that will be performed on the various assemblies.

7.1.3.4 *Validation by Demonstration*

This section describes all validation demonstrations that will be performed on the various assemblies.

7.2 Complete System Integration

This section describes the verification and validation activities that will be performed on the systems after all its assemblies are integrated together to form the complete integrated system. In some cases this will not be practical. Rationale for what cannot be done should be captured.

7.2.1 Developmental/Engineering Unit Evaluations

This section describes the unofficial (not the formal verification/validation) testing/analysis that will be performed on the complete integrated system and the pedigree of the units that will be used. This may include system-level testing of configurations using engineering units, breadboard, simulators, or other forms or combination of forms.

7.2.2 Verification Activities

This section describes the verification activities that will be performed on the completely integrated system

7.2.2.1 *Verification Testing*

This section describes all verification testing that will be performed on the integrated system. The section may be broken up to describe qualification testing performed at the integrated system level and other types of testing.

7.2.2.2 *Verification Analysis*

This section describes all verification analysis that will be performed on the integrated system.

7.2.2.3 *Verification Inspection*

This section describes all verification inspections that will be performed on the integrated system.

7.2.2.4 *Verification Demonstration*

This section describes all verification demonstrations that will be performed on the integrated system.

7.2.3 Validation Activities

This section describes the validation activities that will be performed on the completely integrated system.

7.2.3.1 *Validation by Testing*

This section describes all validation testing that will be performed on the integrated system.

7.2.3.2 *Validation by Analysis*

This section describes all validation analysis that will be performed on the integrated system.

7.2.3.3 *Validation by Inspection*

This section describes the validation inspections that will be performed on the integrated system.

7.2.3.4 *Validation by Demonstration*

This section describes the validation demonstrations that will be performed on the integrated system.

8.0 Program Verification and Validation

This section describes any further testing that the system will be subjected to. For example, if the system is an instrument, the section may include any verification/validation that the system will undergo when integrated into its spacecraft/platform. If the system is a spacecraft, the section may include any verification/validation the system will undergo when integrated with its launch vehicle.

8.1 Vehicle Integration

This section describes any further verification or validation activities that will occur when the system is integrated with its external interfaces.

8.2 End-to-End Integration

This section describes any end-to-end testing that the system may undergo. For example, this configuration would include data being sent from a ground control center through one or more relay satellites to the system and back.

8.3 On-Orbit V&V Activities

This section describes any remaining verification/validation activities that will be performed on a system after it reaches orbit or is placed in the field.

9.0 System Certification Products

This section describes the type of products that will be generated and provided as part of the certification process. This package may include the verification and validation matrix and results, pressure vessel certifications, special forms, materials certifications, test reports or other products as is appropriate for the system being verified and validated.

Appendix A: Acronyms and Abbreviations

This is a list of all the acronyms and abbreviations used in the V&V Plan and their spelled-out meaning.

Appendix B: Definition of Terms

This section is a definition of the key terms that are used in the V&V Plan.

Appendix C: Requirement Verification Matrix

The V&V Plan needs to be baselined after the comments from PDR are incorporated. The information in this section may take various forms. It could be a pointer to another document or model where the matrix and its results may be found. This works well for large projects using a requirements-tracking application. The information in this section could also be the requirements matrix filled out with all but the results information and a pointer to where the results can be found. This allows the key information to be available at the time of baselining. For a smaller project, this may be the completed verification matrix. In this case, the V&V Plan would be filled out as much as possible before. See Appendix D *for an example of a verification matrix.*

Appendix D: Validation Matrix

As with the verification matrix, this product may take various forms from a completed matrix to just a pointer for where the information can be found. Appendix E *provides an example of a validation matrix.*

Appendix J: SEMP Content Outline

J.1 SEMP Content

The Systems Engineering Management Plan (SEMP) is the foundation document for the technical and engineering activities conducted during the project. The SEMP conveys information to all of the personnel on the technical integration methodologies and activities for the project within the scope of the project plan. SEMP content can exist as a stand-alone document or, for smaller projects, in higher-level project documentation.

The SEMP provides the specifics of the technical effort and describes what technical processes will be used, how the processes will be applied using appropriate activities, how the project will be organized to accomplish the activities, and the resources required for accomplishing the activities. The SEMP provides the framework for realizing the appropriate work products that meet the entry and success criteria of the applicable project life cycle phases to provide management with necessary information for assessing technical progress.

Because the SEMP provides the specific technical and management information to understand the technical integration and interfaces, its documentation and approval serve as an agreement within the project of how the technical work will be conducted. The SEMP communicates to the team itself, managers, customers, and other stakeholders the technical effort that will be performed by the assigned technical team.

The technical team, working under the overall program/project plan, develops and updates the SEMP as necessary. The technical team works with the project manager to review the content and obtain concurrence. The SEMP includes the following three general sections:

1. Technical program planning and control, which describe the processes for planning and control of the engineering efforts for the design, development, test, and evaluation of the system.

2. Systems engineering processes, which include specific tailoring of the systems engineering process as described in the NPR, implementation procedures, trade study methodologies, tools, and models to be used.

3. Engineering specialty integration describes the integration of the technical disciplines' efforts into the systems engineering process and summarizes each technical discipline effort and cross references each of the specific and relevant plans.

The SEMP outline in this appendix is guidance to be used in preparing a stand-alone project SEMP. The level of detail in the project SEMP should be adapted based on the size of the project. For a small project, the material in the SEMP can be placed in the project plan's technical summary, and this annotated outline should be used as a topic guide.

Some additional important points on the SEMP:

- The SEMP is a living document. The initial SEMP is used to establish the technical content of the engineering work early in the Formulation Phase for each project and updated as needed throughout the project life cycle. Table J-1 provides some high level guidance on the scope of SEMP content based on the life cycle phase.
- Project requirements that have been tailored or significant customization of SE processes should be described in the SEMP.
- For multi-level projects, the SEMP should be consistent with higher-level SEMPs and the project plan.
- For a technical effort that is contracted, the SEMP should include details on developing requirements for source selection, monitoring performance, and transferring and integrating externally produced products to NASA.

J.2 Terms Used

Terms used in the SEMP should have the same meaning as the terms used in the NPR 7123.1, Systems Engineering Processes and Requirements.

J.3 Annotated Outline

Title Page

Systems Engineering Management Plan

(Provide a title for the candidate program/project and designate a short title or proposed acronym in parenthesis, if appropriate.)

.

.

.

Designated Governing Authority/Technical Authority	Date
Program/Project Manager	Date
Chief Engineer	Date
	Date
	Date

By signing this document, signatories are certifying that the content herein is acceptable as direction for engineering and technical management of this program/project and that they will ensure its implementation by those over whom they have authority.

1.0 Purpose and Scope

This section provides a brief description of the purpose, scope, and content of the SEMP.

- Purpose: *This section should highlight the intent of the SEMP to provide the basis for implementing and communicating the technical effort.*
- Scope: *The scope describes the work that encompasses the SE technical effort required to generate the work products. The plan is used by the technical team to provide personnel the information necessary to successfully accomplish the required task.*
- Content: *This section should briefly describe the organization of the document.*

2.0 Applicable Documents

This section of the SEMP lists the documents applicable to this specific project and its SEMP implementation. This section should list major standards and procedures that this technical effort for this specific project needs to follow. Examples of specific procedures to list could include procedures for hazardous material handling, crew training plans for control room operations, special instrumentation techniques, special interface documentation for vehicles, and maintenance procedures specific to the project.

3.0 Technical Summary

This section contains an executive summary describing the problem to be solved by this technical effort and the purpose, context, and products to be developed and integrated with other interfacing systems identified.

Key Questions

1. *What is the problem we're trying to solve?*
2. *What are the influencing factors?*
3. *What are the critical questions?*
4. *What are the overall project constraints in terms of cost, schedule, and technical performance*
5. *How will we know when we have adequately defined the problem?*
6. *Who are the customers?*
7. *Who are the users?*
8. *What are the customer and user priorities?*
9. *What is the relationship to other projects?*

3.1 *System Description*

This section contains a definition of the purpose of the system being developed and a brief description of the purpose of the products of the product layer of the system structure to which this SEMP applies. Each product layer includes the system end products and their subsystems and the supporting or enabling products and any other work products (plans, baselines) required for the development of the system. The description should include any interfacing systems and system products, including humans with which the system products will interact physically, cognitively, functionally, or electronically.

3.2 *System Structure*

This section contains an explanation of how the technical portion of the product layer (including enabling products, technical cost, and technical schedule) will be developed, how the resulting product layers will be integrated into the project portion of the WBS, and how the overall system structure will be developed. This section contains a description of the relationship of the specification tree and the drawing tree with the products of the system structure and how the relationship and interfaces of the system end products and their life cycle-enabling products will be managed throughout the planned technical effort.

3.3 *Product Integration*

This section contains an explanation of how the products will be integrated and describes clear organizational responsibilities and interdependencies and whether the organizations are geographically dispersed or managed across Centers. This section should also address how products created under a diverse set of contracts are to be integrated, including roles and responsibilities. This includes identifying organizations—intra-and inter-NASA, other Government agencies, contractors, or other partners—and delineating their roles and responsibilities. Product integration includes the integration of analytical products.

When components or elements will be available for integration needs to be clearly understood and identified on the schedule to establish critical schedule issues.

3.4 *Planning Context*

This section contains the programmatic constraints (e.g., NPR 7120.5) that affect the planning and implementation of the common technical processes to be applied in performing the technical effort. The constraints provide a linkage of the technical effort with the applicable product life cycle phases covered by the SEMP including, as applicable, milestone decision gates, major technical reviews, key intermediate events leading to project completion, life cycle phase, event entry and success criteria, and major baseline and other work products to be delivered to the sponsor or customer of the technical effort.

3.5 *Boundary of Technical Effort*

This section contains a description of the boundary of the general problem to be solved by the technical effort, including technical and project constraints that affect the planning. Specifically, it identifies what can be controlled by the technical team (inside the boundary) and what influences the technical effort and is influenced by the technical effort but not controlled by the technical team (outside the boundary). Specific attention should be given to physical, cognitive, functional, and electronic interfaces across the boundary.

A description of the boundary of the system can include the following:

- *Definition of internal and external elements/items involved in realizing the system purpose as well as the system boundaries in terms of space, time, physical, and operational.*
- *Identification of what initiates the transitions of the system to operational status and what initiates its disposal is important. General and functional descriptions of the subsystems inside the boundary.*
- *Current and established subsystem performance characteristics.*
- *Interfaces and interface characteristics.*
- *Functional interface descriptions and functional flow diagrams.*
- *Key performance interface characteristics.*
- *Current integration strategies and architecture.*
- *Documented Human System Integration Plan (HSIP)*

3.6 *Cross References*

This section contains cross references to appropriate nontechnical plans and critical reference material that interface with the technical effort. It contains a summary description of how the technical activities covered in other plans are accomplished as fully integrated parts of the technical effort.

4.0 Technical Effort Integration

This section describes how the various inputs to the technical effort will be integrated into a coordinated effort that meets cost, schedule, and performance objectives.

The section should describe the integration and coordination of the specialty engineering disciplines into the systems engineering process during each iteration of the processes. Where there is potential for overlap of specialty efforts, the SEMP should define the relative responsibilities and authorities of each specialty. This section should contain, as needed, the project's approach to the following:

- *Concurrent engineering*
- *The activity phasing of specialty engineering*
- *The participation of specialty disciplines*
- *The involvement of specialty disciplines,*
- *The role and responsibility of specialty disciplines,*
- *The participation of specialty disciplines in system decomposition and definition*
- *The role of specialty disciplines in verification and validation*
- *Reliability*
- *Maintainability*
- *Quality assurance*
- *Integrated logistics*
- *Human engineering*
- *Safety*
- *Producibility*
- *Survivability/vulnerability*
- *National Environmental Policy Act (NEPA) compliance*
- *Launch approval/flight readiness*

The approach for coordination of diverse technical disciplines and integration of the development tasks should be described. For example, this can include the use of multidiscipline integrated teaming approaches—e.g., an HSI team—or specialized control boards. The scope and timing of the specialty engineering tasks should be described along with how specialty engineering disciplines are represented on all technical teams and during all life cycle phases of the project.

4.1 *Responsibility and Authority*

This section describes the organizing structure for the technical teams assigned to this technical effort and includes how the teams will be staffed and managed.

Key Questions

1. *What organization/panel will serve as the designated governing authority for this project?*
2. *How will multidisciplinary teamwork be achieved?*
3. *What are the roles, responsibilities, and authorities required to perform the activities of each planned common technical process?*
4. *What should be the planned technical staffing by discipline and expertise level?*
5. *What is required for technical staff training?*
6. *How will the assignment of roles, responsibilities, and authorities to appropriate project stakeholders or technical teams be accomplished?*
7. *How are we going to structure the project to enable this problem to be solved on schedule and within cost?*
8. *What does systems engineering management bring to the table?*

The section should provide an organization chart and denote who on the team is responsible for each activity. It should indicate the lines of authority and responsibility. It should define the resolution authority to make decisions/decision process. It should show how the engineers/engineering disciplines relate.

The systems engineering roles and responsibilities need to be addressed for the following: project office, user, Contracting Officer's Representative (COR), systems engineering, design engineering, specialty engineering, and contractor.

4.2 Contractor Integration

This section describes how the technical effort of in-house and external contractors is to be integrated with the NASA technical team efforts. The established technical agreements should be described along with how contractor progress will be monitored against the agreement, how technical work or product requirement change requests will be handled, and how deliverables will be accepted. The section specifically addresses how interfaces between the NASA technical team and the contractor will be implemented for each of the 17 common technical processes. For example, it addresses how the NASA technical team will be involved with reviewing or controlling contractor-generated design solution definition documentation or how the technical team will be involved with product verification and product validation activities.

Key deliverables for the contractor to complete their systems and those required of the contractor for other project participants need to be identified and established on the schedule.

4.3 Analytical Tools that Support Integration

This section describes the methods (such as integrated computer-aided tool sets, integrated work product databases, and technical management information systems) that will be used to support technical effort integration.

5.0 Common Technical Processes Implementation

Each of the 17 common technical processes will have a separate subsection that contains a plan for performing the required process activities as appropriately tailored. (See NPR 7123.1 for the process activities required and tailoring.) Implementation of the 17 common technical processes includes (1) the generation of the outcomes needed to satisfy the entry and success criteria of the applicable product life cycle phase or phases identified in D.4.4.4, and (2) the necessary inputs for other technical processes. These sections contain a description of the approach, methods, and tools for:

- *Identifying and obtaining adequate human and nonhuman resources for performing the planned process, developing the work products, and providing the services of the process.*
- *Assigning responsibility and authority for performing the planned process (e.g., RACI matrix, [*http://en.wikipedia.org/wiki/Responsibility_assignment_matrix*]), developing the work products, and providing the services of the process.*
- *Training the technical staff performing or supporting the process, where training is identified as needed.*
- *Designating and placing designated work products of the process under appropriate levels of configuration management.*
- *Identifying and involving stakeholders of the process.*
- *Monitoring and controlling the systems engineering processes.*
- *Identifying, defining, and tracking metrics and success.*
- *Objectively evaluating adherence of the process and the work products and services of the process to the applicable requirements, objectives, and standards and addressing noncompliance.*
- *Reviewing activities, status, and results of the process with appropriate levels of management and resolving issues.*

This section should also include the project-specific description of each of the 17 processes to be used, including the specific tailoring of the requirements to the system and the project; the procedures to be used in implementing the processes; in-house documentation; trade study methodology; types of mathematical and/or simulation models to be used; and generation of specifications.

Key Questions

1. *What are the systems engineering processes for this project?*
2. *What are the methods that we will apply for each systems engineering task?*
3. *What are the tools we will use to support these methods? How will the tools be integrated?*
4. *How will we control configuration development?*
5. *How and when will we conduct technical reviews?*
6. *How will we establish the need for and manage trade-off studies?*
7. *Who has authorization for technical change control?*
8. *How will we manage requirements? interfaces? documentation?*

6.0 Technology Insertion

This section describes the approach and methods for identifying key technologies and their associated risks and criteria for assessing and inserting technologies, including those for inserting critical technologies from technology development projects. An approach should be developed for appropriate level and timing of technology insertion. This could include alternative approaches to take advantage of new technologies to meet systems needs as well as alternative options if the technologies do not prove appropriate in result or timing. The strategy for an initial technology assessment within the scope of the project requirements should be provided to identify technology constraints for the system.

Key Questions

1. *How and when will we insert new of special technology into the project?*
2. *What is the relationship to research and development efforts? How will they support the project? How will the results be incorporated?*
3. *How will we incorporate system elements provided by others? How will these items be certified for adequacy?*
4. *What facilities are required?*
5. *When and how will these items be transitioned to be part of the configuration?*

7.0 Additional SE Functions and Activities

This section describes other areas not specifically included in previous sections but that are essential for proper planning and conduct of the overall technical effort.

7.1 *System Safety*

This section describes the approach and methods for conducting safety analysis and assessing the risk to operators, the system, the environment, or the public.

7.2 *Engineering Methods and Tools*

This section describes the methods and tools that are not included in the technology insertion sections but are needed to support the overall technical effort. It identifies those tools to be acquired and tool training requirements.

This section defines the development environment for the project, including automation, simulation, and software tools. If required, it describes the tools and facilities that need to be developed or acquired for all disciplines on the project. It describes important enabling strategies such as standardizing tools across the project, or utilizing a common input and output format to support a broad range of tools used on the project. It defines the requirements for information management systems and for using existing elements. It defines and plans for the training required to use the tools and technology across the project.

7.3 Specialty Engineering

This section describes engineering discipline and specialty requirements that apply across projects and the WBS models of the system structure. Examples of these requirement areas would include planning for safety, reliability, human factors, logistics, maintainability, quality, operability, and supportability. It includes estimates of staffing levels for these disciplines and incorporates them with the project requirements.

7.4 Technical Performance Measures

a. *This section describes the TPMs that have been derived from the MOEs and MOPs for the project. The TPMs are used to define and track the technical progress of the systems engineering effort. (The unique identification numbers in red reference the corresponding requirement in NPR 7123.1.) The performance metrics need to address the minimally required TPMs as defined in NPR 7123.1. These include:*

   1. *Mass margins for projects involving hardware* [SE-62].

   2. *Power margins for projects that are powered* [SE-63].

   3. *Review Trends including closure of review action documentation (Request for Action, Review Item Discrepancies, and/or Action Items as established by the project) for all software and hardware projects* [SE-64].

b. Other performance measure that should be considered by the project include:

   » *Requirement trends (percent growth, TBD/TBR closures, number of requirement changes);*
   » *Interface trends (percent ICD approval, TBD/TBR burndown, number of interface requirement changes);*
   » *Verification trends (closure burndown, number of deviations/waivers approved/open);*
   » *Software-unique trends (number of requirements per build/release versus plan);*
   » *Problem report/discrepancy report trends (number open, number closed);*
   » *Cost trends (plan, actual, UFE, EVM, NOA);*
   » *Schedule trends (critical path slack/float, critical milestone dates); and*
   » *Staffing trends (FTE, WYE).*

Key Questions

1. *What metrics will be used to measure technical progress?*
2. *What metrics will be used to identify process improvement opportunities?*
3. *How will we measure progress against the plans and schedules?*
4. *How often will progress be reported? By whom? To whom?*

7.5 Heritage

This section describes the heritage or legacy products that will be used in the project. It should include a discussion of which products are planned to be used, the rationale for their use, and the analysis or testing needed to assure they will perform as intended in the stated use.

7.6 *Other*

This section is reserved to describe any unique SE functions or activities for the project that are not covered in other sections.

8.0 Integration with the Project Plan and Technical Resource Allocation

This section describes how the technical effort will integrate with project management and defines roles and responsibilities. It addresses how technical requirements will be integrated with the project plan to determine the allocation of resources, including cost, schedule, and personnel, and how changes to the allocations will be coordinated.

Key Questions

1. *How will we assess risk? What thresholds are needed for triggering mitigation activities? How will we integrate risk management into the technical decision process?*
2. *How will we communicate across and outside of the project?*
3. *How will we record decisions?*
4. *How do we incorporate lessons learned from other projects?*

This section describes the interface between all of the technical aspects of the project and the overall project management process during the systems engineering planning activities and updates. All activities to coordinate technical efforts with the overall project are included, such as technical interactions with the external stakeholders, users, and contractors.

9.0 Compliance Matrices

Appendix H.2 in NPR 7123.1A is the basis for the compliance matrix for this section of the SEMP. The project will complete this matrix from the point of view of the project and the technical scope. Each requirement will be addressed as compliant, partially compliant, or noncompliant. Compliant requirements should indicate which process or activity addresses the compliance. For example, compliance can be accomplished by using a Center process or by using a project process as described in another section of the SEMP or by reference to another documented process. Noncompliant areas should state the rationale for noncompliance.

Appendices

Appendices are included, as necessary, to provide a glossary, acronyms and abbreviations, and information published separately for convenience in document maintenance. Included are: (1) information that may be pertinent to multiple topic areas (e.g., description of methods or procedures); (2) charts and proprietary data applicable to the technical efforts required in the SEMP; and (3) a summary of technical plans associated with the project. Each appendix should be referenced in one of the sections of the engineering plan where data would normally have been provided.

Templates

Any templates for forms, plans, or reports the technical team will need to fill out, like the format for the verification and validation plan, should be included in the appendices.

References

This section contains all documents referenced in the text of the SEMP.

TABLE J-1 Guidance on SEMP Content per Life-Cycle Phase

SEMP Section	SEMP Subsection	Pre-Phase A KDP A	Phase A KDP B		Phase B KDP C	Phase C KDP D		Phase D KDP E		Phase E KDP F	Phase F
		MCR	SRR	SDR/MDR	PDR	CDR	SIR	ORR	MRR/FRR	DR	DRR
Purpose and Scope		Final	Final	Final	Final	Final	Final	Final	Final	Final	Final
Applicable Documents		Initial	Initial	Initial	Final	Final	Final	Final	Final	Final	Final
Technical Summary		Final	Final	Final	Final	Final	Final	Final	Final	Final	Final
System Description		Initial	Initial	Initial	Final	Final	Final	Final	Final	Final	Final
System Structure	Product Integration	Define thru SDR timeframe	Define thru SDR timeframe	Define thru SDR timeframe	Define thru SIR	Define thru SIR	Define thru SIR	Define sustaining thru end of program	Define sustaining thru end of program	Define sustaining thru end of program	Define sustaining thru end of program
	Planning Context	Define thru SDR timeframe	Define thru SDR timeframe	Define thru SDR timeframe	Define thru SIR	Define thru SIR	Define thru SIR	Define sustaining thru end of program	Define sustaining thru end of program	Define sustaining thru end of program	Define sustaining thru end of program
	Boundary of Technical Effort	Initial	Initial	Initial	Final	Final	Final	Final	Final	Final	Final
	Cross References	Initial	Initial	Initial	Final	Final	Final	Final	Final	Final	Final
Technical Effort Integration	Responsibility and Authority	Define thru SDR timeframe	Define thru SDR timeframe	Define thru SDR timeframe	Define thru SIR timeframe	Define thru SIR timeframe	Define thru SIR timeframe	Define sustaining Roles and Responsibilities through end of program	Define sustaining Roles and Responsibilities through end of program	Define sustaining Roles and Responsibilities through end of program	
	Contractor Integration	Define acquisitions needed		Define insight/oversight through SIR timeframe				Define sustaining insight/oversight through end of program			
	Support Integration	Define acquisitions needed		Define insight/oversight through SIR timeframe				Define sustaining insight/oversight through end of program			

(continued)

SEMP Section	SEMP Subsection	Pre-Phase A KDP A	Phase A KDP B		Phase B KDP C	Phase C KDP D		Phase D KDP E		Phase E KDP F	Phase F
		MCR	SRR	SDR/MDR	PDR	CDR	SIR	ORR	MRR/FRR	DR	DRR
Common Technical Processes Implementation		Processes defined for Concept Development and Formulation		Processes defined for the Design Phase		Processes added for the integration and Operations Phase		Update Operations processes. Define close out processes and sustaining engineering processes			
Technology Insertion		Define technologies to be developed		Define decision process for on ramps and off ramps of technology efforts				Define technology sustaining effort through end of program.			
Additional SE Functions and Activities	System Safety	Define process through CDR						Define sustaining Roles and Responsibilities through end of program			
	Engineering Methods and tools	Define process through CDR						Define sustaining Roles and Responsibilities through end of program			
	Specialty Engineering	Define process through CDR						Define sustaining Roles and Responsibilities through end of program			
Integration with the Project Plan and Technical Resource Allocation		Define through SDR timeframe			Define through SIR	Define through SIR	Define through SIR	Define sustaining through end of program	Define sustaining through end of program	Define sustaining through end of program	Define sustaining through end of program
Compliance Matrix (Appendix H.2 of SE NPR)		Initial	Initial	Initial	Final	Final	Final	Final	Final	Final	Final
Appendices		As required	As required	As required	As required	As required	As required	As required	As required	As required	As required
Templates		As required	As required	As required	As required	As required	As required	As required	As required	As required	As required
References		As required	As required	As required	As required	As required	As required	As required	As required	As required	As required

Appendix K: Technical Plans

The following table represents a typical expectation of maturity of some of the key technical plans developed during the SE processes. This example is for a space flight project. Requirements for work product maturity can be found in the governing PM document (i.e., NPR 7120.5) for the associated type of project.

TABLE K-1 Example of Expected Maturity of Key Technical Plans

Plan	Pre-Phase A	Phase A		Phase B	Phase C		Phase D		Phase E	Phase F	Ref. Page
	MCR	SRR	SDR/MDR	PDR	CDR	SIR	ORR	MRR/FRR	DR	DRR	
Systems Engineering Management Plan	P	B	U	U	U	U	U	U	U	U	
Risk Management Plan	A	B	U	U	U						
Integrated Logistics Support Plan	A	P	P	B	U						
Technology Development Plan	B	U	U	U							
Review Plan	P	B	U	U	U	U	U	U	U	U	
Verification and Validation Plan	A	A	P	B	U						
Integration Plan			P	B	U						
Configuration Management Plan		B	U	U							
Data Management Plan		B	U	U							
Human Systems Integration Plan		B	U	U	U						
Software Management Plan		P	B	U							
Reliability and Maintainability Plan			P	B	U						
Mission Operations Plan						P	B	U			
Project Protection Plan			P	B	U	U	U	U	U	U	
Decommissioning Plan			A					B	U		
Disposal Plan			A					B	U	U	

A = Approach B = Baseline P = Preliminary U = Update

Appendix L: Interface Requirements Document Outline

1.0 Introduction

1.1 Purpose and Scope

State the purpose of this document and briefly identify the interface to be defined. (For example, "This IRD defines and controls the interface(s) requirements between ______ and ______.")

1.2 Precedence

Define the relationship of this document to other program documents and specify which is controlling in the event of a conflict.

1.3 Responsibility and Change Authority

State the responsibilities of the interfacing organizations for development of this document and its contents. Define document approval authority (including change approval authority).

2.0 Documents

2.1 Applicable Documents

List binding documents that are invoked to the extent specified in this IRD. The latest revision or most recent version should be listed. Documents and requirements imposed by higher-level documents (higher order of precedence) should not be repeated.

2.2 Reference Documents

List any document that is referenced in the text in this subsection.

3.0 Interfaces

3.1 General

In the subsections that follow, provide the detailed description, responsibilities, coordinate systems, and numerical requirements as they relate to the interface plane.

3.1.1 *Interface Description*

Describe the interface as defined in the system specification. Use tables, figures, or drawings as appropriate.

3.1.2 *Interface Responsibilities*

Define interface hardware and interface boundary responsibilities to depict the interface plane. Use tables, figures, or drawings as appropriate.

3.1.3 *Coordinate Systems*

Define the coordinate system used for interface requirements on each side of the interface. Use tables, figures, or drawings as appropriate.

3.1.4 *Engineering Units, Tolerances, and Conversion*

Define the measurement units along with tolerances. If required, define the conversion between measurement systems.

3.2 Interface Requirements

In the subsections that follow, define structural limiting values at the interface, such as interface loads, forcing functions, and dynamic conditions. Define the interface requirements on each side of the interface plane.

3.2.1 Mass Properties

Define the derived interface requirements based on the allocated requirements contained in the applicable specification pertaining to that side of the interface. For example, this subsection should cover the mass of the element.

3.2.2 Structural/Mechanical

Define the derived interface requirements based on the allocated requirements contained in the applicable specification pertaining to that side of the interface. For example, this subsection should cover attachment, stiffness, latching, and mechanisms.

3.2.3 Fluid

Define the derived interface requirements based on the allocated requirements contained in the applicable specification pertaining to that side of the interface. For example, this subsection should cover fluid areas such as thermal control, O_2 and N_2, potable and waste water, fuel cell water, and atmospheric sampling.

3.2.4 Electrical (Power)

Define the derived interface requirements based on the allocated requirements contained in the applicable specification pertaining to that side of the interface. For example, this subsection should cover various electric current, voltage, wattage, and resistance levels.

3.2.5 Electronic (Signal)

Define the derived interface requirements based on the allocated requirements contained in the applicable specification pertaining to that side of the interface. For example, this subsection should cover various signal types such as audio, video, command data handling, and navigation.

3.2.6 Software and Data

Define the derived interface requirements based on the allocated requirements contained in the applicable specification pertaining to that side of the interface. For example, this subsection should cover various data standards, message timing, protocols, error detection/correction, functions, initialization, and status.

3.2.7 Environments

Define the derived interface requirements based on the allocated requirements contained in the applicable specification pertaining to that side of the interface. For example, cover the dynamic envelope measures of the element in English units or the metric equivalent on this side of the interface.

3.2.7.1 *Electromagnetic Effects*

3.2.7.1.a Electromagnetic Compatibility

Define the appropriate electromagnetic compatibility requirements. For example, the end-item-1-to-end-item-2 interface shall meet the requirements [to be determined] of systems requirements for electromagnetic compatibility.

3.2.7.1.b Electromagnetic Interference

Define the appropriate electromagnetic interference requirements. For example, the end-item-1-to-end-item-2 interface shall meet the requirements [to be determined] of electromagnetic emission and susceptibility requirements for electromagnetic compatibility.

3.2.7.1.c Grounding

Define the appropriate grounding requirements. For example, the end-item-1-to-end-item-2 interface shall meet the requirements [to be determined] of grounding requirements.

3.2.7.1.d Bonding

Define the appropriate bonding requirements. For example, the end-item-1-to-end-item-2 structural/mechanical interface shall meet the requirements [to be determined] of electrical bonding requirements.

3.2.7.1.e Cable and Wire Design

Define the appropriate cable and wire design requirements. For example, the end-item-1-to-end-item-2 cable and wire interface shall meet the requirements [to be determined] of cable/wire design and control requirements for electromagnetic compatibility.

3.2.7.2 *Acoustic*

Define the appropriate acoustics requirements. Define the acoustic noise levels on each side of the interface in accordance with program or project requirements.

3.2.7.3 *Structural Loads*

Define the appropriate structural loads requirements. Define the mated loads that each end item should accommodate.

3.2.7.4 *Vibroacoustics*

Define the appropriate vibroacoustics requirements. Define the vibroacoustic loads that each end item should accommodate.

3.2.7.5 *Human Operability*

Define the appropriate human interface requirements. Define the human-centered design considerations, constraints, and capabilities that each end item should accommodate.

3.2.8 Other Types of Interface Requirements

Define other types of unique interface requirements that may be applicable.

Appendix M: CM Plan Outline

A comprehensive Configuration Management (CM) Plan that reflects efficient application of configuration management principles and practices would normally include the following topics:

- General product definition and scope
- Description of CM activities and procedures for each major CM function
- Organization, roles, responsibilities, and resources
- Definitions of terms
- Programmatic and organizational interfaces
- Deliverables, milestones, and schedules
- Subcontract flow down requirements

The documented CM planning should be reevaluated following any significant change affecting the context and environment, e.g., changes in suppliers or supplier responsibilities, changes in diminishing manufacturing sources/part obsolescence, changes in resource availabilities, changes in customer contract, and changes in the product. CM planning should also be reviewed on a periodic basis to make sure that an organization's application of CM functions is current.

For additional information regarding a CM Plan, refer to SAE EIA-649, Rev. B.

Appendix N: Guidance on Technical Peer Reviews/ Inspections

This appendix has been removed. For additional guidance on how to perform technical peer reviews refer to Appendix N in the NASA Expanded Guidance for Systems Engineering at *https://nen.nasa.gov/web/se/doc-repository*.

Appendix O: Reserved

Appendix P: SOW Review Checklist

This appendix has been removed. For additional guidance on checklists for editorial and content review questions refer to Appendix P in the NASA Expanded Guidance for Systems Engineering at *https://nen.nasa.gov/web/se/doc-repository*.

Appendix Q: Reserved

Appendix R: HSI Plan Content Outline

R.1 HSI Plan Overview

The Human Systems Integration (HSI) Plan documents the strategy for and planned implementation of HSI through a particular program's/project's life cycle. The intent of HSI is:

- To ensure the human elements of the total system are effectively integrated with hardware and software elements,
- To ensure all human capital required to develop and operate the system is accounted for in life cycle costing, and
- To ensure that the system is built to accommodate the characteristics of the user population that will operate, maintain, and support the system.

The HSI Plan is specific to a program or project and applies to NASA systems engineering per NPR 7123.1, NASA Systems Engineering Processes and Requirements. The HSI Plan should address the following:

- Roles and responsibilities for integration across HSI domains;
- Roles and responsibilities for coordinating integrated HSI domain inputs with the program team and stakeholders;
- HSI goals and deliverables for each phase of the life cycle;
- Entry and exit criteria with defined metrics for each phase, review, and milestone;
- Planned methods, tools, requirements, processes, and standards for conducting HSI;
- Strategies for identifying and resolving HSI risks; and
- Alignment strategy with the SEMP.

The party or parties responsible for program/project HSI implementation—e.g., an HSI integrator (or team)—should be identified by the program/project manager. The HSI integrator or team develops and maintains the HSI Plan with support from and coordination with the project manager and systems engineer.

Implementation of HSI on a program/project utilizes many of the tools and products already required by systems engineering; e.g., development of a ConOps, clear functional allocation across the elements of a system (hardware, software, and human), and the use of key performance measurements through the life cycle to validate and verify HSI's effectiveness. It is not the intent of the HSI Plan or its implementation to duplicate other systems engineering plans or processes, but rather to define the uniquely HSI effort being made to ensure the human element is given equal consideration to hardware/software elements of a program/project.

R.2 HSI Plan Content Outline

Each program/project-specific HSI Plan should be tailored to fit the program/project's size, scope, and purpose. The following is an example outline for a major program; e.g., space flight or aeronautics.

1.0 Introduction

1.1 Purpose

This section briefly identifies the ultimate objectives for this program/project's HSI Plan. This section also introduces the intended implementers and users of this HSI Plan.

1.2 Scope

This section describes the overall scope of the HSI Plan's role in documenting the strategy for and implementation of HSI. Overall, this section describes that the HSI Plan:

- *Is a dynamic document that will be updated at key life cycle milestones.*
- *Is a planning and management guide that describes how HSI will be relevant to the program/project's goals.*
- *Describes planned HSI methodology, tools, schedules, and deliverables.*
- *Identifies known program/project HSI issues and concerns and how their resolutions will be addressed.*
- *Defines program/project HSI organizational elements, roles, and responsibilities.*
- *May serve as an audit trail that documents HSI data sources, analyses, activities, trade studies, and decisions not captured in other program/project documentation.*

1.3 Definitions

This section defines key HSI terms and references relevant program/project-specific terms.

2.0 Applicable Documents

This section lists all documents, references, and data sources that are invoked by HSI's implementation on the program/project, that have a direct impact on HSI outcomes, and/or are impacted by the HSI effort.

3.0 HSI Objectives

3.1 System Description

This section describes the system, missions to be performed, expected operational environment(s), predecessor and/or legacy systems (and lessons learned), capability gaps, stage of development, etc. Additionally, reference should be made to the acquisition strategy for the system; e.g., if it is developed in-house within NASA or if major systems are intended for external procurement. The overall strategy for program integration should be referenced.

Note that this information is likely captured in other program/project documentation and can be referenced in the HSI Plan rather than repeated.

3.2 HSI Relevance

At a high level, this section describes HSI's relevance to the program/project; i.e., how the HSI strategy will improve the program/project's outcome. Known HSI challenges should be described along with mention of areas where human performance in the system's operations is predicted to directly impact the probability of overall system performance and mission success.

4.0 HSI Strategy

4.1 HSI Strategy Summary

This section summarizes the HSI approaches, planning, management, and strategies for the program/project. It should describe how HSI products will be integrated across all HSI domains and how HSI inputs to program/project systems engineering and management processes

HSI RELEVANCE

Key Points

- *Describe performance characteristics of the human elements known to be key drivers to a desired total system performance outcome.*
- *Describe the total system performance goals that require HSI support.*
- *Identify HSI concerns with legacy systems; e.g., if operations and logistics, manpower, skill selection, required training, logistics support, operators' time, maintenance, and/or risks to safety and success exceeded expectations.*
- *Identify potential cost, schedule, risk, and trade-off concerns with the integration of human elements; e.g., quantity and skills of operators, maintainers, ground controllers, etc.*

HSI STRATEGY

Key Points

- *Identify critical program/project-specific HSI key decision points that will be used to track HSI implementation and success.*
- *Identify key enabling (and particularly, emerging) technologies and methodologies that may be overlooked in hardware/software systems trade studies but that may positively contribute to HSI implementation; e.g., in the areas of human performance, workload, personnel management, training, safety, and survivability.*
- *Describe HSI products that will be integrated with program/project systems engineering products, analyses, risks, trade studies, and activities.*
- *Describe efforts to ensure HSI will contribute in critically important Phase A and Pre-Phase A cost-effective design concept studies.*
- *Describe the plan and schedule for updating the HSI Plan through the program/project life cycle.*

contribute to system performance and help contain life cycle cost. This section (or Implementation Summary, Section 6 *of this outline) should include a top-level schedule showing key HSI milestones.*

4.2 *HSI Domains*

This section identifies the HSI domains applicable to the program/project including rationale for their relevance.

5.0 HSI Requirements, Organization, and Risk Management

5.1 *HSI Requirements*

This section references HSI requirements and standards applicable to the program/project and identifies the authority that invokes them; e.g., the NASA Procedural Requirements (NPR) document(s) that invoke applicability.

HSI DOMAINS

Key Points

- *Identify any domain(s) associated with human performance capabilities and limitations whose integration into the program/project is likely to directly affect the probability of successful program/project outcome.*
- *An overview of processes to apply, document, validate, evaluate, and mitigate HSI domain knowledge and to integrate domain knowledge into integrated HSI inputs to program/project and systems engineering processes.*

HSI REQUIREMENTS

Key Points

- *Describe how HSI requirements that are invoked on the program/project contribute to mission success, affordability, operational effectiveness, and safety.*
- *HSI should include requirements that influence the system design to moderate manpower (operators, maintainers, system administrative, and support personnel), required skill sets (occupational specialties with high aptitude or skill requirements), and training requirements.*
- *Define the program/project-specific HSI strategy derived from NASA-STD-3001, NASA Space Flight Human-System Standard, Volume 2: Human Factors, Habitability, and Environmental Health, Standard 3.5 [V2 3005], "Human-Centered Design Process", if applicable.*
- *Capture the development process and rationale for any program/project-specific requirements not derived from existing NASA standards. In particular, manpower, skill set, and training HSI requirements/goals may be so program/project-specific as to not have NASA parent standards or requirements.*
- *Identify functional connections between HSI measures of effectiveness used to verify requirements and key performance measures used throughout the life cycle as indicators of overall HSI effectiveness.*

5.2 *HSI Organization, Roles, and Responsibilities*

In this section, roles and responsibilities for program/project personnel assigned to facilitate and/or manage HSI tasks are defined; e.g., the HSI integrator (and/or team if required by NPR 8705.2). HSI integrator/team functional responsibilities to the program are described in addition to identification of organizational elements with HSI responsibilities. Describe the relationships between HSI integrator/team, stakeholders, engineering technical teams, and governing bodies (control boards).

5.2.1 *HSI Organization*

- *Describe the HSI management structure for the program/project and identify its leaders and membership.*
- *Reference the organizational structure of the program (including industry partners) and describe the roles and responsibilities of the HSI integrator/team within that structure. Describe the HSI responsible party's relationship to other teams, including those*

for systems engineering, logistics, risk management, test and evaluation, and requirements verification.

- *Provide the relationship of responsible HSI personnel to NASA Technical Authorities (Engineering, Safety, and Health/Medical).*

- *Identify if the program/project requires NASA-(Government) and/or contractor-issued HSI Plans, and identify the responsible author(s). Describe how NASA's HSI personnel will monitor and assess contractor HSI activities. For contractor-issued HSI Plans, identify requirements and processes for NASA oversight and evaluation of HSI efforts by subcontractors.*

5.2.2 *HSI Roles & Responsibilities*

- *Describe the HSI responsible personnel's functional responsibilities to the program/project, addressing (as examples) the following:*
    - » *developing HSI program documentation;*
    - » *validating human performance requirements;*
    - » *conducting HSI analyses;*
    - » *designing human machine interfaces to provide the level of human performance required for operations, maintenance, and support, including conduct of training;*
    - » *describing the role of HSI experts in documenting and reporting the results from tests and evaluations.*

- *Define how collaboration will be performed within the HSI team, across program/project integrated product teams and with the program/project manager and systems engineer.*

- *Define how the HSI Plan and the SEMP will be kept aligned with each other.*

- *Define responsibility for maintaining and updating the HSI Plan through the program/project's life cycle.*

5.3 *HSI Issue and Risk Processing*

This section describes any HSI-unique processes for identifying and mitigating human system risks. HSI risks should be processed in the same manner and system as other program/project risks (technical, programmatic, schedule). However, human system risks may only be recognized by HSI domain and integration experts. Therefore, it may be important to document any unique procedures by which the program/project HSI integrator/team identifies, validates, prioritizes, and tracks the status of HSI-specific risks through the program/project risk management system. Management of HSI risks may be deemed the responsibility of the program's/project's HSI integrator/team in coordination with overall program/project risk management.

- *Ensure that potential cost, schedule, risk, and trade-off concerns with the integration of human elements (operators, maintainers, ground controllers, etc.) with the total system are identified and mitigated.*

- *Ensure that safety, health, or survivability concerns that arise as the system design and implementation emerge are identified, tracked, and managed.*

- *Identify and describe any risks created by limitations on the overall program/project HSI effort (time, funding, insufficient availability of information, availability of expertise, etc.).*

- *Describe any unique attributes of the process by which the HSI integrator/team elevates HSI risks to program/project risks.*

- *Describe any HSI-unique aspects of how human system risk mitigation strategies are deemed effective.*

6.0 HSI Implementation

6.1 *HSI Implementation Summary*

This section summarizes the HSI implementation approach by program/project phase. This section shows how an HSI strategy for the particular program/project is planned to be tactically enabled; i.e., establishment of HSI priorities; description of specific activities, tools, and products planned to ensure HSI objectives are met; application of technology in the achievement of HSI objectives; and an HSI risk processing strategy that identifies and mitigates technical and schedule concerns when they first arise.

6.2 *HSI Activities and Products*

In this section, map activities, resources, and products associated with planned HSI technical implementation to each systems engineering phase of the program/project. Consideration might be given to mapping the needs and products of each HSI domain by program/project phase. Examples of HSI activities include analyses, mockup/prototype human-in-the-loop evaluations, simulation/modeling, participation in design and design reviews, formative evaluations, technical interchanges, and trade studies. Examples of HSI resources include acquisition of unique/specific HSI skill sets and domain expertise, facilities, equipment, test articles, specific time allocations, etc.

When activities, products, or risks are tied to life cycle reviews, they should include a description of the HSI entrance and exit criteria to clearly define the boundaries of each phase, as well as resource limitations that may be associated with each activity or product (time, funding, data availability, etc.). A high-level, summary example listing of HSI activities, products, and known risk mitigations by life cycle phase is provided in TABLE R.2-1.

6.3 *HSI Plan Update*

The HSI Plan should be updated throughout the program/project's life cycle management and systems engineering processes at key milestones. Milestones recommended for HSI Plan updates are listed in appendix G of NPR 7123.1, NASA Systems Engineering Processes and Requirements.

HSI IMPLEMENTATION

Key Points

- *Relate HSI strategic objectives to the technical approaches planned for accomplishing these objectives.*
- *Overlay HSI milestones—e.g., requirements definition, verification, known trade studies, etc.—on the program/project schedule and highlight any inconsistencies, conflicts, or other expected schedule challenges.*
- *Describe how critical HSI key decision points will be dealt with as the program/project progresses through its life cycle. Indicate the plan to trace HSI key performance measures through the life cycle; i.e., from requirements to human/system functional performance allocations, through design, test, and operational readiness assessment.*
- *Identify HSI-unique systems engineering processes—e.g., verification using human-in-the-loop evaluations—that may require special coordination with program/project processes.*
- *As the system emerges, indicate plans to identify HSI lessons learned from the application of HSI on the program/project.*
- *Include a high-level summary of the resources required.*

TABLE R.2-1 HSI Activity, Product, or Risk Mitigation by Program/Project Phase

Life-Cycle Phase	Phase Description	Activity, Product, or Risk Mitigation
Pre-Phase A	*Concept Studies*	*ConOps (Preliminary—to include training, maintenance, logistics, etc.)*
Phase A	*Concept & Technology Development*	*HSI Plan (baseline)* *ConOps (initial)* *HSI responsible party(ies) and/or team identified before SRR* *Develop mockup(s) for HSI evaluations* *Crew Workload Evaluation Plan* *Functional allocation, crew task lists* *Validation of ConOps (planning)*
Phase B	*Preliminary Design & Technology Completion*	*HSI Plan (update)* *ConOps (baseline)* *Develop engineering-level mockup(s) for HSI evaluations* *Define crew environmental and crew health support needs (e.g., aircraft flight decks, human space flight missions)* *Assess operator interfaces through task analyses (for, e.g., aircraft cockpit operations, air traffic management, spacecraft environments, mission control for human space flight missions)* *Human-in-the-loop usability plan* *Human-rating report for PDR*
Phase C	*Final Design & Fabrication*	*HSI Plan (update)* *First Article HSI Tests* *Human-rating report for CDR*
Phase D	*System Assembly, Integ. & Test, Launch & Checkout*	*Human-rating report for ORR* *Validation of human-centered design activities* *Validation of ConOps*
Phase E	*Operations & Sustainment*	*Monitoring of human-centered design performance*
Phase F	*Closeout*	*Lessons learned report*

HSI PLAN UPDATES

Key points to be addressed in each update

- *Identify the current program/project phase, the publication date of the last iteration of the HSI Plan, and the HSI Plan version number. Update the HSI Plan revision history.*
- *Describe the HSI entrance criteria for the current phase and describe any unfinished work prior to the current phase.*
- *Describe the HSI exit criteria for the current program/project phase and the work that must be accomplished to successfully complete the current program/project phase.*

Appendix S: Concept of Operations Annotated Outline

This Concept of Operations (ConOps) annotated outline describes the type and sequence of information that should be contained in a ConOps, although the exact content and sequence will be a function of the type, size, and complexity of the project. The text in italics describes the type of information that would be provided in the associated subsection. Additional subsections should be added as necessary to fully describe the envisioned system.

Cover Page

Table of Contents

1.0 Introduction

1.1 Project Description

This section will provide a brief overview of the development activity and system context as delineated in the following two subsections.

1.1.1 *Background*

Summarize the conditions that created the need for the new system. Provide the high-level mission goals and objective of the system operation. Provide the rationale for the development of the system.

1.1.2 *Assumptions and Constraints*

State the basic assumptions and constraints in the development of the concept. For example, that some technology will be matured enough by the time the system is ready to be fielded, or that the system has to be provided by a certain date in order to accomplish the mission.

1.2 Overview of the Envisioned System

This section provides an executive summary overview of the envisioned system. A more detailed description will be provided in Section 3.0

1.2.1 *Overview*

This subsection provides a high-level overview of the system and its operation. Pictorials, graphics, videos, models, or other means may be used to provide this basic understanding of the concept.

1.2.2 *System Scope*

This section gives an estimate of the size and complexity of the system. It defines the system's external interfaces and enabling systems. It describes what the project will encompass and what will lie outside of the project's development.

2.0 Documents

2.1 Applicable Documents

This section lists all the documents, models, standards or other material that are applicable and some or all of which will form part of the requirements of the project.

2.2 Reference Documents

This section provides supplemental information that might be useful in understanding the system or its scenarios.

3.0 Description of Envisioned System

This section provides a more detailed description of the envisioned system and its operation as contained in the following subsections.

3.1 Needs, Goals and Objectives of Envisioned System

This section describes the needs, goals, and objectives as expectations for the system capabilities, behavior, and operations. It may also point to a separate document or model that contains the current up-to-date agreed-to expectations.

3.2 Overview of System and Key Elements

This section describes at a functional level the various elements that will make up the system, including the users and operators. These descriptions should be implementation free; that is, not specific to any implementation or design but rather a general description of what the system and its elements will be expected to do. Graphics, pictorials, videos, and models may be used to aid this description.

3.3 Interfaces

This section describes the interfaces of the system with any other systems that are external to the project. It may also include high-level interfaces between the major envisioned elements of the system. Interfaces may include mechanical, electrical, human user/operator, fluid, radio frequency, data, or other types of interactions.

3.4 Modes of Operations

This section describes the various modes or configurations that the system may need in order to accomplish its intended purpose throughout its life cycle. This may include modes needed in the development of the system, such as for testing or training, as well as various modes that will be needed during it operational and disposal phases.

3.5 Proposed Capabilities

This section describes the various capabilities that the envisioned system will provide. These capabilities cover the entire life cycle of the system's operation, including special capabilities needed for the verification/validation of the system, its capabilities during its intended operations, and any special capabilities needed during the decommissioning or disposal process.

4.0 Physical Environment

This section should describe the environment that the system will be expected to perform in throughout its life cycle, including integration, tests, and transportation. This may include expected and off-nominal temperatures, pressures, radiation, winds, and other atmospheric, space, or aquatic conditions. A description of whether the system needs to operate, tolerate with degraded performance, or just survive in these conditions should be noted.

5.0 Support Environment

This section describes how the envisioned system will be supported after being fielded. This includes how operational planning will be performed and how commanding or other uploads will be determined and provided, as required. Discussions may include ***how*** *the envisioned system would be maintained, repaired, replaced, it's sparing philosophy, and how future upgrades may be performed. It may also include assumptions on the level of continued support from the design teams.*

6.0 Operational Scenarios, Use Cases and/or Design Reference Missions

This section takes key scenarios, use cases, or DRM and discusses what the envisioned system provides or how it functions throughout that single-thread timeline.

The number of scenarios, use cases, or DRMs discussed should cover both nominal and off-nominal conditions and cover all expected functions and capabilities. A good practice is to label each of these scenarios to facilitate requirements traceability; e.g., [DRM-0100], [DRM-0200], etc.

6.1 Nominal Conditions

These scenarios, use cases, or DRMs cover how the envisioned system will operate under normal circumstances where there are no problems or anomalies taking place.

6.2 Off-Nominal Conditions

These scenarios cover cases where some condition has occurred that will need the system to perform in a way that is different from normal. This would cover failures, low performance, unexpected environmental conditions, or operator errors. These scenarios should reveal any additional capabilities or safeguards that are needed in the system.

7.0 Impact Considerations

This section describes the potential impacts, both positive and negative, on the environment and other areas.

7.1 Environmental Impacts

Describes how the envisioned system could impact the environment of the local area, state, country, worldwide, space, and other planetary bodies as appropriate for the systems intended purpose. This includes the possibility of the generation of any orbital debris, potential contamination of other planetary bodies or atmosphere, and generation of hazardous wastes that will need disposal on earth and other factors. Impacts should cover the entire life cycle of the system from development through disposal.

7.2 Organizational Impacts

Describes how the envisioned system could impact existing or future organizational aspects. This would include the need for hiring specialists or operators, specialized or widespread training or retraining, and use of multiple organizations.

7.3 Scientific/Technical Impacts

This subsection describes the anticipated scientific or technical impact of a successful mission or deployment, what scientific questions will be answered, what knowledge gaps will be filled, and what services will be provided. If the purpose of this system is to improve operations or logistics instead of science, describe the anticipated impact of the system in those terms.

8.0 Risks and Potential Issues

This section describes any risks and potential issues associated with the development, operations or disposal of the envisioned system. Also includes concerns/risks with the project schedule, staffing support, or implementation approach. Allocate subsections as needed for each risk or issue consideration. Pay special attention to closeout issues at the end of the project.

Appendix A: Acronyms

This part lists each acronym used in the ConOps and spells it out.

Appendix B: Glossary of Terms

The part lists key terms used in the ConOps and provides a description of their meaning.

Appendix T: Systems Engineering in Phase E

T.1 Overview

In general, normal Phase E activities reflect a reduced emphasis on system design processes but a continued focus on product realization and technical management. Product realization process execution in Phase E takes the form of continued mission plan generation (and update), response to changing flight conditions (and occurrence of in-flight anomalies), and update of mission operations techniques, procedures, and guidelines based on operational experience gained. Technical management processes ensure that appropriate rigor and risk management practices are applied in the execution of the product realization processes.

Successful Phase E execution requires the prior establishment of mission operations capabilities in four (4) distinct categories: tools, processes, products, and trained personnel. These capabilities may be developed as separate entities, but need to be fused together in Phase E to form an end-to-end operational capability.

Although systems engineering activities and processes are constrained throughout the entire project life cycle, additional pressures exist in Phase E:

- **Increased resource constraints:** Even when additional funding or staffing can be secured, building new capabilities or training new personnel may require more time or effort than is available. Project budget and staffing profiles generally decrease at or before entry into Phase E, and the remaining personnel are typically focused on mission execution.
- **Unforgiving schedule:** Unlike pre-flight test activities, it may be difficult or even impossible to pause mission execution to deal with technical issues of a spacecraft in operation. It is typically difficult or impossible to truly pause mission execution after launch.

These factors must be addressed when considering activities that introduce change and risk during Phase E.

NOTE: When significant hardware or software changes are required in Phase E, the logical decomposition process may more closely resemble that exercised in earlier project phases. In such cases, it may be more appropriate to identify the modification as a new project executing in parallel—and coordinated with—the operating project.

T.2 Transition from Development to Operations

An effective transition from development to operations phases requires prior planning and coordination among stakeholders. This planning should focus not only on the effective transition of hardware and software systems into service but also on the effective transfer of knowledge, skills, experience, and processes into roles that support the needs of flight operations.

Development phase activities need to clearly and concisely document system knowledge in the form of operational techniques, characteristics, limits, and constraints—these are key inputs used by flight operations personnel in building operations tools and techniques. Phase D Integration and Test (I&T) activities share many common needs with Phase E operations activities. Without prior planning and agreement, however, similar products used in these two phases may be formatted so differently that one set cannot be used for both purposes. The associated product duplication is often unexpected and results in increased cost and schedule risk. Instead, system engineers should identify opportunities for product reuse early in the development process and establish common standards, formats, and content expectations to enable transition and reuse.

Similarly, the transfer of skills and experience should be managed through careful planning and placement of key personnel. In some cases, key design, integration, and test personnel may be transitioned into the mission operations team roles. In other cases, dedicated mission operations personnel may be assigned to shadow or assist other teams during Phase A–D activities. In both cases, assignees bring knowledge, skills, and experience into the flight operations environment. Management of this transition process can, however, be complex as these personnel may be considered key to both ongoing I&T and preparation for upcoming operations. Careful and early planning of personnel assignments and transitions is key to success in transferring skills and experience.

T.3 System Engineering Processes in Phase E

T.3.1 System Design Processes

In general, system design processes are complete well before the start of Phase E. However, events during operations may require that these processes be revisited in Phase E.

T.3.1.1 *Stakeholder Expectations Definition*

Stakeholder expectations should have been identified during development phase activities, including the definition of operations concepts and design reference missions. Central to this definition is a consensus on mission success criteria and the priority of all intended operations. The mission operations plan should state and address these stakeholder expectations with regard to risk management practices, planning flexibility and frequency of opportunities to update the plan, time to respond and time/scope of status communication, and other key parameters of mission execution. Additional detail in the form of operational guidelines and constraints should be incorporated in mission operations procedures and flight rules.

The Operations Readiness Review (ORR) should confirm that stakeholders accept the mission operations plan and operations implementation products.

However, it is possible for events in Phase E to require a reassessment of stakeholder expectations. Significant in-flight anomalies or scientific discoveries during flight operations may change the nature and goals of a mission. Mission systems engineers, mission operations managers, and program management need to remain engaged with stakeholders throughout Phase E to identify potential changes in expectations and to manage the acceptance or rejection of such changes during operations.

T.3.1.2 *Technical Requirements Definition*

New technical requirements and changes to existing requirements may be identified during operations as a result of:

- New understanding of system characteristics through flight experience;
- The occurrence of in-flight anomalies; or

- Changing mission goals or parameters (such as mission extension).

These changes or additions are generally handled as change requests to an operations baseline already under configuration management and possibly in use as part of ongoing flight operations. Such changes are more commonly directed to the ground segment or operations products (operational constraints, procedures, etc.). Flight software changes may also be considered, but flight hardware changes for anything other than human-tended spacecraft are rarely possible.

Technical requirement change review can be more challenging in Phase E as fewer resources are available to perform comprehensive review. Early and close involvement of Safety and Mission Assurance (SMA) representatives can be key in ensuring that proposed changes are appropriate and within the project's allowable risk tolerance.

T.3.1.3 *Logical Decomposition*

In general, logical decomposition of mission operations functions is performed during development phases. Additional logical decomposition during operations is more often applied to the operations products: procedures, user interfaces, and operational constraints. The authors and users of these products are often the most qualified people to judge the appropriate decomposition of new or changed functionality as a series of procedures or similar products.

T.3.1.4 *Design Solution Definition*

Similar to logical decomposition, design solution definition tasks may be better addressed by those who develop and use the products. Minor modifications may be handled entirely within an operations team (with internal reviews), while larger changes or additions may warrant the involvement of program-level system engineers and Safety and Mission Assurance (SMA) personnel.

Scarcity of time and resources during Phase E can make implementation of these design solutions challenging. The design solution needs to take into account the availability of and constraints to resources.

T.3.1.5 *Product Implementation*

Personnel who implement mission operations products such as procedures and spacecraft command scripts should be trained and certified to the appropriate level of skill as defined by the project. Processes governing the update and creation of operations products should be in place and exercised prior to Phase E.

T.3.2 Product Realization Processes

Product realization processes in Phase E are typically executed by Configuration Management (CM) and test personnel. It is common for these people to be "shared resources;" i.e., personnel who fulfil other roles in addition to CM and test roles.

T.3.2.1 *Product Integration*

Product integration in Phase E generally involves bringing together multiple operations products—some pre-existing and others new or modified—into a proposed update to the baseline mission operations capability.

The degree to which a set of products is integrated may vary based on the size and complexity of the project. Small projects may define a baseline—and update to that baseline—that spans the entire set of all operations products. Larger or more complex projects may choose to create logical baseline subsets divided along practical boundaries. In a geographically disperse set of separate mission operations Centers, for example, each Center may be initially integrated as a separate product. Similarly, the different functions within a single large control Center—planning, flight dynamics, command and control, etc.—may be established as separately baselined products. Ultimately, however, some method needs to be established to ensure that the product realization processes identify and assess all potential impacts of system changes.

T.3.2.2 *Product Verification*

Product verification in Phase E generally takes the form of unit tests of tools, data sets, procedures, and other items under simulated conditions. Such "thread tests" may exercise single specific tasks or functions. The fidelity of simulation required for verification varies with the nature and criticality of the product. Key characteristics to consider include:

- **Runtime:** Verification of products during flight operations may be significantly time constrained. Greater simulation fidelity can result in slower simulation performance. This slower performance may be acceptable for some verification activities but may be too constraining for others.

- **Level of detail:** Testing of simple plans and procedures may not require high-fidelity simulation of a system's dynamics. For example, simple state change processes may be tested on relatively low-fidelity simulations. However, operational activities that involve dynamic system attributes – such as changes in pressure, temperature, or other physical properties may require testing with much higher-fidelity simulations.

- **Level of integration:** Some operations may impact only a single subsystem, while others can affect multiple systems or even the entire spacecraft.

- **Environmental effects:** Some operations products and procedures may be highly sensitive to environmental conditions, while others may not. For example, event sequences for atmospheric entry and deceleration may require accurate weather data. In contrast, simple system reconfiguration procedures may not be impacted by environmental conditions at all.

T.3.2.3 *Product Validation*

Product validation is generally executed through the use of products in integrated operational scenarios such as mission simulations, operational readiness tests, and/or spacecraft end-to-end tests. In these environments, a collection of products is used by a team of operators to simulate an operational activity or set of activities such as launch, activation, rendezvous, science operations, or Entry, Descent, and Landing (EDL). The integration of multiple team members and operations products provides the context necessary to determine if the product is appropriate and meets the true operations need.

T.3.2.4 *Product Transition*

Transition of new operational capabilities in Phase E is generally overseen by the mission operations manager or a Configuration Control Board (CCB) chaired by the mission operations manager or the project manager.

Proper transition management includes the inspection of product test (verification and validation) results as well as the readiness of the currently operating operations system to accept changes. Transition during Phase E can be particularly challenging as the personnel using these capabilities also need to change techniques, daily practices, or other behaviors as a result. Careful attention should be paid to planned operations, such as spacecraft maneuvers or other mission critical events and risks associated with performing product transition at times near such events.

T.3.3 Technical Management Processes

Technical management processes are generally a shared responsibility of the project manager and the mission operations manager. Clear agreement between these two parties is essential in ensuring that Phase E efforts are managed effectively.

T.3.3.1 *Technical Planning*

Technical planning in Phase E generally focuses on the management of scarce product development resources during mission execution. Key decision-makers, including the mission operations manager and lower operations team leads, need to review the benefits of a change against the resource cost to implement changes. Many resources are shared in Phase E – for example, product developers may also serve other real-time operations roles– and the additional workload placed on these resources should be viewed as a risk to be mitigated during operations.

T.3.3.2 *Requirements Management*

Requirements management during Phase E is similar in nature to pre-Phase E efforts. Although some streamlining may be implemented to reduce process overhead in Phase E, the core need to review and validate requirements remains. As most Phase E changes are derived from a clearly demonstrated need, program management may reduce or waive the need for complete requirements traceability analysis and documentation.

T.3.3.3 *Interface Management*

It is relatively uncommon for interfaces to change in Phase E, but this can occur when a software tool is modified or a new need is uncovered. Interface definitions should be managed in a manner similar to that used in other project phases.

T.3.3.4 *Technical Risk Management*

Managing technical risks during operations can be more challenging during Phase E than during other phases. New risks discovered during operations may be the result of system failures or changes in the surrounding environment. Where additional time may be available to assess and mitigate risk in other project phases, the nature of flight operations may limit the time over which risk management can be executed. For this reason, every project should develop a formal process for handling anomalies and managing risk during operations. This process should be exercised before flight, and decision-makers should be well versed in the process details.

T.3.3.5 *Configuration Management*

Effective and efficient Configuration Management (CM) is essential during operations. Critical operations materials, including procedures, plans, flight datasets, and technical reference material need to be secure, up to date, and easily accessed by those who make and enact mission critical decisions. CM systems—in their intended flight configuration—should be exercised as part of operational readiness tests to ensure that the systems, processes, and participants are flight-ready.

Access to such operations products is generally time-critical, and CM systems supporting that access should be managed accordingly. Scheduled maintenance or other "downtime" periods should be coordinated with flight operations plans to minimize the risk of data being inaccessible during critical activities.

T.3.3.6 *Technical Data Management*

Tools, procedures, and other infrastructure for Technical Data Management must be baselined, implemented, and verified prior to flight operations. Changes to these capabilities are rarely made during Phase E due to the high risk of data loss or reduction in operations efficiency when changing during operations.

Mandatory Technical Data Management infrastructure changes, when they occur, should be carefully reviewed by those who interact with the data on a regular basis. This includes not only operations personnel, but also engineering and science customers of that data.

T.3.3.7 *Technical Assessment*

Formal technical assessments during Phase E are typically focused on the upcoming execution of a specific operational activity such as launch, orbit entry, or decommissioning. Reviews executed while flight operations are in progress should be scoped to answer critical questions while not overburdening the project or operations team.

Technical Performance Measures (TPMs) in Phase E may differ significantly from those in other project phases. Phase E TPMs may focus on the accomplishment of mission events, the performance of the system in operation, and the ability of the operations team to support upcoming events.

T.3.3.8 *Decision Analysis*

The Phase E Decision Analysis Process is similar to that in other project phases but may emphasize different criteria. For example, the ability to change a schedule may be limited by the absolute timing of events such as an orbit entry or landing on a planetary surface. Cost trades may be more constrained by the inability to add trained personnel to support an activity. Technical trades may be limited by the inability to modify hardware in operation.

References Cited

This appendix contains references that were cited in the sections of the handbook.

Preface

NPR 7123.1, Systems Engineering Processes and Requirements

NASA Chief Engineer and the NASA Integrated Action Team (NIAT) report, *Enhancing Mission Success – A Framework for the Future*, December 21, 2000. Authors: McBrayer, Robert O and Thomas, Dale, NASA Marshall Space Flight Center, Huntsville, AL United States.

NASA. *Columbia Accident Investigation Board (CAIB) Report*, 6 volumes: Aug. 26, Oct. 2003. *http://www.nasa.gov/columbia/caib/html/report.html*

NASA. Diaz Report, *A Renewed Commitment to Excellence: An Assessment of the NASA Agency-wide Applicability of the Columbia Accident Investigation Board Report*, January 30, 2004. Mr. Al Diaz, Director, Goddard Space Flight Center, and team.

International Organization for Standardization (ISO) 9000:2015, *Quality management systems – Fundamentals and vocabulary*. Geneva: International Organization for Standardization, 2015.

Section 1.1 Purpose

NPR 7123.1. Systems Engineering Processes and Requirements

Section 1.2 Scope and Depth

NASA Office of Chief Information Officer (OCIO), *Information Technology Systems Engineering Handbook Version 2.0*

NASA-HDBK-2203, *NASA Software Engineering Handbook* (February 28, 2013)

Section 2.0 Fundamentals of Systems Engineering

NPR 7120.5, NASA Space Flight Program and Project Management Requirements

NPR 7120.7, NASA Information Technology and Institutional Infrastructure Program and Project Management Requirements

NPR 7120.8, NASA Research and Technology Program and Project Management Requirements

NPR 7123.1, NASA Systems Engineering Processes and Requirements

NASA Engineering Network (NEN) Systems Engineering Community of Practice (SECoP), located at *https://nen.nasa.gov/web/se*

Griffin, Michael D., NASA Administrator. "System Engineering and the Two Cultures of Engineering." Boeing Lecture, Purdue University, March 28, 2007.

Rechtin, Eberhardt. *Systems Architecting of Organizations: Why Eagles Can't Swim*. Boca Raton: CRC Press, 2000.

Section 2.1 The Common Technical Processes and the SE Engine

NPR 7123.1, NASA Systems Engineering Processes and Requirements

Society of Automotive Engineers (SAE) and the European Association of Aerospace Industries (EAAI). *AS9100C Quality Management Systems (QMS) – Requirements for Aviation, Space, and Defense Organizations* Revision C: January 15, 2009.

Section 2.3 Example of Using the SE Engine

NPD 1001.0, 2006 NASA Strategic Plan

NPR 7120.5, NASA Space Flight Program and Project Management Requirements

Section 2.5 Cost Effectiveness Considerations

Department of Defense (DOD) Defense Acquisition University (DAU). *Systems Engineering Fundamentals Guide*. Fort Belvoir, VA, 2001.

INCOSE-TP-2003-002-04, *Systems Engineering Handbook: A Guide for System Life Cycle Processes and Activities*, Version 4, edited by Walden, David D., et al., 2015

Section 2.6 Human Systems Integration (HSI) in the SE Process

NPR 7120.5, NASA Space Flight Program and Project Management Requirements

NPR 7123.1, NASA Systems Engineering Processes and Requirements

Section 3.0 NASA Program/Project Life Cycle

NPR 7120.5, NASA Space Flight Program and Project Management Requirements

NPR 7120.8, NASA Research and Technology Program and Project Management Requirements

NASA Office of the Chief Information Officer (OCIO), *Information Technology Systems Engineering Handbook* Version 2.0

NASA/SP-2014-3705, *NASA Space Flight Program and Project Management Handbook*

Section 3.1 Program Formulation

NPR 7120.5, NASA Space Flight Program and Project Management Requirements

NPR 7120.7, NASA Information Technology and Institutional Infrastructure Program and Project Management Requirements

NPR 7120.8, NASA Research and Technology Program and Project Management Requirements

NPR 7123.1, NASA Systems Engineering Processes and Requirements

Section 3.2 Program Implementation

NPR 7120.5, NASA Space Flight Program and Project Management Requirements

NPR 7123.1, NASA Systems Engineering Processes and Requirements

Section 3.3 Project Pre-Phase A: Concept Studies

NPR 7120.5, NASA Space Flight Program and Project Management Requirements

NPR 7123.1, NASA Systems Engineering Processes and Requirements

Section 3.4 Project Phase A: Concept and Technology Development

NPD 1001.0, 2014 NASA Strategic Plan

NPR 2810.1, Security of Information Technology

NPR 7120.5, NASA Space Flight Program and Project Management Requirements

NPR 7123.1, NASA Systems Engineering Processes and Requirements

NPR 7150.2, NASA Software Engineering Requirements

NASA-STD-8719.14, *Handbook for Limiting Orbital Debris*. Rev A with Change 1. December 8, 2011.

National Institute of Standards and Technology (NIST), Federal Information Processing Standard Publication (FIPS PUB) 199, *Standards for Security Categorization of Federal Information and Information Systems*, February 2004.

Section 3.5 Project Phase B: Preliminary Design and Technology Completion

NPR 7120.5, NASA Space Flight Program and Project Management Requirements

NPR 7123.1, NASA Systems Engineering Processes and Requirements

Section 3.6 Project Phase C: Final Design and Fabrication

NPR 7120.5, NASA Space Flight Program and Project Management Requirements

NPR 7123.1, NASA Systems Engineering Processes and Requirements

Section 3.7 Project Phase D: System Assembly, Integration and Test, Launch

NPR 7120.5, NASA Space Flight Program and Project Management Requirements

NPR 7123.1, NASA Systems Engineering Processes and Requirements

NASA Office of the Chief Information Officer (OCIO), *Information Technology Systems Engineering Handbook* Version 2.0

Section 3.8 Project Phase E: Operations and Sustainment

NPR 7120.5, NASA Space Flight Program and Project Management Requirements

NPR 7123.1, NASA Systems Engineering Processes and Requirements

Section 3.9 Project Phase F: Closeout

NPR 7120.5, NASA Space Flight Program and Project Management Requirements

NPR 7123.1, NASA Systems Engineering Processes and Requirements

NPD 8010.3, Notification of Intent to Decommission or Terminate Operating Space Systems and Terminate Missions

NPR 8715.6, NASA Procedural Requirements for Limiting Orbital Debris

Section 3.10 Funding: The Budget Cycle

NASA's *Financial Management Requirements (FMR)* Volume 4

Section 3.11 Tailoring and Customization of NPR 7123.1 Requirements

NPD 1001.0, 2014 NASA Strategic Plan

NPR 7120.5, NASA Space Flight Program and Project Management Requirements

NPR 7120.7, NASA Information Technology and Institutional Infrastructure Program and Project Management Requirements

NPR 7120.8, NASA Research and Technology Program and Project Management Requirements

NPR 7123.1, NASA Systems Engineering Processes and Requirements

NPR 7150.2, NASA Software Engineering Requirements

NPR 8705.4, Risk Classification for NASA Payloads

NASA-HDBK-2203, *NASA Software Engineering Handbook* (February 28, 2013)

NASA Engineering Network (NEN) Systems Engineering Community of Practice (SECoP), located at *https://nen.nasa.gov/web/se*

Section 4.1 Stakeholder Expectations Definition

NPR 7120.5, NASA Space Flight Program and Project Management Requirements

NASA Science Mission Directorate strategic plans

Presidential Policy Directive PPD-4 (2010), National Space Policy

Presidential Policy Directive PPD-21 (2013), Critical Infrastructure Security and Resilience

Ball, Robert E. (Naval Postgraduate School), *The Fundamentals of Aircraft Combat Survivability Analysis and Design*, 2nd Edition, AIAA Education Series, 2003

Larson (Wiley J.), Kirkpatrick, Sellers, Thomas, and Verma. *Applied Space Systems Engineering: A Practical Approach to Achieving Technical Baselines.* 2nd Edition, Boston, MA: McGraw-Hill Learning Solutions, CEI Publications, 2009.

Section 4.2 Technical Requirements Definition

NPR 7120.10, Technical Standards for NASA Programs and Projects

NPR 8705.2, Human-Rating Requirements for Space Systems

NPR 8715.3, NASA General Safety Program Requirements

NASA-STD-3001, *NASA Space Flight Human System Standard* – 2 volumes

NASA-STD-8719.13, *Software Safety Standard*, Rev C. Washington, DC, May 7, 2013.

NASA/SP-2010-3407, *Human Integration Design Handbook (HIDH)*

Section 4.3 Logical Decomposition

Department of Defense (DOD) Architecture Framework (DODAF) Version 2.02 Change 1, January 2015

Institute of Electrical and Electronics Engineers (IEEE) STD 610.12-1990, *IEEE Standard Glossary of Software Engineering Terminology.* Reaffirmed 2002. Superseded by ISO/IEC/IEEE 24765:2010, *Systems and Software Engineering – Vocabulary*

Section 4.4 Design Solution Definition

NPD 8730.5, NASA Quality Assurance Program Policy

NPR 8735.2, Management of Government Quality Assurance Functions for NASA Contracts

NASA-HDBK-1002, *Fault Management (FM) Handbook*, Draft 2, April 2012.

NASA-STD-3001, *NASA Space Flight Human System Standard* – 2 volumes

NASA-STD-8729.1, *Planning, Developing, and Maintaining an Effective Reliability and Maintainability (R&M) Program.* Washington, DC, December 1, 1998.

Code of Federal Regulations (CFR), Title 48 – Federal Acquisition Regulation (FAR) System, Part 46.4 Government Contract Quality Assurance (48 CFR 46.4)

International Organization for Standardization, ISO 9001:2015 *Quality Management Systems (QMS)*

Society of Automotive Engineers and the European Association of Aerospace Industries. AS9100C *Quality Management Systems (QMS)—Requirements for Aviation, Space, and Defense Organizations* Revision C: 2009-01-15

Blanchard, Benjamin S., *System Engineering Management.* 4th Edition, Hoboken, NJ: John Wiley & Sons, Inc., 2008

Section 5.1 Product Implementation

NPR 7150.2, NASA Software Engineering Requirements

NASA Engineering Network (NEN) Systems Engineering Community of Practice (SECoP), located at https://nen.nasa.gov/web/se

NASA Engineering Network (NEN) V&V Community of Practice, located at https://nen.nasa.gov/web/se

American Institute of Aeronautics and Astronautics (AIAA) G-118-2006e. *AIAA Guide for Managing the Use of Commercial Off the Shelf (COTS) Software Components for Mission-Critical Systems.* Reston, VA, 2006

Section 5.2 Product Integration

NASA Lyndon B. Johnson Space Center (JSC-60576), National Space Transportation System (NSTS), Space Shuttle Program, Transition Management Plan, May 9, 2007

Section 5.3 Product Verification

NPR 7120.5, NASA Space Flight Program and Project Management Requirements

NPR 7120.8, NASA Research and Technology Program and Project Management Requirements

NPR 7123.1, NASA Systems Engineering Processes and Requirements

NPR 8705.4, Risk Classification for NASA Payloads

NASA-STD-7009, *Standard for Models and Simulations*. Washington, DC, October 18, 2013

NASA GSFC-STD-7000, Goddard Technical Standard: *General Environmental Verification Standard (GEVS) for GSFC Flight Programs and Projects*. Goddard Space Flight Center. April 2005

Department of Defense (DOD). MIL-STD-1540D, *Product Verification Requirements for Launch, Upper Stage, and Space Vehicles*. January 15, 1999

Section 5.4 Product Validation

NPD 7120.4, NASA Engineering and Program/Project Management Policy

NPR 7150.2, NASA Software Engineering Requirements

Section 5.5 Product Transition

(The) National Environmental Policy Act of 1969 (NEPA). See 42 U.S.C. 4321-4347. *https://ceq.doe.gov/welcome.html*

Section 6.1 Technical Planning

NPR 7120.5, NASA Space Flight Program and Project Management Requirements

NPD 7120.6, Knowledge Policy on Programs and Projects

NPR 7123.1, NASA Systems Engineering Processes and Requirements

NASA-SP-2010-3403, *NASA Schedule Management Handbook*

NASA-SP-2010-3404, *NASA Work Breakdown Structure Handbook*

NASA Cost Estimating Handbook (CEH), Version 4, February 2015.

DOD. MIL-STD-881C, *Work Breakdown Structure (WBS) for Defense Materiel Items*. Washington, DC, October 3, 2011.

Institute of Electrical and Electronics Engineers (IEEE) STD 1220-2005. *IEEE Standard for Application and Management of the Systems Engineering Process*, Washington, DC, 2005.

Office of Management and Budget (OMB) Circular A-94, "Guidelines and Discount Rates for Benefit-Cost Analysis of Federal Programs" (10/29/1992)

Joint (cost and schedule) Confidence Level (JCL). Frequently asked questions (FAQs) can be found at: *http://www.nasa.gov/pdf/394931main_JCL_FAQ_10_12_09.pdf*

The U. S. Chemical Safety Board (CSB) case study reports on mishaps found at: *http://www.csb.gov/*

Section 6.3 Interface Management

NPR 7120.5, NASA Space Flight Program and Project Management Requirements

Section 6.4 Technical Risk Management

NPR 8000.4, Agency Risk Management Procedural Requirements

NASA/SP-2010-576, *NASA Risk-Informed Decision Making Handbook*

NASA/SP-2011-3421, *Probabilistic Risk Assessment Procedures Guide for NASA Managers and Practitioners*

NASA/SP-2011-3422, *NASA Risk Management Handbook*

Code of Federal Regulations (CFR) Title 22 – Foreign Relations, Parts 120-130 Department of State: International Traffic in Arms Regulations (ITAR) (22 CFR 120-130). Implements 22 U.S.C. 2778 of the Arms Export Control Act (AECA) of 1976 and Executive Order 13637, "Administration of Reformed Export Controls," March 8, 2013

Section 6.5 Configuration Management

NPR 7120.5, NASA Space Flight Program and Project Management Requirements

NASA. *Columbia Accident Investigation Board (CAIB) Report*, 6 volumes: Aug. 26, Oct. 2003. *http://www.nasa.gov/columbia/caib/html/report.html*

NASA. *NOAA N-Prime Mishap Investigation Final Report*, Sept. 13, 2004 *http://www.nasa.gov/pdf/65776main_noaa_np_mishap.pdf*

SAE International (SAE)/Electronic Industries Alliance (EIA) 649B-2011, *Configuration Management Standard (Aerospace Sector)* April 1, 2011

American National Standards Institute (ANSI)/Electronic Industries Alliance (EIA). ANSI/EIA-649, *National Consensus Standard for Configuration Management*, 1998–1999

Section 6.6 Technical Data Management

NPR 1441.1, NASA Records Retention Schedules

NPR 1600.1, NASA Security Program Procedural Requirements

NID 1600.55, Sensitive But Unclassified (SBU) Controlled Information

NPR 7120.5, NASA Space Flight Program and Project Management Requirements

NPR 7123.1, NASA Systems Engineering Processes and Requirements

NASA Form (NF) 1686, NASA Scientific and Technical Document Availability Authorization (DAA) for Administratively Controlled Information.

Code of Federal Regulations (CFR) Title 22 – Foreign Relations, Parts 120-130 Department of State: International Traffic in Arms Regulations (ITAR) (22 CFR 120-130). Implements 22 U.S.C. 2778 of the Arms Export Control Act (AECA) of 1976 and Executive Order 13637, "Administration of Reformed Export Controls," March 8, 2013

The Invention Secrecy Act of 1951, 35 U.S.C. §181–§188. Secrecy of Certain Inventions and Filing Applications in Foreign Country; §181 – Secrecy of Certain Inventions and Withholding of Patent.

Code of Federal Regulations (CFR) Title 37 – Patents, Trademarks, and Copyrights; Part 5 Secrecy of Certain Inventions and Licenses to Export and File Applications in Foreign Countries; Part 5.2 Secrecy Order. (37 CFR 5.2)

Section 6.7 Technical Assessment

NPR 1080.1, Requirements for the Conduct of NASA Research and Technology (R&T)

NPR 7120.5, NASA Space Flight Program and Project Management Requirements

NPR 7120.7, NASA Information Technology and Institutional Infrastructure Program and Project Management Requirements

NPR 7120.8, NASA Research and Technology Program and Project Management Requirements

NPR 7123.1, NASA Systems Engineering Processes and Requirements

NPR 8705.4, Risk Classification for NASA Payloads

NPR 8705.6, Safety and Mission Assurance (SMA) Audits, Reviews, and Assessments

NPR 8715.3, NASA General Safety Program Requirements

NASA-HDBK-2203, *NASA Software Engineering Handbook*. February 28, 2013

NASA/SP-2012-599, *NASA's Earned Value Management (EVM) Implementation Handbook*

NASA Federal Acquisition Regulation (FAR) Supplement (NFS) 1834.201, Earned Value Management System Policy.

NASA EVM website *http://evm.nasa.gov/index.html*

NASA Engineering Network (NEN) EVM Community of Practice located at *https://nen.nasa.gov/web/pm/evm*

NASA Engineering Network (NEN) Systems Engineering Community of Practice (SECoP) under Tools and Methods at *https://nen.nasa.gov/web/se/tools/* and then NASA Tools & Methods

American National Standards Institute/Electronic Industries Alliance (ANSI-EIA), Standard 748-C *Earned Value Management Systems.* March, 2013.

International Council on Systems Engineering (INCOSE). INCOSE-TP-2003-020-01, *Technical Measurement*, Version 1.0, 27 December 2005. Prepared by Garry J. Roedler (Lockheed Martin) and Cheryl Jones (U.S. Army).

Section 6.8 Decision Analysis

NPR 7120.5, NASA Space Flight Program and Project Management Requirements

NPR 7123.1, NASA Systems Engineering Processes and Requirements

Brughelli, Kevin (Lockheed Martin), Deborah Carstens (Florida Institute of Technology), and Tim Barth (Kennedy Space Center), "Simulation Model Analysis Techniques," Lockheed Martin presentation to KSC, November 2003

Saaty, Thomas L. *The Analytic Hierarchy Process.* New York: McGraw-Hill, 1980

Appendix B: Glossary

NPR 2210.1, Release of NASA Software

NPD 7120.4, NASA Engineering and Program/Project Management Policy

NPR 7120.5, NASA Space Flight Program and Project Management Requirements

NPR 7123.1, NASA Systems Engineering Processes and Requirements

NPR 7150.2, NASA Software Engineering Requirements

NPR 8000.4, Agency Risk Management Procedural Requirements

NPR 8705.2, Human-Rating Requirements for Space Systems

NPR 8715.3, NASA General Safety Program Requirements

International Organization for Standardization (ISO). ISO/IEC/IEEE 42010:2011. *Systems and Software Engineering – Architecture Description*. Geneva: International Organization for Standardization, 2011. (*http://www.iso-architecture.org/ieee-1471/index.html*)

Avizienis, A., J.C. Laprie, B. Randell, C. Landwehr, "Basic concepts and taxonomy of dependable and secure computing," *IEEE Transactions on Dependable and Secure Computing* 1 (1), 11–33, 2004

Appendix F: Functional, Timing, and State Analysis

NASA Reference Publication 1370, *Training Manual for Elements of Interface Definition and Control*. 1997

Defense Acquisition University. *Systems Engineering Fundamentals Guide*. Fort Belvoir, VA, 2001

Buede, Dennis. *The Engineering Design of Systems: Models and Methods*. New York: Wiley & Sons, 2000

Long, James E. *Relationships Between Common Graphical Representations in Systems Engineering*. Vienna, VA: Vitech Corporation, 2002

Sage, Andrew, and William Rouse. *The Handbook of Systems Engineering and Management*. New York: Wiley & Sons, 1999

Appendix G: Technology Assessment/Insertion

NPR 7120.5, NASA Space Flight Program and Project Management Requirements

NPR 7123.1, NASA Systems Engineering Processes and Requirements

Appendix H: Integration Plan Outline

Federal Highway Administration and CalTrans, *Systems Engineering Guidebook for ITS*, Version 2.0. Washington, DC: U.S. Department of Transportation, 2007

Appendix J: SEMP Content Outline

NPR 7120.5, NASA Space Flight Program and Project Management Requirements

NPR 7123.1, Systems Engineering Processes and Requirements

Appendix K: Technical Plans

NPR 7120.5, NASA Space Flight Program and Project Management Requirements

Appendix M: CM Plan Outline

SAE International (SAE)/Electronic Industries Alliance (EIA) 649B-2011, *Configuration Management Standard (Aerospace Sector)* April 1, 2011

Appendix N: Guidance on Technical Peer Reviews/Inspections

NPR 7123.1, Systems Engineering Processes and Requirements

NPR 7150.2, NASA Software Engineering Requirements

NASA Langley Research Center (LARC), *Instructional Handbook for Formal Inspections.* http://sw-eng.larc.nasa.gov/files/2013/05/Instructional-Handbook-for-Formal-Inspections.pdf

Appendix P: SOW Review Checklist

NASA Langley Research Center (LaRC) Procedural Requirements (LPR) 5000.2 Procurement Initiator's Guide

NASA Langley Research Center (LaRC) *Guidance on System and Software Metrics for Performance-Based Contracting* sites-e.larc.nasa.gov/sweng/files/2013/05/Guidance_on_Metrics_for_PBC_R1V01.doc

Appendix R: HSI Plan Content Outline

NPR 7123.1, NASA Systems Engineering Processes and Requirements

NPR 8705.2, Human-Rating Requirements for Space Systems

NASA-STD-3001, *Space Flight Human-System Standard*, Volume 2: Human Factors, Habitability, and Environmental Health, Section 3.5 [V2 3005], "Human-Centered Design Process." February 10, 2015

Bibliography

The bibliography contains sources cited in sections of the document and additional sources for developing the material in the document.

AIAA	American Institute of Aeronautics and Astronautics
ANSI	American National Standards Institute
ASME	American Society of Mechanical Engineers
ASQ	American Society for Quality
CCSDS	Consultative Committee for Space Data Systems
CFR	(U.S.) Code of Federal Regulations
COSPAR	The Committee on Space Research
DOD	(U.S.) Department of Defense
EIA	Electronic Industries Alliance
GEIA	Government Electronics Information Technology Association
IEEE	Institute of Electrical and Electronics Engineers
INCOSE	International Council on Systems Engineering
ISO	International Organization for Standardization
NIST	National Institute of Standards and Technology
SAE	Society of Automotive Engineers
TOR	Technical Operating Report
U.S.C.	United States Code

A

Adams, R. J., et al. *Software Development Standard for Space Systems, Aerospace Corporation Report* No. TOR-2004(3909)3537, Revision B. March 11, 2005. Prepared for the U.S. Air Force.

AIAA G-118-2006e, *AIAA Guide for Managing the Use of Commercial Off the Shelf (COTS) Software Components for Mission-Critical Systems*, Reston, VA, 2006

AIAA S-120-2006, *Mass Properties Control for Space Systems*. Reston, VA, 2006

AIAA S-122-2007, *Electrical Power Systems for Unmanned Spacecraft*, Reston, VA, 2007

ANSI/AIAA G-043-1992, *Guide for the Preparation of Operational Concept Documents*, Washington, DC, 1992

ANSI/EIA-632, *Processes for Engineering a System*, Arlington, VA, 1999

ANSI/EIA-649, *National Consensus Standard for Configuration Management*, 1998-1999

ANSI/GEIA-649, *National Consensus Standard for Configuration Management*, National Defense Industrial Association (NDIA), Arlington, VA 1998

ANSI/EIA-748-C *Standard: Earned Value Management Systems*, March, 2013

ANSI/GEIA GEIA-859, *Data Management*, National Defense Industrial Association (NDIA), Arlington, VA 2004

ANSI/IEEE STD 1042. *IEEE Guide to Software Configuration Management*. Washington, DC, 1987

Architecture Analysis & Design Language (AADL): https://wiki.sei.cmu.edu/aadl/index.php/Main_Page

(The) Arms Export Control Act (AECA) of 1976, see 22 U.S.C. 2778

ASME Y14.24, *Types and Applications of Engineering Drawings*, New York, 1999

ASME Y14.100, *Engineering Drawing Practices*, New York, 2004

ASQ, Statistics Division, Statistical Engineering, http://asq.org/statistics/quality-information/statistical-engineering

Avizienis, A., J.C. Laprie, B. Randell, C. Landwehr, "Basic concepts and taxonomy of dependable and secure computing," *IEEE Transactions on Dependable and Secure Computing* 1 (1), 11–33, 2004

B

Ball, Robert E. *The Fundamentals of Aircraft Combat Survivability Analysis and Design*. 2nd Edition, AIAA Education Series, 2003

Bayer, T.J., M. Bennett, C. L. Delp, D. Dvorak, J. S. Jenkins, and S. Mandutianu. "Update: Concept of Operations for Integrated Model-Centric Engineering at JPL," paper #1122, *IEEE Aerospace Conference 2011*

Blanchard, Benjamin S., *System Engineering Management*. 4th Edition, Hoboken, NJ: John Wiley & Sons, Inc., 2008

Blanchard, Benjamin S., and Wolter J. Fabrycky. *Systems Engineering and Analysis*, 5th Edition Prentice Hall International Series in Industrial & Systems Engineering; February 6, 2010

Brown, Barclay. "Model-based systems engineering: Revolution or Evolution," IBM Software, Thought Leadership White Paper, *IBM Rational*, December 2011

Brughelli, Kevin (Lockheed Martin), Deborah Carstens (Florida Institute of Technology), and Tim Barth (Kennedy Space Center), "Simulation Model Analysis Techniques," Lockheed Martin presentation to KSC, November 2003

Buede, Dennis. *The Engineering Design of Systems: Models and Methods*. New York: Wiley & Sons, 2000.

Business Process Modeling Notation (BPMN) http://www.bpmn.org/

C

CCSDS 311.0-M-1, *Reference Architecture for Space Data Systems,* Recommended Practice (Magenta), Sept 2008. http://public.ccsds.org/publications/MagentaBooks.aspx

CCSDS 901-0-G-1, *Space Communications Cross Support Architecture Description Document*, Informational Report (Green) Sept 2013. http://public.ccsds.org/publications/GreenBooks.aspx

Chapanis, A. "The Error-Provocative Situation: A Central Measurement Problem in Human

Factors Engineering." In *The Measurement of Safety Performance*. Edited by W. E. Tarrants. New York: Garland STPM Press, 1980

Chattopadhyay, Debarati, Adam M. Ross, and Donna H. Rhodes, "A Method for Tradespace Exploration of Systems of Systems," presentation in Track 34-SSEE-3: Space Economic Cost Modeling, *AIAA Space 2009*, September 15, 2009. © 2009 Massachusetts Institute of Technology (MIT), SEARI: Systems Engineering Advancement Research Initiative, seari.mit.edu

Chung, Seung H., Todd J. Bayer, Bjorn Cole, Brian Cooke, Frank Dekens, Christopher Delp, Doris Lam. "Model-Based Systems Engineering Approach to Managing Mass Margin," in *Proceedings of the 5th International Workshop on Systems & Concurrent Engineering for Space Applications (SECESA)*, Lisbon, Portugal, October, 2012

Clark, J.O. "System of Systems Engineering and Family of Systems Engineering From a Standards, V-Model, and Dual-V Model Perspective," *3rd Annual IEEE International Systems Conference*, Vancouver, Canada, March 23–26, 2009

Clemen, R., and T. Reilly. *Making Hard Decisions with DecisionTools Suite*. Pacific Grove, CA: Duxbury Resource Center, 2002

CFR, Title 14 – Aeronautics and Space, Part 1214 NASA Space Flight (14 CFR 1214)

CFR, Title 14 – Aeronautics and Space, Part 1216.3 NASA Environmental Quality: Procedures for Implementing the National Environmental Policy Act (NEPA) (14 CFR 1216.3)

CFR Title 22 – Foreign Relations, Parts 120-130 Department of State: International Traffic in Arms Regulations (ITAR) (22 CFR 120-130). Implements 22 U.S.C. 2778 of the Arms Export Control Act (AECA) of 1976 and Executive Order 13637, "Administration of Reformed Export Controls," March 8, 2013

CFR Title 37 – Patents, Trademarks, and Copyrights; Part 5 Secrecy of Certain Inventions and Licenses to Export and File Applications in Foreign Countries; Part 5.2 Secrecy Order. (37 CFR 5.2)

CFR Title 40 – Protection of Environment, Part 1508.27 Council on Environmental Quality: Terminology "significantly." (40 CFR 1508.27)

CFR Title 48 – Federal Acquisition Regulation (FAR) System, Part 1214 NASA Acquisition Planning: Acquisition of Commercial Items: Space Flight. (48 CFR 1214)

CFR Title 48 – Federal Acquisition Regulation (FAR) System, Part 46.103 Government Contract Quality Assurance: Contracting office responsibilities. (48 CFR 46.103)

CFR Title 48 – Federal Acquisition Regulation (FAR) System, Part 46.4 Government Contract Quality Assurance (48 CFR 46.4)

CFR Title 48 – Federal Acquisition Regulation (FAR) System, Part 46.407 Government Contract Quality Assurance: Nonconforming Supplies or Services (48 CFR 46.407)

COSPAR, *Planetary Protection Policy*. March 24, 2005. *http://w.astro.berkeley.edu/~kalas/ethics/documents/environment/COSPAR%20Planetary%20Protection%20Policy.pdf*

D

Deming, W. Edwards, see https://www.deming.org/

Dezfuli, H. "Role of System Safety in Risk-informed Decisionmaking." In *Proceedings, the NASA Risk Management Conference 2005*. Orlando, December 7, 2005

DOD Architecture Framework (DODAF) Version 2.02 Change 1, January 2015 http://dodcio.defense.gov/Library/DoDArchitectureFramework.aspx

DOD. *Defense Acquisition Guidebook (DAG)*. 2014

DOD Defense Acquisition University (DAU). *Systems Engineering Fundamentals Guide*. Fort Belvoir, VA, 2001

DOD Defense Logistics Agency (DLA). *Cataloging Handbook, H4/H8 Series*. Washington, DC, February 2003

DOD Defense Technical Information Center (DTIC). *Directory of Design Support Methods (DDSM)*. 2007. http://www.dtic.mil/dtic/tr/fulltext/u2/a437106.pdf

DOD MIL-HDBK-727 (Validation Notice 1). *Military Handbook: Design Guidance for Producibility*, U.S. Army Research Laboratory, Weapons and Materials Research Directorate: Adelphi,MD, 1990

DOD. MIL-HDBK-965. *Acquisition Practices for Parts Management*. Washington, DC, September 26, 1996. Notice 1: October 2000

DOD. MIL-STD-881C. *Work Breakdown Structure (WBS) for Defense Materiel Items*. Washington, DC, October 3, 2011

DOD. MIL-STD-1472G, *DOD Design Criteria Standard: Human Engineering*. Washington, DC, January 11, 2012

DOD. MIL-STD-1540D, *Product Verification Requirements for Launch, Upper Stage, and Space Vehicles*. January 15, 1999

DOD. MIL-STD-46855A, *Human Engineering Requirements for Military Systems, Equipment, and Facilities*. May 24, 2011. Replacement for DOD HDBK 763 and DOD MIL-HDBK-46855A, which have been cancelled.

DOD Office of the Under Secretary of Defense, Acquisition, Technology, & Logistics. SD-10. *Defense Standardization Program: Guide for Identification and Development of Metric Standards*. Washington, DC, April, 2010

DOD Systems Management College. *Systems Engineering Fundamentals*. Defense Acquisition University Press: Fort Belvoir, VA 22060-5565, 2001 http://ocw.mit.edu/courses/aeronautics-and-astronautics/16-885j-aircraft-systems-engineering-fall-2005/readings/sefguide_01_01.pdf

Duren, R. et al., "Systems Engineering for the Kepler Mission: A Search for Terrestrial Planets," *IEEE Aerospace Conference*, 2006

E

Eggemeier, F. T., and G. F. Wilson. "Performance and Subjective Measures of Workload in Multitask Environments." In *Multiple-Task Performance*. Edited by D. Damos. London: Taylor and Francis, 1991

Endsley, M. R., and M. D. Rogers. "Situation Awareness Information Requirements

Analysis for En Route Air Traffic Control." In *Proceedings of the Human Factors and Ergonomics Society 38th Annual Meeting.* Santa Monica: Human Factors and Ergonomics Society, 1994

Eslinger, Suellen. *Software Acquisition Best Practices for the Early Acquisition Phases.* El Segundo, CA: The Aerospace Corporation, 2004

Estefan, Jeff, *Survey of Model-Based Systems Engineering (MBSE) Methodologies*, Rev B, Section 3.2. NASA Jet Propulsion Laboratory (JPL), June 10, 2008. The document was originally authored as an internal JPL report, and then modified for public release and submitted to INCOSE to support the INCOSE MBSE Initiative.

Executive Order (EO) 12114, *Environmental Effects Abroad of Major Federal Actions.* January 4, 1979.

Executive Order (EO) 12770, *Metric Usage in Federal Government Programs*, July 25, 1991.

Executive Order (EO) 13637, *Administration of Reformed Export Controls*, March 8, 2013.

Extensible Markup Language (XML) http://www.w3.org/TR/REC-xml/

Extensible Markup Language (XML) Metadata Interchange (XMI) http://www.omg.org/spec/XMI/

F

Federal Acquisition Regulation (FAR). See: Code of Federal Regulations (CFR), Title 48.

Federal Aviation Administration (FAA), HF-STD-001, *Human Factors Design Standard (HFDS).* Washington, DC, May 2003. Updated: May 03, 2012. hf.tc.faa.gov/hfds

Federal Highway Administration, and CalTrans. *Systems Engineering Guidebook for ITS*, Version 2.0. Washington, DC: U.S. Department of Transportation, 2007

Friedenthal, Sanford, Alan Moore, and Rick Steiner. *A Practical Guide to SysML: Systems Modeling Language*, Morgan Kaufmann Publishers, Inc., July 2008

Fuld, R. B. "The Fiction of Function Allocation." *Ergonomics in Design* (January 1993): 20–24

G

Garlan, D., W. Reinholtz, B. Schmerl, N. Sherman, T. Tseng. "Bridging the Gap between Systems Design and Space Systems Software," *Proceedings of the 29th IEEE/NASA Software Engineering Workshop*, 6-7 April 2005, Greenbelt, MD, USA

Glass, J. T., V. Zaloom, and D. Gates. "A Micro-Computer-Aided Link Analysis Tool." *Computers in Industry* 16, (1991): 179–87

Gopher, D., and E. Donchin. "Workload: An Examination of the Concept." In *Handbook of Perception and Human Performance: Vol. II. Cognitive Processes and Performance.* Edited by K. R. Boff, L. Kaufman, and J. P. Thomas. New York: John Wiley & Sons, 1986

Griffin, Michael D., NASA Administrator. "System Engineering and the Two Cultures of Engineering." Boeing Lecture, Purdue University, March 28, 2007

H

Hart, S. G., and C. D. Wickens. "Workload Assessment and Prediction." In *MANPRINT: An Approach to Systems Integration*. Edited by H. R. Booher. New York: Van Nostrand Reinhold, 1990

Hoerl, R.W. and R.S. Snee, *Statistical Thinking – Improving Business Performance*, John Wiley & Sons. 2012

Hoffmann, Hans-Peter, "Harmony-SE/SysML Deskbook: Model-Based Systems Engineering with Rhapsody," Rev. 1.51, Telelogic/I-Logix white paper, Telelogic AB, May 24, 2006

Hofmann, Hubert F., Kathryn M. Dodson, Gowri S. Ramani, and Deborah K. Yedlin. *Adapting CMMI® for Acquisition Organizations: A Preliminary Report*, CMU/SEI-2006-SR-005. Pittsburgh: Software Engineering Institute, Carnegie Mellon University, 2006, pp. 338–40

Huey, B. M., and C. D. Wickens, eds. *Workload Transition*. Washington, DC: National Academy Press, 1993

I

IEEE STD 610.12-1990. *IEEE Standard Glossary of Software Engineering Terminology*. 1999, superceded by ISO/IEC/IEEE 24765:2010, *Systems and Software Engineering – Vocabulary*. Washington, DC, 2010

IEEE STD 828. *IEEE Standard for Software Configuration Management Plans*. Washington, DC, 1998

IEEE STD 1076-2008 *IEEE Standard VHDL Language Reference Manual*, 03 February 2009

IEEE STD 1220-2005. *IEEE Standard for Application and Management of the Systems Engineering Process*, Washington, DC, 2005

IEEE Standard12207.1, *EIA Guide for Information Technology Software Life Cycle Processes—Life Cycle Data*, Washington, DC, 1997

INCOSE. *Systems Engineering Handbook*, Version 3.2.2. Seattle, 2011

INCOSE-TP-2003-002-04, *Systems Engineering Handbook: A Guide for System Life Cycle Processes and Activities*, Version 4, Edited by Walden, David D., et al., 2015

INCOSE-TP-2003-020-01, *Technical Measurement*, Version 1.0, 27 December 2005. Prepared by Garry J. Roedler (Lockheed Martin) and Cheryl Jones (U.S. Army).

INCOSE-TP-2004-004-02, *Systems Engineering Vision 2020*, Version 2.03, September 2007, *http://www.incose.org/ProductsPubs/pdf/SEVision2020_20071003_v2_03.pdf*

INCOSE-TP-2005-001-03, *Systems Engineering Leading Indicators Guide*, Version 2.0, January 29, 2010; available at *http://seari.mit.edu/documents/SELI-Guide-Rev2.pdf*. Edited by Garry J. Roedler and Howard Schimmoller (Lockheed Martin), Cheryl Jones (U.S. Army), and Donna H. Rhodes (Massachusetts Institute of Technology)

ISO 9000:2015, *Quality management systems – Fundamentals and vocabulary*. Geneva: International Organization for Standardization, 2015

ISO 9001:2015 *Quality Management Systems (QMS)*. Geneva: International Organization for Standardization, September 2015

ISO 9100/AS9100, *Quality Systems Aerospace—Model for Quality Assurance in Design, Development, Production, Installation, and Servicing*. Geneva: International Organization for Standardization, 1999

ISO 10007: 1995(E). *Quality Management—Guidelines for Configuration Management*, Geneva: International Organization for Standardization, 1995

ISO 10303-AP233, *Application Protocol (AP) for Systems Engineering Data Exchange (AP-233)* Working Draft 2 published July 2006

ISO/TS 10303-433:2011 *Industrial automation systems and integration – Product data representation and exchange – Part 433: Application module: AP233 systems engineering*. ISO: Geneva, 2011

ISO/IEC 10746-1 to 10746-4, *ITU-T Specifications X.901 to x.904, Reference Model of Open distributed Processing (RM-ODP)*, Geneva: International Organization for Standardization, 1998. http://www.rm-odp.net

ISO 13374-1, *Condition monitoring and diagnostics of machines—Data processing, communication and presentation – Part 1: General guidelines*. Geneva: International Organization for Standardization, 2002

ISO/IEC 15288:2002. *Systems Engineering—System Life Cycle Processes*. Geneva: International Organization for Standardization, 2002

ISO/TR 15846. *Information Technology—Software Life Cycle Processes Configuration Management*, Geneva: International Organization for Standardization, 1998

ISO/IEC TR 19760:2003. Systems Engineering—A Guide for the Application of ISO/IEC 15288. Geneva: International Organization for Standardization, 2003

ISO/IEC/IEEE 24765:2010, Systems and Software Engineering – Vocabulary. Geneva: International Organization for Standardization, 2010

ISO/IEC/IEEE 42010:2011. Systems and Software Engineering—Architecture Description. Geneva: International Organization for Standardization, 2011 http://www.iso-architecture.org/ieee-1471/index.html

(The) Invention Secrecy Act of 1951, see 35 U.S.C. §181–§188. Secrecy of Certain Inventions and Filing Applications in Foreign Country; §181 – Secrecy of Certain Inventions and Withholding of Patent

J

Joint (cost and schedule) Confidence Level (JCL). Frequently asked questions (FAQs) can be found at: http://www.nasa.gov/pdf/394931main_JCL_FAQ_10_12_09.pdf

Jennions, Ian K. editor. *Integrated Vehicle Health Management (IVHM): Perspectives on an Emerging Field*. SAE International, Warrendale PA (IVHM Book) September 27, 2011

Jennions, Ian K. editor. *Integrated Vehicle Health Management (IVHM): Business Case Theory and*

Practice. SAE International, Warrendale PA (IVHM Book) November 12, 2012

Jennions, Ian K. editor. *Integrated Vehicle Health Management (IVHM): The Technology*. SAE International, Warrendale PA (IVHM Book) September 5, 2013

Johnson, Stephen B. et al., editors. *System Health Management with Aerospace Applications*. John Wiley & Sons, Ltd, West Sussex, UK, 2011

Jones, E. R., R. T. Hennessy, and S. Deutsch, eds. *Human Factors Aspects of Simulation*. Washington, DC: National Academy Press, 1985

K

Kaplan, S., and B. John Garrick. "On the Quantitative Definition of Risk." *Risk Analysis* 1(1). 1981

Karpati, G., Martin, J., Steiner, M., Reinhardt, K., "The Integrated Mission Design Center (IMDC) at NASA Goddard Space Flight Center," *IEEE Aerospace Conference 2003 Proceedings*, Volume 8, Page(s): 8_3657–8_3667, 2003

Keeney, Ralph L. *Value-Focused Thinking: A Path to Creative Decisionmaking*. Cambridge, MA: Harvard University Press, 1992

Keeney, Ralph L., and Timothy L. McDaniels. "A Framework to Guide Thinking and Analysis Regarding Climate Change Policies." *Risk Analysis* 21(6): 989–1000. 2001

Keeney, Ralph L., and Howard Raiffa. *Decisions with Multiple Objectives: Preferences and Value Tradeoffs*. Cambridge, UK: Cambridge University Press, 1993

Kirwin, B., and L. K. Ainsworth. *A Guide to Task Analysis*. London: Taylor and Francis, 1992

Kluger, Jeffrey with Dan Cray, "Management Tips from the Real Rocket Scientists," *Time Magazine*, November 2005

Knowledge Based Systems, Inc. (KBSI), *Integration Definition for functional modeling (IDEF0) ISF0 Function Modeling Method*, found at http://www.idef.com/idef0.htm

Kruchten, Philippe B. *The Rational Unified Process: An Introduction*, Third Edition, Addison-Wesley Professional: Reading, MA, 2003

Kruchten, Philippe B. "A 4+1 view model of software architecture," *IEEE Software Magazine* 12(6) (November 1995), 42–50

Kurke, M. I. "Operational Sequence Diagrams in System Design." *Human Factors* 3: 66–73. 1961

L

Larson, Wiley J.et al.. *Applied Space Systems Engineering: A Practical Approach to Achieving Technical Baselines*. 2nd Edition, Boston, MA: McGraw-Hill Learning Solutions, CEI Publications, 2009

Long, James E., *Relationships Between Common Graphical Representations in Systems Engineering*. Vienna, VA: Vitech Corporation, 2002

Long, James E., "Systems Engineering (SE) 101," *CORE®: Product & Process Engineering Solutions*, Vitech training materials. Vienna, VA: Vitech Corporation, 2000

M

Maier, M.W. "Architecting Principles for Systems-of-Systems," *Systems Engineering* 1(1998), 267-284, John Wiley & Sons, Inc.

Maier, M.W., D. Emery, and R. Hillard, "ANSI/IEEE 1471 and Systems Engineering," *Systems Engineering* 7 (2004), 257–270, Wiley InterScience, http://www.interscience.wiley.com

Maier, M.W. "System and Software Architecture Reconciliation," *Systems Engineering* 9 (2006), 146–159, Wiley InterScience, http://www.interscience.wiley.com

Maier, M.W., and E. Rechtin, *The Art of Systems Architecting*, 3rd Edition, CRC Press, Boca Raton, FL, 2009

Martin, James N., *Processes for Engineering a System: An Overview of the ANSI/GEIA EIA-632 Standard and Its Heritage.* New York: Wiley & Sons, 2000

Martin, James N., *Systems Engineering Guidebook: A Process for Developing Systems and Products.* Boca Raton: CRC Press, 1996.

Mathworks: Matlab http://www.mathworks.com/

McGuire, M., Oleson, S., Babula, M., and Sarver-Verhey, T., "Concurrent Mission and Systems Design at NASA Glenn Research Center: The origins of the COMPASS Team," *AIAA Space 2011 Proceedings*, September 27-29, 2011, Long Beach, CA

Meister, David, *Behavioral Analysis and Measurement Methods.* New York: John Wiley & Sons, 1985

Meister, David, *Human Factors: Theory and Practice.* New York: John Wiley & Sons, 1971

(The) Metric Conversion Act of 1975 (Public Law 94-168) amended by the Omnibus Trade and Competitiveness Act of 1988 (Public Law 100-418), the Savings in Construction Act of 1996 (Public Law 104-289), and the Department of Energy High-End Computing Revitalization Act of 2004 (Public Law 108-423). See 15 U.S.C. §205a et seq.

Miao, Y., and J. M. Haake. "Supporting Concurrent Design by Integrating Information Sharing and Activity Synchronization." In *Proceedings of the 5th ISPE International Conference on Concurrent Engineering Research and Applications (CE98).* Tokyo, 1998, pp. 165–74

The Mitre Corporation, *Common Risks and Risk Mitigation Actions for a COTS-based System.* McLean, VA. http://www2.mitre.org/.../files/CommonRisksCOTS.doc (no date)

MODAF http://www.modaf.com/

Moeller, Robert C., Chester Borden, Thomas Spilker, William Smythe, Robert Lock , "Space Missions Trade Space Generation and Assessment using the JPL Rapid Mission Architecture (RMA) Team Approach," *IEEE Aerospace Conference*, Big Sky, Montana, March 2011

Morgan, M. Granger, and M. Henrion, *Uncertainty: A Guide to Dealing with Uncertainty in Quantitative Risk and Policy Analysis.* Cambridge, UK: Cambridge University Press, 1990

M. Moshir, et al., "Systems engineering and application of system performance modeling in SIM Lite mission," *Proceedings. SPIE* 7734, 2010

Mulqueen, J.; R. Hopkins; D. Jones, "The MSFC Collaborative Engineering Process for Preliminary Design and Concept Definition Studies." 2012 *http://ntrs.nasa.gov/archive/nasa/casi.ntrs.nasa.gov/20120001572.pdf*

NASA Publications

NASA Federal Acquisition Regulation (FAR) Supplement (NFS) 1834.201, Earned Value Management System Policy

NASA Form (NF) 1686, NASA Scientific and Technical Document Availability Authorization (DAA) for Administratively Controlled Information

Reports

NASA Chief Engineer and the NASA Integrated Action Team (NIAT) report, "Enhancing Mission Success—A Framework for the Future," December 21, 2000. Authors: McBrayer, Robert O and Thomas, Dale, NASA Marshall Space Flight Center, Huntsville, AL United States

NASA. *Columbia Accident Investigation Board (CAIB) Report*, 6 volumes: Aug. 26, Oct. 2003. *http://www.nasa.gov/columbia/caib/html/report.html*

NASA. *NOAA N-Prime Mishap Investigation Final Report*, Sept. 13, 2004. *http://www.nasa.gov/pdf/65776main_noaa_np_mishap.pdf*

NASA. Diaz Report, *A Renewed Commitment to Excellence: An Assessment of the NASA Agency-wide Applicability of the Columbia Accident Investigation Board Report*, January 30, 2004. Mr. Al Diaz, Director, Goddard Space Flight Center, and team

NASA JPL D-71990, *Europa Study 2012 Full Report*. May 1 2012, publicly available here: *http://solarsystem.nasa.gov/europa/2012study.cfm*

NASA Office of Inspector General. *Final Memorandum on NASA's Acquisition Approach Regarding Requirements for Certain Software Engineering Tools to Support NASA Programs*, Assignment No. S06012. Washington, DC, 2006

NASA Office of Inspector General. *Performance-Based Contracting* *https://oig.nasa.gov/august/report/FY06/s06012*

Specialty Web Sites

NASA Engineering Network (NEN) Systems Engineering Community of Practice (SECoP) located at *https://nen.nasa.gov/web/se*

NASA Engineering Network (NEN) Systems Engineering Community of Practice (SECoP) under Tools and Methods at *https://nen.nasa.gov/web/se/tools/* and then NASA Tools & Methods

NASA Engineering Network (NEN) V&V Community of Practice, located at *https://nen.nasa.gov/web/se*

NASA Engineering Network (NEN) EVM Community of Practice, *https://nen.nasa.gov/web/pm/evm*

NASA EVM website *http://evm.nasa.gov/index.html*

NASA Procurement Library found at *http://www.hq.nasa.gov/office/procurement/*

Conference Publications

NASA 2011 Statistical Engineering Symposium, Proceedings. *http://engineering.larc.nasa.gov/2011_NASA_Statistical_Engineering_Symposium.html*

Aerospace Conference, 2007 IEEE Big Sky, MT 3–10 March 2007. NASA/Aerospace Corp. paper: "Using Historical NASA Cost and Schedule Growth to Set Future Program and Project Reserve Guidelines," by Emmons, D. L., R.E. Bitten, and C.W. Freaner. IEEE Conference Publication pages: 1–16, 2008. Also presented at the NASA Cost Symposium, Denver CO, July 17–19, 2007

NASA Cost Symposium 2014, NASA "Mass Growth Analysis: Spacecraft & Subsystems." LaRC, August 14th, 2014. Presenter: Vincent Larouche – Tecolote Research, also James K. Johnson, NASA HQ Study Point of Contact

Planetary Science Subcommittee, NASA Advisory Council, 23 June, 2008, NASA GSFC. NASA/Aerospace Corp. presentation; "An Assessment of the Inherent Optimism in Early Conceptual Designs and its Effect on Cost and Schedule Growth," by Freaner, Claude, Bob Bitten, Dave Bearden, and Debra Emmons

Technical Documents

NASA Office of Chief Information Officer (OCIO). *Information Technology Systems Engineering Handbook* Version 2.0

NASA Science Mission Directorate, *Risk Communication Plan for Planetary and Deep Space Missions*, 1999

NASA PD-EC-1243, *Preferred Reliability Practices for Fault Protection*, October 1995

NASA-CR-192656, *Contractor Report: Research and technology goals and objectives for Integrated Vehicle Health Management (IVHM).* October 10, 1992

NASA Jet Propulsion Laboratory (JPL), JPL-D-17868 (REV.1), *JPL Guideline: Design, Verification/Validation and Operations Principles for Flight Systems.* February 16, 2001

NASA Lyndon B. Johnson Space Center (JSC-65995), *Commercial Human Systems Integration Processes (CHSIP)*, May 2011

NASA/TP-2014-218556, *Technical Publication: Human Integration Design Processes (HIDP).* NASA ISS Program, Lyndon B. Johnson Space Center, Houston TX, September 2014. *http://ston.jsc.nasa.gov/collections/TRS/_techrep/TP-2014-218556.pdf*

NASA Lyndon B. Johnson Space Center (JSC-60576), *National Space Transportation System (NSTS), Space Shuttle Program, Transition Management Plan*, May 9, 2007

NASA Langley Research Center (LARC) *Guidance on System and Software Metrics for Performance-Based Contracting.* 2013 *sites-e.larc.nasa.gov/sweng/files/2013/05/Guidance_on_Metrics_for_PBC_R1V01.doc*

NASA Langley Research Center (LARC), *Instructional Handbook for Formal Inspections.* 2013 *http://sw-eng.larc.nasa.gov/files/2013/05/Instructional-Handbook-for-Formal-Inspections.pdf*

NASA/TM-2008-215126/Volume II (NESC-RP-06-108/05-173-E/Part 2), *Technical Memorandum: Design Development Test and Evaluation (DDT&E) Considerations for Safe and Reliable Human-Rated Spacecraft Systems.* April 2008.Volume II: *Technical Consultation Report.* James Miller, Jay Leggett, and Julie Kramer-White, NASA Langley Research Center, Hampton VA, June 14, 2007

NASA Reference Publication 1370. *Training Manual for Elements of Interface Definition and Control.* Vincent R. Lalli, Robert E. Kastner, and Henry N. Hartt. NASA Lewis Research Center, Cleveland OH, January 1997

NASA. *Systems Engineering Leading Indicators Guide,* http://seari.mit.edu/

NASA *Cost Estimating Handbook (CEH),* Version 4, February 2015

NASA *Financial Management Requirements (FMR)* Volume 4

Special Publications

NASA/SP-2010-576 *NASA Risk-Informed Decision Making Handbook*

NASA/SP-2012-599, *NASA's Earned Value Management (EVM) Implementation Handbook*

NASA/SP-2010-3403, *NASA Schedule Management Handbook*

NASA/SP-2010-3404, *NASA Work Breakdown Structure Handbook*

NASA/SP-2010-3406, *Integrated Baseline Review (IBR) Handbook*

NASA/SP-2010-3407, *Human Integration Design Handbook (HIDH)*

NASA/SP-2011-3421, *Probabilistic Risk Assessment Procedures Guide for NASA Managers and Practitioners*

NASA/SP-2011-3422, *NASA Risk Management Handbook*

NASA/SP-2013-3704, *Earned Value Management (EVM) System Description*

NASA/SP-2014-3705, *NASA Space Flight Program and Project Management Handbook*

NASA/SP-2015-3709, *Human Systems Integration Practitioners Guide*

Handbooks and Standards

NASA-HDBK-1002, *Fault Management (FM) Handbook,* Draft 2, April 2012

NASA-HDBK-2203, *NASA Software Engineering Handbook,* February 28, 2013

NASA Safety Standard (NSS) 1740.14, *Guidelines and Assessment Procedures for Limiting Orbital Debris.* Washington, DC, 1995 http://www.hq.nasa.gov/office/codeq/doctree/174014.htm NASA-STD 8719.14 should be used in place of NSS 1740.14 to implement NPR 8715.6. See NPR 8715.6 for restrictions on the use of NSS 1740.14.

NASA GSFC-STD-1000, *Rules for the Design, Development, Verification, and Operation of Flight Systems.* NASA Goddard Space Flight Center, February 8, 2013

NASA-STD-3001, *Space Flight Human System Standard.* Volume 1: *Crew Health.* Rev. A, July 30, 2014

NASA-STD-3001, *Space Flight Human System Standard.* Volume 2: *Human Factors, Habitability, and Environmental Health.* Rev. A, February 10, 2015

NASA GSFC-STD-7000, *Goddard Technical Standard: General Environmental Verification Standard (GEVS) for GSFC Flight Programs and Projects.* Goddard Space Flight Center, April 2005

NASA KSC-NE-9439 *Kennedy Space Center Design Engineering Handbook, Best Practices for Design and Development of Ground Systems.* Kennedy Space Center, November 20 2009

NASA-STD-7009, *Standard for Models and Simulations.* Washington, DC, October 18, 2013

NASA-STD-8719.13, *Software Safety Standard,* Rev C. Washington, DC, May 7, 2013

NASA-STD-8719.14, *Handbook for Limiting Orbital Debris.* Rev A with Change 1. December 8, 2011

NASA-STD-8729.1, *Planning, Developing, and Maintaining an Effective Reliability and Maintainability (R&M) Program.* Washington, DC, December 1, 1998

Policy Directives

NPD 1001.0, 2014 NASA Strategic Plan

NID 1600.55, Sensitive But Unclassified (SBU) Controlled Information

NPD 2820.1, NASA Software Policy

NPD 7120.4, NASA Engineering and Program/ Project Management Policy

NPD 7120.6, Knowledge Policy on Programs and Projects

NPD 8010.2, Use of the SI (Metric) System of Measurement in NASA Programs

NPD 8010.3, Notification of Intent to Decommission or Terminate Operating Space Systems and Terminate Missions

NPD 8020.7, Biological Contamination Control for Outbound and Inbound Planetary Spacecraft

NPD 8730.5, NASA Quality Assurance Program Policy

Procedural Requirements

NPR 1080.1, Requirements for the Conduct of NASA Research and Technology (R&T)

NPR 1441.1, NASA Records Retention Schedules

NPR 1600.1, NASA Security Program Procedural Requirements

NPR 2210.1, Release of NASA Software

NPR 2810.1, Security of Information Technology

LPR 5000.2, Procurement Initiator's Guide. NASA Langley Research Center (LARC)

JPR 7120.3, Project Management: Systems Engineering & Project Control Processes and Requirements. NASA Lyndon B. Johnson Space Center (JSC)

NPR 7120.5, NASA Space Flight Program and Project Management Processes and Requirements

NPR 7120.7, NASA Information Technology and Institutional Infrastructure Program and Project Management Requirements

NPR 7120.8, NASA Research and Technology Program and Project Management Requirements

NPR 7120.10, Technical Standards for NASA Programs and Projects

NPR 7120.11, NASA Health and Medical Technical Authority (HMTA) Implementation

NPR 7123.1, Systems Engineering Processes and Requirements

NPR 7150.2, NASA Software Engineering Requirements

NPR 8000.4, Risk Management Procedural Requirements

NPI 8020.7, NASA Policy on Planetary Protection Requirements for Human Extraterrestrial Missions

NPR 8020.12, Planetary Protection Provisions for Robotic Extraterrestrial Missions

APR 8070.2, EMI/EMC Class D Design and Environmental Test Requirements. NASA Ames Research Center (ARC)

NPR 8580.1, Implementing the National Environmental Policy Act and Executive Order 12114

NPR 8705.2, Human-Rating Requirements for Space Systems

NPR 8705.3, Probabilistic Risk Assessment Procedures Guide for NASA Managers and Practitioners

NPR 8705.4, Risk Classification for NASA Payloads

NPR 8705.5, Probabilistic Risk Assessment (PRA) Procedures for NASA Programs and Projects

NPR 8705.6, Safety and Mission Assurance (SMA) Audits, Reviews, and Assessments

NPR 8710.1, Emergency Preparedness Program

NPR 8715.2, NASA Emergency Preparedness Plan Procedural Requirements

NPR 8715.3, NASA General Safety Program Requirements

NPR 8715.6, NASA Procedural Requirements for Limiting Orbital Debris

NPR 8735.2, Management of Government Quality Assurance Functions for NASA Contracts

NPR 8900.1, NASA Health and Medical Requirements for Human Space Exploration

Work Instructions

MSFC NASA MWI 8060.1, Off-the-Shelf Hardware Utilization in Flight Hardware Development. NASA Marshall Space Flight Center.

JSC Work Instruction EA-WI-016, Off-the-Shelf Hardware Utilization in Flight Hardware Development. NASA Lyndon B. Johnson Space Center.

Acquisition Documents

NASA. *The SEB Source Evaluation Process.* Washington, DC, 2001

NASA. *Solicitation to Contract Award.* Washington, DC, NASA Procurement Library, 2007

NASA. *Statement of Work Checklist.* Washington, DC. See: *Appendix P* in this handbook.

N

(The) National Environmental Policy Act of 1969 (NEPA). See 42 U.S.C. 4321-4347. *https://ceq.doe.gov/welcome.html*

National Research Council (NRC) of the National Academy of Sciences (NAS), The Planetary Decadal Survey 2013–2022, *Vision and Voyagers for Planetary Science in the Decade 2013–2022*, The National Academies Press: Washington, D.C., 2011. *http://www.nap.edu*

NIST Special Publication 330: *The International System of Units (SI)* Barry N. Taylor and Ambler Thompson, Editors, March 2008. The United States version of the English text of the eighth edition (2006) of the International Bureau of Weights and Measures publication *Le Système International d' Unités (SI)*

NIST Special Publication 811: *NIST Guide for the Use of the International System of Units (SI)* A. Thompson and B. N. Taylor, Editors. Created July 2, 2009; Last updated January 28, 2016

NIST, Federal Information Processing Standard Publication (FIPS PUB) 199, *Standards for Security Categorization of Federal Information and Information Systems,* February 2004

O

Oberto, R.E., Nilsen, E., Cohen, R., Wheeler, R., DeFlorio, P., and Borden, C., "The NASA Exploration Design Team; Blueprint for a New Design Paradigm", *2005 IEEE Aerospace Conference*, Big Sky, Montana, March 2005

Object Constraint Language (OCL) *http://www.omg.org/spec/OCL/*

Office of Management and Budget (OMB) Circular A-94, *Guidelines and Discount Rates for Benefit-Cost Analysis of Federal Programs*, October 29, 1992

Oliver, D., T. Kelliher, and J. Keegan, *Engineering Complex Systems with Models and Objects*, New York, NY, USA: McGraw-Hill 1997

OOSEM Working Group, *Object-Oriented Systems Engineering Method (OOSEM) Tutorial*, Version 03.00, Lockheed Martin Corporation and INCOSE, October 2008

OWL, Web Ontology Language (OWL) *http://www.w3.org/2001/sw/wiki/OWL*

P

Paredis, C., Y. Bernard, R. Burkhart, H.P. Koning, S. Friedenthal, P. Fritzon, N.F. Rouquette, W. Schamai. "Systems Modeling Language (SysML)-Modelica Transformation." *INCOSE 2010*

Pennell, J. and Winner, R., "Concurrent Engineering: Practices and Prospects," Global Telecommunications Conference, *GLOBECOM '89*, 1989

Presidential Directive/National Security Council Memorandum No. 25 (PD/NSC-25), "Scientific or Technological Experiments with Possible Large-Scale Adverse Environmental Effects and Launch of Nuclear Systems into Space," as amended May 8, 1996

Presidential Policy Directive PPD-4 (2010), *National Space Policy*

Presidential Policy Directive PPD-21 (2013), *Critical Infrastructure Security and Resilience*

Price, H. E. "The Allocation of Functions in Systems." *Human Factors* 27: 33–45. 1985

The Project Management Institute® (PMI). *Practice Standards for Work Breakdown Structures.* Newtown Square, PA, 2001

Q

Query View Transformation (QVT) http://www.omg.org/spec/QVT/1.0/

R

Rasmussen, Robert. "Session 1: Overview of State Analysis," (internal document), *State Analysis Lite Course*, Jet Propulsion Laboratory, California Institute of Technology, Pasadena, CA, 2005

R. Rasmussen, B. Muirhead, *Abridged Edition: A Case for Model-Based Architecting in NASA*, California Institute of Technology, August 2012

Rechtin, Eberhardt. *Systems Architecting of Organizations: Why Eagles Can't Swim.* Boca Raton: CRC Press, 2000

S

Saaty, Thomas L. *The Analytic Hierarchy Process.* New York: McGraw-Hill, 1980

SAE Standard AS5506B, *Architecture Analysis & Design Language (AADL)*, SAE International, September 10, 2012

SAE International and the European Association of Aerospace Industries (EAAI) AS9100C, *Quality Management Systems (QMS): Requirements for Aviation, Space, and Defense Organizations* Revision C, January 15, 2009

SAE International/Electronic Industries Alliance (EIA) 649B-2011, *Configuration Management Standard (Aerospace Sector)*, April 1, 2011

Sage, Andrew, and William Rouse. *The Handbook of Systems Engineering and Management*, New York: Wiley & Sons, 1999

Shafer, J. B. "Practical Workload Assessment in the Development Process." In *Proceedings of the Human Factors Society 31st Annual Meeting*, Santa Monica: Human Factors Society, 1987

Shames, P., and J. Skipper. "Toward a Framework for Modeling Space Systems Architectures," *SpaceOps 2006 Conference*, AIAA 2006-5581, 2006

Shaprio, J., "George H. Heilmeier," *IEEE Spectrum*, 31(6), 1994, pg. 56–59 http://ieeexplore.ieee.org/iel3/6/7047/00284787.pdf?arnumber=284787

Software Engineering Institute (SEI). *A Framework for Software Product Line Practice*, Version 5.0. Carnegie Mellon University, http://www.sei.cmu.edu/productlines/frame_report/arch_def.htm

Stamelatos, M., H. Dezfuli, and G. Apostolakis. "A Proposed Risk-Informed Decision making Framework for NASA." In *Proceedings of the 8th International Conference on Probabilistic Safety*

Assessment and Management. New Orleans, LA, May 14–18, 2006

Stern, Paul C., and Harvey V. Fineberg, eds. *Understanding Risk: Informing Decisions in a Democratic Society*. Washington, DC: National Academies Press, 1996

Systems Modeling Language (SysML) http://www.omgsysml.org/

T

Taylor, Barry. *Guide for the Use of the International System of Units (SI)*, Special Publication 811. Gaithersburg, MD: NIST, Physics Laboratory, 2007

U

Unified Modeling Language (UML) http://www.uml.org/

UPDM: Unified Profile for the (US) Department of Defense Architecture Framework (DoDAF) and the (UK) Ministry Of Defense Architecture Framework (MODAF) http://www.omg.org/spec/UPDM/

U.S. Air Force. *SMC Systems Engineering Primer and Handbook*, 3rd ed. Los Angeles: Space and Missile Systems Center, 2005

U. S. Chemical Safety Board (CSB) case study reports on mishaps found at: http://www.csb.gov/

U.S. Navy. Naval Air Systems Command, *Systems Engineering Guide*: 2003 (based on requirements of ANSI/EIA 632:1998). Patuxent River, MD, 2003

U.S. Nuclear Regulatory Commission. SECY-98-144, *White Paper on Risk-Informed and Performance-Based Regulation*,Washington, DC, 1998

U.S. Nuclear Regulatory Commission. NUREG-0700, *Human-System Interface Design Review Guidelines*, Rev.2. Washington, DC, Office of Nuclear Regulatory Research, 2002

United Nations, Office for Outer Space Affairs. *Treaty of Principles Governing the Activities of States in the Exploration and Use of Outer Space, Including the Moon and Other Celestial Bodies.* Known as the "Outer Space Treaty of 1967"

W

Wall, S., "Use of Concurrent Engineering in Space Mission Design," *Proceedings of EuSEC 2000*, Munich, Germany, September 2000

Warfield, K., "Addressing Concept Maturity in the Early Formulation of Unmanned Spacecraft," *Proceedings of the 4th International Workshop on System and Concurrent Engineering for Space Applications*, October 13–15, 2010, Lausanne, Switzerland

Web Ontology Language (OWL) http://www.w3.org/2001/sw/wiki/OWL

Wessen, Randii R., Chester Borden, John Ziemer, and Johnny Kwok. "Space Mission Concept Development Using Concept Maturity Levels," Conference paper presented at the American Institute of Aeronautics and Astronautics (AIAA) Space 2013 Conference and Exposition; September 10–12, 2013; San Diego, CA. Published in the *AIAA Space 2013 Proceedings*

Winner, R., Pennell, J., Bertrand, H., and Slusarczuk, M., *The Role Of Concurrent Engineering In Weapons System Acquisition*,

Institute of Defense Analyses (IDA) Report R-338, Dec 1988

Wolfram, *Mathematica* http://www.wolfram.com/mathematica/

X

XMI: Extensible Markup Language (XML) Metadata Interchange (XMI) http://www.omg.org/spec/XMI/

XML: Extensible Markup Language (XML) http://www.w3.org/TR/REC-xml/

Z

Ziemer, J., Ervin, J., Lang, J., "Exploring Mission Concepts with the JPL Innovation Foundry A-Team," *AIAA Space 2013 Proceedings*, September 10–12, 2013, San Diego, CA

Made in the USA
Las Vegas, NV
09 March 2022